70674

NUTRITIONAL ASPECTS
and
CLINICAL MANAGEMENT
of
CHRONIC DISORDERS
and
DISEASES

CRC SERIES IN MODERN NUTRITION
Edited by Ira Wolinsky and James F. Hickson, Jr.

Published Titles

Manganese in Health and Disease, Dorothy J. Klimis-Tavantzis

Nutrition and AIDS: Effects and Treatments, Ronald R. Watson

Nutrition Care for HIV-Positive Persons: A Manual for Individuals and Their Caregivers,
Saroj M. Bahl and James F. Hickson, Jr.

Calcium and Phosphorus in Health and Disease, John J.B. Anderson and
Sanford C. Garner

Edited by Ira Wolinsky

Published Titles

Practical Handbook of Nutrition in Clinical Practice, Donald F. Kirby
and Stanley J. Dudrick

Handbook of Dairy Foods and Nutrition, Gregory D. Miller, Judith K. Jarvis,
and Lois D. McBean

Advanced Nutrition: Macronutrients, Carolyn D. Berdanier

Childhood Nutrition, Fima Lifschitz

Nutrition and Health: Topics and Controversies, Felix Bronner

Nutrition and Cancer Prevention, Ronald R. Watson and Siraj I. Mufti

Nutritional Concerns of Women, Ira Wolinsky and Dorothy J. Klimis-Tavantzis

Nutrients and Gene Expression: Clinical Aspects, Carolyn D. Berdanier

Antioxidants and Disease Prevention, Harinda S. Garewal

Advanced Nutrition: Micronutrients, Carolyn D. Berdanier

Nutrition and Women's Cancers, Barbara Pence and Dale M. Dunn

Nutrients and Foods in AIDS, Ronald R. Watson

Nutrition: Chemistry and Biology, Second Edition, Julian E. Spallholz,
L. Mallory Boylan, and Judy A. Driskell

Melatonin in the Promotion of Health, Ronald R. Watson

Nutritional and Environmental Influences on the Eye, Allen Taylor

Laboratory Tests for the Assessment of Nutritional Status, Second Edition,
H.E. Sauberlich

Advanced Human Nutrition, Robert E.C. Wildman and Denis M. Medeiros

Handbook of Dairy Foods and Nutrition, Second Edition, Gregory D. Miller,
Judith K. Jarvis, and Lois D. McBean

Nutrition in Space Flight and Weightlessness Models, Helen W. Lane
and Dale A. Schoeller

*Eating Disorders in Women and Children: Prevention, Stress Management,
  and Treatment*, Jacalyn J. Robert-McComb
Childhood Obesity: Prevention and Treatment, Jana Pařízková and Andrew Hills
Alcohol and Coffee Use in the Aging, Ronald R. Watson
Handbook of Nutrition in the Aged, Third Edition, Ronald R. Watson
Vegetables, Fruits, and Herbs in Health Promotion, Ronald R. Watson
Nutrition and AIDS, Second Edition, Ronald R. Watson
Advances in Isotope Methods for the Analysis of Trace Elements in Man,
  Nicola Lowe and Malcolm Jackson
Nutritional Anemias, Usha Ramakrishnan
Handbook of Nutraceuticals and Functional Foods, Robert E. C. Wildman
The Mediterranean Diet: Constituents and Health Promotion, Antonia-Leda Matalas,
  Antonis Zampelas, Vassilis Stavrinos, and Ira Wolinsky
Vegetarian Nutrition, Joan Sabaté
Nutrient–Gene Interactions in Health and Disease, Naïma Moustaïd-Moussa
  and Carolyn D. Berdanier
Micronutrients and HIV Infection, Henrik Friis
Tryptophan: Biochemicals and Health Implications, Herschel Sidransky
Nutritional Aspects and Clinical Management of Chronic Disorders and Diseases,
  Felix Bronner

Forthcoming Titles

Handbook of Nutraceuticals and Nutritional Supplements and Pharmaceuticals,
  Robert E. C. Wildman
Insulin and Oligofructose: Functional Food Ingredients, Marcel B. Roberfroid

NUTRITIONAL ASPECTS
and
CLINICAL MANAGEMENT
of
CHRONIC DISORDERS
and
DISEASES

Edited by
Felix Bronner

CRC PRESS

Boca Raton   London   New York   Washington, D.C.

Library of Congress Cataloging-in-Publication Data

Nutritional aspects and clinical management of chronic disorders and diseases / edited by
Felix Bronner.
    p. cm — (CRC series in modern nutrition)
    Includes bibliographical references and index.
    ISBN 0-8493-0945-X
    1. Diet therapy. 2. Cookery for the sick. 3. Diet in disease. 4. Chronic
diseases—Nutritional aspects. I. Bronner, Felix. II. Modern nutrition (Boca Raton, Fla.)
    [DNLM: 1. Chronic Disease—therapy. 2. Diet Therapy. 3. Nutrition. WT 500 N976 2002]
    RM216 .N886 2002
    616′.044—dc21
                                                    2002023353
                                                       CIP

Visit the CRC Press Web site at www.crcpress.com

© 2003 by CRC Press LLC

No claim to original U.S. Government works
International Standard Book Number 0-8493-0945-X
Library of Congress Card Number 2002023353
Printed in the United States of America 1 2 3 4 5 6 7 8 9 0
Printed on acid-free paper

Preface

Nutritional counseling and management are becoming important in health care, particularly in the management of a number of chronic conditions and diseases. The publication of this book is timely, because it aims to help physicians and their staffs identify conditions and diseases that can be treated effectively with nutritional intervention, and provides specifics on appropriate counseling and management.

The first of the 13 chapters discusses nutritional support for children, with emphasis on premature infants, cystic fibrosis, and bronchopulmonary dysplasia. As in all subsequent chapters, the authors, Valentine, Griffin, and Abrams, emphasize the need for good general nutrition to ensure that an individual attains full genetic potential. Malnutrition may be the result of inadequate nutrient intake — a possibility that even in wealthy societies cannot be neglected — or may be due to illness or a condition that magnifies the need for one or several nutrients. Diagnosis and assessment thus become the first and essential steps in an evaluation of nutritional status that is then followed by counseling, intervention, and clinical management. Convincing the patient, including young children, and caretakers of the rationale for the proposed approach is essential. The authors here, as authors of other chapters, deal with the question of enteral vs. parenteral and intravenous feeding, appropriately emphasizing the importance of milk or modified milks for the nutrition of children. In cystic fibrosis, the principal nutritional defect is fat malabsorption, and the use of supplementary fats, pancreatic enzymes, and vitamin supplements is discussed. An adequate energy supply, often accomplished by increasing the fat-to-carbohydrate ratio in the diet, is needed in bronchopulmonary dysplasia, and the authors discuss monitoring the patient to ensure reasonable weight gain over the long term.

The next several chapters discuss nutritional support and therapy in major organ systems. Kotchen and Kotchen, in their discussion of cardiovascular health (Chapter 2), enumerate the risk factors that contribute to coronary heart disease, the impact of diet and caloric balance, the dietary guidelines that have evolved, and the strategies that should be followed by patients and physicians to implement the guidelines. The authors point out how general guidelines — avoiding high fat intake, limiting caloric intake, generous consumption of fruits and vegetables — can be individualized for a specific patient, taking into account the social and cultural environment that may, on occasion, make acceptance of some recommendations difficult.

In Chapter 3, Navder and Lieber point out that even though in the popular view most diseases are thought to have originated in something eaten, diet has little to do with causing gastrointestinal disorders. Yet GI diseases and their treatment can often have serious nutritional consequences. The chapter systematically reviews and discusses disorders of the esophagus, stomach, small and large intestines, and liver, as well as inherited diseases and diseases of the gall bladder and pancreas. Each section includes treatment and nutritional management, with tables providing summary recommendations. In many situations, obesity is a complicating factor and weight reduction is recommended, sometimes facilitated by the need to avoid certain foods or food constituents. An example is peptic ulcer disease, where strong gastric stimulants should be avoided, even though antibiotics now constitute the principal treatment, whereas for a long time previously, diet therapy played a major role. Another example is Wilson's Disease, where penicillamine treatment reduces the body copper content and where copper-rich foods, e.g., beef liver, roasted cashew nuts, and chocolate chips, should be avoided.

Diabetes mellitus is a condition that illustrates the complexity of medical and nutritional management. Preuss and Bagchi, in Chapter 4, point out that glucose–insulin perturbations, a category that includes but is broader than diabetes mellitus, have increased in incidence, probably because of changes in lifestyle brought about by industrialization, urbanization, and increased longevity. One consequence of changes in lifestyle is the increase in obesity, which appears to exacerbate disturbances of insulin homeostasis. A primary goal of therapy is the return to normal metabolism of the three major nutrient groups — carbohydrates, lipids, and proteins — and the avoidance of later complications such as vascular disease and insufficiency that may eventuate the need for amputation. To achieve that requires intervention and the often substantial modification of social and cultural habits. This represents a major challenge to the physician, as well to the patient, and the chapter discusses these challenges as well as less classical approaches, such as the use of botanical supplements.

Chapter 5 by Utermohlen discusses endocrine control of metabolism, with emphasis on thyroid and glucocorticoid disease and, in Part 2, deals with diseases of carbohydrate intolerance, specifically galactosemia and lactose intolerance. Suppression of the excess hormone secretion in thyroiditis or Graves' disease or thyroxine replacement in hypothyroidism needs to be accompanied by management of the nutritional consequences of each condition, e.g., severe nutritional depletion of persons with hypermetabolism due to thyrotoxicosis, or the difficulty of maintaining a normal body weight for the hypothyroid patient. Special tables list the nutritional problems posed by hyper- and hypocorticalism, with medical management discussed in detail. Carbohydrate intolerance is defined and discussed, as is management, including nutritional management and complications resulting from galactosemia and lactose intolerance.

Nutritional support and management of musculoskeletal diseases are the topics of Chapter 6 by Favus, Utset, and Lee. The first topic is osteoporosis, a condition that has become a major public health problem as a result of the increase in longevity. Still more acutely a problem of women, it is becoming a problem for men in their seventh and eighth decades as well. Nutritional rickets in children is fortunately now very rare in countries that supplement their milk supply with vitamin D. Still, vitamin D deficiency and osteomalacia, as well as phosphate depletion, do occur, sometimes slowly over years, but need recognition and treatment. The chapter also discusses primary hyperparathyroidism, osteoarthritis, rheumatoid cachexia, scleroderma, and gout. All have nutritional implications or complications that need to be part of the overall treatment and management, often requiring substantial changes in lifestyle.

Renal failure, its nutrient metabolism, nutritional status, and requirements and medical and nutritional management are the topics of Chapter 7 by Ikizler. Renal failure is a chronic, progressive disease that requires different approaches at its several stages. For example, in early renal failure, phosphate control can usually be achieved by moderate phosphorus restriction, but when the glomerular filtration is below 20ml/min, it becomes necessary to stimulate calcitriol production and to employ phosphate binders. In malnourished patients who require dialysis, intradialytic parenteral nutrition may need to be considered. Clearly, nutritional management of renal failure patients is an integral component of therapy. Readers of this chapter will be helped by tables summarizing the various points made in the text.

The importance of vitamin A for proper night vision has been known for some time, but the overall importance of nutritional strategies to reduce the risk of eye diseases and their use to treat such diseases is perhaps less well known and is the subject of Chapter 8 by Trevithick and Mitton. In addition to dealing with specific eye diseases — cataracts, macular degeneration, retinopathy, retinitis pigmentosa, glaucoma, and keratoconus — the authors describe the rod visual cycle and give a lesson in genetics and oxidation stress. They also devote attention to the various herbal nutritional supplements that have been proposed, some of which may be harmful. Their reference list is particularly extensive for a field that may be unfamiliar.

Patients with cancer frequently suffer from protein-calorie malnutrition, with weight loss the most common manifestation. Yet, as discussed by Mason and Choi, in Chapter 9, evaluating this kind of malnutrition quantitatively is difficult, because it depends entirely on the assessment tool employed. Still, physicians need to be aware that malnutrition of a degree that worsens clinical outcome is common among cancer patients. The chapter deals with the mechanisms of body weight loss in cancer, the effects of the various major nutrients on cancer wasting, discusses the efficacy of nutritional support, and provides specifics on how to accomplish this in a variety of cancers, as well as in patients on chemotherapy or radiation therapy. Targeted nutrient therapy, i.e., the administration of specific nutrients in more than the usual quantities, e.g., omega-3 polyunsaturated fatty acids, or certain amino acids, is critically discussed, as is the advisability of aggressive nutritional support, either prophylactically or concurrent with treatment. Because many cancer patients tend to seek and use "alternative" treatments, physicians and their staffs must try to know this in order to better manage the patient's treatment.

Smith and Souba deal in considerable detail with the nutritional aspects of trauma and postsurgical care in Chapter 10. They analyze stress and the stress response in relation to surgical stress, the determinants of the host response to stress, and the role of cytokines as mediators of the stress response. Whereas most patients undergoing elective surgery are reasonably well nourished, there are specific endocrine and neuroendocrine responses to surgery that have nutritional consequences and need to be taken into account. In trauma patients, responses and consequences tend to be more dramatic, and the increased metabolic demands following injury can readily lead to malnutrition if adequate nutritional support is not provided. The chapter lists how nutritional requirements of the trauma patient can be determined. It also deals with sepsis, the choice of feeding routes — enteral or parenteral — and how to maintain gut function in the perioperative period.

Immunocompromised patients present a special challenge because of the intricate and complicated relationship between immunity and nutritional status. In

Chapter 11, Richter, Teuber, and Gershwin address this question with special reference to the human immunodeficiency virus and people with acquired human immune deficiency. Specific macro- and micronutrients are considered, as are deficiencies in zinc and selenium, fairly often encountered in patients with AIDS. Infections with Candida, Pneumocystis, or Mycobacterium are associated with high iron stores which contribute to morbidity. Treatment with an iron chelator may therefore be desirable. The authors discuss AIDS progression and vertical transmission in relation to nutrition, as well as the use of specific supplements as part of treatment.

Because of the important role played by food socially and culturally, it is not surprising that food intake management plays an important role in the treatment of psychiatric disorders. Lucas, Olson, and Olson, in Chapter 12, provide practical guidelines to physicians managing the dietary requirements of patients with psychiatric disorders. They deal with feeding disorders of infants and young children, pointing out that most problems can be prevented by routine support and education about feeding and parenting. Obviously, the role of the parent or caretaker is crucial for infants and young children. Older children who are hyperactive may have feeding problems, due either to their condition or their medication. Eating disorders, more common in girls than boys, often start at puberty and may continue throughout early adulthood. Following assessment, they are best treated by a weight-maintenance program. In major psychiatric disorders, weight gain is often a primary concern, and the remainder of the chapter deals with approaches to be taken in mood and psychotic disorders. Patients receiving monoamine oxidase inhibitors need to be on a tyramine-restricted diet, which is described and discussed in detail.

For more than 3000 years, alcoholic beverages have been desirable drinks, yet excessive drinking, leading to drunkenness and ultimately alcoholism, has been known just as long. In the last chapter, Navder and Lieber discuss alcoholic beverages, their place and effect on nutrition and nutritional status, the process of intoxication, and alcoholic liver disease. Potential treatment with polyunsaturated phosphatidylcholine, s-adenosylmethionine, or silymarin is discussed, and the effects of alcoholism on the brain and other tissues besides the liver are described, as are drug interactions. The authors thus deal with the correction of medical and nutritional problems of alcoholism; more direct approaches, focusing on medication-induced prevention, are emerging and when combined with the correction of nutritional deficiencies, may, in the words of the authors, alleviate the suffering of the alcoholic and reduce the public health impact of alcoholism.

In developing this book I was aware that most readers will read some, but not all chapters. Repetition of nutritional principles and of applications therefore seemed desirable. I thank the authors for their effort, patience, and willingness to accept editorial suggestions, and CRC Press for bringing this project to fruition.

Felix Bronner
Farmington, Connecticut
March 2002

The Editor

Felix Bronner, Ph.D., is professor emeritus of biostructure and function and of nutritional science at the University of Connecticut Health Center. A doctoral thesis on calcium metabolism in adolescent boys, under the supervision of Professor R.S. Harris at MIT, was the beginning of a long research career in bone and calcium metabolism and nutrition. A major aspect of Dr. Bronner's research has been the quantitative elucidation of active and passive calcium absorption in human and experimental animals, a definition of renal calcium movement, and quantitative analysis of plasma calcium homeostasis.

Author of more than 100 research papers, 77 book chapters, and editor of 54 books and treatises, including serving as founding editor of *Current Topics in Membranes and Transport*, Dr. Bronner has organized many scientific meetings and symposia, has trained graduate students and many postdoctoral fellows, and is the recipient of the 1975 Andre Lichtwitz Prize awarded by the French National Institute for Health and Medical Research (INSERM) for excellence in calcium and phosphate research. In 1996 he was awarded an honorary doctorate by the École Pratique des Hautes Études under the auspices of the French Ministry of Higher Education. He has been a visiting professor at the Universities of Cape Town, Tel Aviv, and Lyon, and at the Pasteur and Weizmann Institutes. He is a fellow of the American Association for the Advancement of Science and of the American Society for Nutritional Sciences. He is a member of numerous professional societies and has been on the editorial boards of *The American Journal of Physiology, The Journal of Nutrition*, and *The American Journal of Clinical Nutrition*. Currently he is principal editor of the Bone Biology domain of *TheScientific World Journal*. He was the founder and first chair of the Gordon Research Conference on Bones and Teeth.

Contributors

Steven A. Abrams, MD
Children's Nutrition Research Center
Baylor College of Medicine
Houston, TX

Debasis Bagchi, PhD
Creighton University School of
  Pharmacy and Allied Health
  Professions
Omaha, NE

Sang-Woon Choi, MD, PhD
Jean Mayer USDA Human Nutrition
  Research Center on Aging
Tufts University
Boston, MA

Murray J. Favus, MD
Pritzker School of Medicine
University of Chicago
Chicago, IL

M. Eric Gershwin, MD
School of Medicine
University of California at Davis
Davis, CA

Ian J. Griffin, MB, ChB
Section of Neonatology
Baylor College of Medicine
Houston, TX

T. Alp Ikizler, MD
Vanderbilt University Medical Center
Nashville, TN

Jane Morley Kotchen, MD, MPH
Medical College of Wisconsin
Milwaukee, WI

Theodore A. Kotchen, MD
Medical College of Wisconsin
Milwaukee, WI

Chin Lee, MD
Pritzker School of Medicine
University of Chicago
Chicago, IL

Charles S. Lieber, MD
Bronx VA Medical Center and Mt. Sinai
  School of Medicine
Bronx, NY

Alexander R. Lucas, MD
Mayo Clinic
Rochester, MN

Joel B. Mason, MD
Jean Mayer USDA Human Nutrition
  Research Center on Aging
Tufts University
Boston, MA

Kenneth P. Mitton, PhD
Eye Research Institute
Oakland University
Rochester, MI

Khursheed P. Navder, PhD, RD
Hunter College of the City University of
  New York and Bronx VA Medical
  Center
New York, NY

Diane L. Olson, RD, LD
Mayo Clinic
Rochester, MN

F. Karen Olson, RD, LD
Mayo Clinic
Rochester, MN

Harry G. Preuss, MD
Georgetown University Medical
  Center
Washington, DC

Sarah S. Richter, BS
School of Medicine
University of California at Davis
Davis, CA

J. Stanley Smith, Jr., MD
Penn State College of Medicine
M.S. Hershey Medical Center
Hershey, PA

Wiley W. Souba, MD, ScD, MBA
Penn State College of Medicine
M.S. Hershey Medical Center
Hershey, PA

Suzanne S. Teuber, MD
School of Medicine
University of California at Davis
Davis, CA

John R. Trevithick, PhD
Faculty of Medicine and Dentistry
University of Western Ontario
London, Ontario, Canada

Virginia Utermohlen, MD
Cornell University
Ithaca, NY

Tammy O. Utset, MD
Pritzker School of Medicine
University of Chicago
Chicago, IL

Christina J. Valentine, RD, MD
Department of Pediatrics
Baylor College of Medicine
Houston, TX

Contents

1 Nutritional Support in Children

Christina J. Valentine, Ian J. Griffin,
and Steven A. Abrams

CONTENTS

INTRODUCTION

The care of children is increasingly driven by modern technology. With the use of artificial surfactants and new ventilatory methods, the survival of the majority of infants born at greater than 700 grams birth weight and a substantial portion of those 500 to 700 g at birth can be ensured. This achievement, however, has made it necessary for nutritional support for premature infants and for the many children with acute and chronic illnesses to constitute an integral part of clinical management.

Complete nutritional support includes nutritional assessment, management, and surveillance strategies. The goal of this integrated approach is to avoid malnutrition and its resulting adverse effects on growth,[110] immune function,[11] mental development, and school performance.[97]

Malnutrition can be primary (i.e., due to inadequate nutrient intake) or secondary (i.e., due to illness or disease that increases nutritional needs or leads to poor nutrient absorption or increased nutritional losses).[105] Malnourished pediatric patients have greater morbidity and require longer hospitalizations than their well-nourished counterparts.[68] The clinical responsibility therefore exists to optimize a child's diet and ensure each child attains his/her genetic potential. Systematic nutritional assessment, performed as soon as possible after hospital admission or during outpatient visits, should initiate the nutritional support of the ill or prematurely delivered child.

Nutrition assessment classically uses clinical signs, diet adequacy, growth, and biochemical values.[13,105] Clinically, physical signs can reflect nutrient deficiencies but are late indicators of nutritional status[105] (Table 1.1). Subtle nutrient deficiencies can be identified earlier with the use of tools such as 24-hour dietary recalls or 3- to 5-day food diaries, which are then compared with the U.S. recommended dietary intakes (RDA). These intakes for most nutrients have been updated recently but a single revised RDA is not yet available.

Indirect calorimetry is a more sophisticated method to measure caloric requirements,[20] but it is technically difficult and not yet widely used clinically in children. Body composition is routinely evaluated with the aid of measurements of weight, length (or height for children over 2 years of age), and head circumference and is plotted for age using the new Centers for Disease Control (CDC) growth charts (http//www.cdc.gov/growthcharts). Generally, percentiles of normative values are between the 5th and 95th percentile. Body stores of protein and calories are indirectly measured using anthropometric measures such as mid-arm circumference and triceps skinfold, respectively.[37] Mid-arm circumference is measured by relaxing the arm and measuring the arm circumference midway between the olecranon and the tip of the acromion.[34] A nomogram can then be utilized to calculate muscle circumference[43] and the result compared to standardized values[34] (Tables 1.2 and 1.3). Triceps skinfold can also be measured at the same site by holding the skinfold parallel to the longitudinal axis of the upper arm with a pressure of $10g/mm^2$ using skinfold calipers.[34]

Biochemical estimates of visceral protein stores are often assessed by means of the serum albumin concentration. Serum albumin has a half-life of 18 to 20 days[13] and can reflect the severity of malnutrition. Prealbumin has a shorter half-life (2 days) and is often useful to monitor acute nutritional changes.[13] Additional methods, such as hydrodensitometry, total body potassium, total body water, neutron activation, photon and x-ray absorptiometry, bioelectrical impedance, and total body electrical conductivity, can be used to estimate the proportion of lean and fat tissue but have limited clinical use in hospitalized children.[2]

Effective management and surveillance begins after classification of nutritional status is obtained. Waterlow[110] has established a triad approach to the determination of the level of malnutrition by dividing subjects into three groups: 1) normal;

TABLE 1.1
Clinical Signs Associated with Nutritional Deficiencies

Nutritional Deficiency	Clinical Signs
Skeletal & Muscle Systems	
Protein-calorie	Muscle wasting
Vitamin D	Craniotabes, frontal and parietal bossing, persistently open anterior fontanel, pigeon chest, Rachitic rosary, genu varum, genu valgum
Vitamin D, vitamin C	Epiphyseal enlargement
Vitamin C	Hemorrhages
Skin	
Vitamin A or essential fatty acids	Xerosis, follicular hyperkeratosis
Vitamin C, vitamin K	Petechiae
Niacin	Pellagrous dermatitis
Nails	
Iron	Spoon-shaped
Hair	
Protein-calorie	Lackluster, thin or sparse, straight, flaccid, easy to pluck
Face	
Protein-calorie	Diffuse pigmentation
Protein	Moon face
Eyes	
Iron, folate, vitamin B_{12}	Pale conjunctiva
Vitamin A	Bitot's spots, corneal
Xerosis, keratomalacia	
Lips	
Riboflavin, niacin, iron, pyridoxine	Angular stomatitis, cheilosis
Mouth	
Zinc	Absence or impairment of sense of taste
Tongue	
Riboflavin	Magenta tongue
Folate, niacin, riboflavin, iron, vitamin B_{12}	Atrophic filiform papillae
Niacin, folate, riboflavin, iron, vitamin B_{12}	
Pyridoxine	Glossitis
Teeth	
Fluorine	Caries
Gums	
Vitamin C	Swollen, bleeding
Glands	
Iodine	Thyroid enlarged

From Suskind RM and Varma RN. Assessment of nutritional status of children. *Pediatr. Rev.* 5:195–202, 1984. With permission.

TABLE 1.2
Percentiles for Triceps Skinfold for Whites of the United States Health and Nutrition Examination Survey I of 1971 to 1974

Age Group	Triceps Skinfold Percentiles (mm²)							
	5	25	50	95	5	25	50	95
	Males				Females			
1–1.9	6	8	10	16	6	8	10	16
2–2.9	6	8	10	15	6	9	10	16
3–3.9	6	8	10	15	7	9	11	15
4–4.9	6	8	9	14	7	8	10	16
5–5.9	6	8	9	15	6	8	10	18
6–6.9	5	7	8	16	6	8	10	16
7–7.9	5	7	9	17	6	9	11	18
8–8.9	5	7	8	16	6	9	12	24
9–9.9	6	7	10	18	8	10	13	22
10–10.9	6	8	10	21	7	10	12	27
11–11.9	6	8	11	24	7	10	13	28
12–12.9	6	8	11	28	8	11	14	27
13–13.9	5	7	10	26	8	12	15	30
14–14.9	4	7	9	24	9	13	16	28
15–15.9	4	6	8	24	8	12	17	32
16–16.9	4	6	8	22	10	15	18	31
17–17.9	5	6	8	19	10	13	19	37

Data from Frisancho AR. New norms of upper limb fat and muscle areas for assessment of nutritional status. *Am. J. Clin. Nutr*. 34:2540–2545, 1981. With permission.

2) stunted (length-for-age deficiency); and 3) wasted (weight-for-height deficiency). In general, acute malnutrition is primarily associated with wasting, and chronic malnutrition with linear growth retardation.

After the level of nutritional status is determined, an effective plan can then be developed based on the age, sex, and nutritional status of the child, with adjustments for circumstances such as illness, disease, and ongoing losses from diarrhea or ostomy secretion. Extreme caution is necessary when nutritionally rehabilitating a severely malnourished child in order to avoid the "re-feeding syndrome," which consists of hypophosphatemia and hypomagnesemia with subsequent cardiac and/or neuromuscular dysfunction.[98] In children who are malnourished, close surveillance is necessary, including monitoring daily caloric intakes, body weight, and feeding tolerance. Additionally, growth assessment laboratory values should be monitored at routine intervals to avoid re-feeding and malnutrition problems. The assistance of a registered dietitian in this situation can be invaluable.[2,37] Nutrition standards established by the Joint Commission on Hospital Accreditation (JCHO) or perinatal services should be updated and incorporated into the nutrition support policy.

In the balance of this chapter we will consider a few of the more problematic nutritional problems faced in pediatrics and discuss the etiology, assessment,

TABLE 1.3
Percentiles of Upper Arm Circumference (mm) for Whites of the United States Health and Nutrition Examination Survey I of 1971 to 1974

Age Group	5	25	50	95	5	25	50	95
		Male				Female		
1–1.9	142	150	159	183	138	148	156	177
2–2.9	141	153	162	185	142	152	160	184
3–3.9	150	160	167	190	143	158	167	189
4–4.9	149	162	171	192	149	160	169	191
5–5.9	153	167	175	204	153	165	175	211
6–6.9	155	167	179	228	156	170	176	211
7–7.9	162	177	187	230	164	174	183	231
8–8.9	162	177	190	245	168	183	195	261
9–9.9	175	187	200	257	178	194	211	260
10–10.9	181	196	210	274	174	193	210	265
11–11.9	186	202	223	280	185	208	224	303
12–12.9	193	214	232	303	194	216	237	294
13–13.9	194	228	247	301	202	223	243	338
14–14.9	220	237	253	322	214	237	252	322
15–15.9	222	244	264	320	208	239	254	322
16–16.9	244	262	278	343	218	241	258	334
17–17.9	246	267	285	347	220	241	264	350

Data from Frisancho AR. New norms of upper limb fat and muscle areas for assessment of nutritional status. *Am. J. Clin. Nutr.* 34:2540–2545, 1981. With permission.

management, and outlook for these conditions. These conditions are prematurity, cystic fibrosis, and bronchopulmonary dysplasia.

PREMATURE INFANTS

INTRODUCTION

By definition, any infant delivered at less than 37 completed weeks of gestation is considered premature. The care of very small premature infants is complex and includes the use of artificial surfactants, mechanical ventilation, and medications which frequently affect growth. Premature infants less than 1000 g birth weight usually receive parenteral nutrition and are uncommonly discharged after less than 2 to 3 months of hospitalization.

The nutritional consequences of premature delivery are especially important for infants born at less than 30-weeks' gestation or less than 1500 g birth weight (defined as "very low birth weight," VLBW). These infants have missed much or all of the last trimester's accretion of nutrients, which places them at risk for multinutrient deficiencies.[5] VLBW infants have a higher percentage of body water (up to 80%) and decreased protein and fat stores at birth than do full-term infants.[116] These deficiencies make the VLBW infant acutely susceptible to inadequate nutrient intake postnatally. In addition, VLBW infants have immature enzymatic,[82] hormonal,[9] motor,[14] and digestive[71] functions, so standard approaches to feeding may be inappropriate. Consequently, VLBW infants are susceptible to malnutrition that may lead to poor growth,[47] chronic lung disease,[99] rickets,[101] and cholestasis.[10] Unique and timely nutrition support is therefore essential in the immediate newborn period.

NUTRITION ASSESSMENT AND GROWTH GOALS

Nutrition assessment begins with determination of the gestational age, maturity, weight, length, and head circumference of the infant.[108] Standard growth charts can be used to plot weight and classify infants as appropriate for gestational age (AGA), small for gestational age (SGA), or large for gestational age (LGA). Appropriate for gestational (AGA) infants are in the 10th to 95th percentile of weight for their gestational age; small for gestational age infants (SGA) are below the 10th percentile; and large for gestational age (LGA) are above the 90th percentile. The risk of growth failure is increased in SGA infants, and they often have greater nutrient requirements than AGA infants.[78] Large for gestational age infants (LGA) also have health risks such as hypoglycemia, requiring greater glucose intakes initially.[48] In addition, difficulty in labor and delivery, as evidenced by a low Apgar score at 5 minutes of age, has been associated with a higher incidence of necrotizing enterocolitis. The clinical diagnosis, maternal and perinatal history, human milk availability, and medications often influence feeding plans and should be noted.

The usual goal of nutrition support for premature infants is to attempt to achieve intrauterine nutrient accretion rates.[5] Defining these rates is not always simple, because good recent data for normal *in utero* accretion of most nutrients are not readily available. Early data utilized a "reference fetus"[116] and formed the basis for many of the current recommendations regarding growth and nutrition. In general, a desirable weight gain after growth has begun is 15 g/kg/day, length gain of at least 0.8 to 1.1 cm/week, and head circumference growth of 0.5 to 1.0 cm/week. Knee-to-heel length may be a useful measure of linear growth, particularly in very ill infants unable to stretch out on a length board. A kneemometer can be used to measure knee-to-heel acceleration.[69] The goal is an increment of at least 0.4 mm/day. Although this measurement may be useful in extremely sick infants, it appears to be less reliable than crown-to-heel length measurement in healthy, growing, preterm infants.[41]

Mid-arm circumference standards have also been developed for the preterm and are available,[88] but routine body composition measurements are not that practical in the neonatal intensive care unit (NICU). Many nurseries document growth using extrauterine growth grids, many of which have been described.[24,30] It must be emphasized that these should be used as guidelines and not standards because of differences in birth weights, illness patterns, feeding practices, and the management styles of the clinical settings in which they were developed. Optimal nutrition should then provide nutrients of proper quantity and quality to avoid catabolism and provide accretion without adding a toxic overload to such an immature infant.

NUTRIENT NEEDS AND MANAGEMENT

The premature infant, because of his/her immaturity and limited body stores, has unique requirements for fluids, electrolytes, energy, carbohydrates, proteins, lipids, vitamins, minerals, and trace elements.[2,116] Furthermore, infants are unable to suck and swallow in a coordinated fashion before approximately 33 weeks' gestation.[42] Nutrients must therefore initially be given intravenously or via feeding tube until the infant is more mature. In general, for the infant born before 33 weeks' gestation, intravenous feeding begins on day one of life and tube feeding sometime during the first week.

Fluids and Electrolytes

Fluids are essential after delivery because the infant is born with a larger percentage of body water relative to older infants. Newborn infants have high insensible fluid losses and can be expected to lose 5 to 15% body weight during the first week of life.[31] Careful management of fluids is necessary, however, to avoid potential complications from fluid overload and exacerbation of patent ductus arteriosus,[1] intraventricular hemorrhages,[36] or necrotizing enterocolitis.[12] Recommended intravenous volumes include initially 60 to 80 ml/kg/day on day one, advancing to 150 ml/kg/day by day five. Fluid losses can increase significantly in very immature infants, those receiving phototherapy, and those nursed under radiant warmers.[50] Electrolytes needed include sodium, 2 to 4 meq/kg/day, and potassium, 1 to 2 meq/kg/day, to maintain normal serum electrolytes and urine output. Electrolyte supplementation must be guided by frequent monitoring of the serum electrolyte concentrations. If the infant has additional fluid and electrolyte losses, as for example in the case of intra-abdominal infections or losses from chest tubes or ostomies, then frequent monitoring is needed to ensure appropriate fluid and electrolyte replacement.

Parenteral Nutrition

Energy Needs

An energy source should be started rapidly because of the infant's limited glycogen and fat stores. Glucose is the primary energy source *in utero*[112] and should be provided intravenously at an initial rate of 6 to 8 mg/kg per minute, advancing to a maximum of 12 mg/kg/minute.[106] The immature infant may have episodes of hyperglycemia from stress or inability to suppress glucose production during infusion.[104] High glucose levels can be treated with an infusion of approximately 0.06 to 0.10 units insulin/hour,[22] based on careful monitoring of the blood glucose levels. Hypoglycemia caused by insulin infusion is especially dangerous because alternative metabolic fuels (such as free fatty acids and ketones) are also suppressed. The use of insulin in neonatal care remains controversial, as does the degree of hyperglycemia that merits therapy.

Lipids

A lipid source is needed to avoid essential fatty acid deficiency[33] because stores of linoleic acid (C18:2w6) are low in premature compared to term infants.[33] A minimum of 0.5 to 1 g/kg/day of lipid is needed to avoid essential fatty acid deficiency.[3] A continuous 24-hour infusion[54] of 20% Intralipid® (KabiVitrum) intravenous emulsion is well tolerated by most infants.[46] The infusion can be initiated at 0.5 to 1.0 g/kg per day and is typically advanced to about 3 g/kg/day.[5] Even higher infusion rates, up to 4 g/kg/d, may be well tolerated[17] without any adverse effects on alveolar–arteriolar diffusing capacity.

Preterm infants have lower lipid clearance abilities due to limited activity of post-heparin lipoprotein lipase.[27] Infusion rates should therefore be adjusted to keep serum triglyceride concentrations below 150 mg/dl.[5] Premature infants also have reduced tissue carnitine concentrations,[89] and it has been speculated that long-term parenteral nutrition without carnitine may limit efficient fatty acid oxidation.[58] Carnitine can, if desired, be given as an intravenous supplement of 10mg/kg/day, if no source of enteral milk is given for prolonged periods.

Protein

Protein can be begun on the first day of life at 1 to 2 g/kg/day and should be advanced to at least 3g/kg/day as soon as feasible to promote positive nitrogen balance.[85] Cystathionase activity is minimal in the liver of infants,[103] as are cystathionine γ-lyase and cysteinesulfinic acid decarboxylase.[114] Cysteine and taurine, therefore, become conditionally essential amino acids in preterm infants. Amino acid blends should be evaluated to ensure the solution contains tolerable quantities of phenylalanine and methionine and sufficient cysteine and taurine.

The parenteral solution of Trophamine® (Kendall McGaw) currently meets these requirements and has been shown to normalize the amino acid profile and to promote weight gain and a positive nitrogen balance in premature infants.[49] In addition, when Trophamine is supplemented with cysteine hydrochloride at 40 mg/g amino acids, the pH of the solution is optimal to maintain calcium and phosphorus in suspension.[32] Infants with necrotizing enterocolitis (NEC),[95] intestinal surgery,[28] ostomies, or protein-losing enteropathies may require up to 3.5 to 4 grams protein/kg/day.

Vitamins and Minerals

Vitamins in parenteral nutrition mixtures are provided in the United States using MVI-Pediatric (Armour) at a dose of 40% of the 5 ml vial if the infant weighs less than 2kg (38) and at 100% if the infant weighs more than 2kg.

Preterm infants are born with low reserves of vitamin A.[93] Parenteral nutrition sources are low in vitamin A and a substantial amount, up to 50 to 75%, is lost in the intravenous tubing.[94] Several investigators have reported that infants with chronic lung disease have lower plasma vitamin A levels than healthy premature infants,[93,109] and some clinical trials have suggested that supplementation with parenteral vitamin A may decrease the incidence of bronchopulmonary dysplasia in at-risk infants. Delivery of vitamin A can be markedly enhanced by the direct addition of multivitamins[39] or of a retinyl palmitate source directly in the lipid infusion.[111]

Vitamin K is needed for the blood factors II, VII, IX, and X. Vitamin K deficiency can lead to hemorrhages from the umbilical cord and venipuncture sites, or, most dramatically, can cause intracranial bleeding. This can be avoided by giving 0.5 to 1 mg IM dose at delivery.[2] Phylloquinone, the plant source of vitamin K, is a component of MVI-Pediatric. Preterm infants should receive 100 mcg/kg/day of phylloquinone.[38]

Calcium and phosphorus are required for bone mineralization. The VLBW infant misses much of the *in utero* bone mineralization that takes place in the last trimester when calcium accretion rates are about 100 to 120 mg/kg/d and phosphorus is accreted at a rate of 65 to 80 mg/kg/day.[112,116] It is difficult to meet this level of intake using parenteral nutrition. The amino acid preparation, dextrose concentration, temperature, and type of calcium salt used all affect the solubility of calcium and phosphorus in the parenteral solution.[29] Using 1.7 mmol/dL of calcium and 2.0 mmol/dL of phosphorus in a parenteral solution containing Trophamine and cysteine, investigators were able to attain 65% of intrauterine calcium retention rates and 85% of the phosphorus retention rates, thus leading to enhanced bone mineral content.[81] Serum calcium, phosphorus, and alkaline phosphatase should be monitored routinely.[66]

Zinc is important in many enzymatic reactions and zinc deficiencies result in poor growth and impaired T lymphocyte function.[51] Zlotkin and Buchanan[117] found that an intravenous zinc intake of 438 mcg/kg/day resulted in intrauterine accretion rates. This is similar to the current recommendation of 400 mcg/kg/day.[38] It has been suggested that unless parenteral nutrition is required for longer than 4 weeks, zinc is the only trace element that needs to be added.[2,38]

Enteral Feeding, Advancement, Method, and Milk Type

Enteral feeding should begin in premature infants as soon as feasible, preferably in the first week of life if the infant is medically stable. The benefits of enteral nutrition include trophic effects on the intestinal tract,[70] hormonal stimulation,[63,67] improved liver function, enhanced maturation of the gastrointestinal tract,[15] and fewer days to achieve full enteral nutrition. Potential reasons to delay initiating enteral feeds may include unstable blood pressure, severe acidosis, or evidence of severe respiratory depression at birth minutes. Traditionally, infants were not fed enterally while they had umbilical catheters in place. However, this restriction is not well supported by controlled trials.[67]

Currently, for infants less than 1500 g birth weight, we recommend initiating enteral feeding at 10 to 20 ml/kg/day and increasing it by not more than 20 ml/kg/day. It may be useful to provide several days of priming feeds at a rate of 20 ml/kg/day or less prior to advancing to greater volumes, although this practice remains under evaluation. Daily volume increases greater than 20 ml/kg/day have been associated with an increased incidence of NEC.[6] Other data do not support this association, however.[83]

The method of feeding depends on the infant's suck/swallow/breathing maturity. Infants between 33 and 35 weeks' gestation may be allowed to attempt breast-feeding.[4] Bottle-feeding should be initiated cautiously, and each feeding should be limited to 20 minutes in duration in the initial feeding period. If the infant has a grooved palate from long-term intubation or oral gastric tubes,[7] then the Haberman feeder (Medela, Inc., McHenry, IL) may be useful because of its long, shaft-like nipple that allows for improved nipple seal and improved intra-oral pressure.

Infants who cannot exclusively nipple feed require tube feedings, either by intermittent bolus or slow infusion. This allows direct delivery to the stomach and provides a physiologic stimulus to gut hormone production.[9] Bolus feeding has also been reported to result in less feeding intolerance and improved weight gain in preterm infants.[90] However, in infants with intestinal[73] or cardiac disease,[92] the continuous method is reported to enhance tolerance. To ensure adequate delivery, an automated syringe pump should be used with short tubing, and syringes of milk should be changed every 3 to 4 hours.[40] If human milk is being delivered by this method, the syringe pump should be pointed upward to avoid fat layering at the bottom of the syringe.

Transpyloric feeding should be reserved only for those infants who are intolerant to gastric feeding or who have severe reflux. This feeding method may not be ideal because it bypasses the stomach's hormonal, enzymatic, and anti-infective functions.[64,77]

Human milk is the feeding of choice for virtually all infants.[4] The unique components of breast milk which benefit the immature infant include anti-infective components, hormonal, growth factors, fatty acids, nucleotides, glutamine, and enzymes.[90] Preterm infants fed fortified human milk may have less sepsis and NEC.[90] There is controversy regarding potential developmental benefits relating to the use of human milk for preterm infants, but some data support an improved long-term outcome during childhood comparing human milk to formula-fed infants.[62]

Adapted formulas designed for full-term infants, in addition to unfortified human milk, are nutritionally inadequate for exclusive feeding of preterm infants whose birth weight is below 1800 g. To achieve nutrient intakes required for growth and bone mineralization, human milk should be supplemented with specialized, powdered human milk fortifier. These fortifiers provide additional protein, calories, vitamins, and minerals. If weight gain is inadequate on 150 to 160 ml/kg/d of the preterm formula, or on 180 to 200 ml/kg/d fortified human milk, then infants should be evaluated for causes of growth failure, such as hyponatremia, hypokalemia, metabolic acidosis, or urinary tract infections. In some infants, feeding volume should be increased once these conditions are excluded or treated. Alternatively, in human milk-fed infants, hindmilk fractionation and feeding should be considered. Hindmilk is milk expressed after the first 2 minutes of pumping or feeding.[72] It should be bottled separately from the foremilk, fortified as usual, and then fed to the infant. This practice has been reported to significantly improve fat and calorie intake and subsequent weight gain.[107]

When human milk is not available, or in rare cases such as maternal therapy with radio-pharmaceuticals, special formulas designed for premature infants should be utilized.[2,44] These formulas use whey-predominant protein, mixtures of medium- and long-chain fats, glucose polymers, and lactose sugars. Vitamins, minerals, and trace elements are added to provide recommended intakes when fed at 150 to 160 ml/kg/day.[44] Soy-protein-based formulas are not appropriate for use by premature infants.[5] Likewise, other specialized infant formulas, such as casein–hydolysate formulas are not recommended for the preterm infant.

It is necessary to supplement premature infants with iron once active erythropoiesis has begun.[5] The optimal age for starting supplementation is unclear, but may be begun as soon as enteral feedings of nearly full volume are tolerated. Ferrous sulfate given to provide 2 mg/kg/day should be started in infants receiving fortified expressed breast milk. Iron-fortified formulas should be used for formula-fed infants. Unless they are receiving erythropoietin, there is no evidence that infants receiving preterm formulas benefit from additional iron supplementation beyond that provided by iron-fortified formulas.

Modular additives to provide carbohydrate (Polycose®, Ross; Moducal®, Mead Johnson), protein (Casec®, Ross; ProMod®, MJ), or fat calories (MCT or LCT oil) should be used sparingly because of concerns with nutrient ratios, sedimentation, and delivery. These products, however, may be needed in specialized situations when intake is severely limited or the energy requirement is very high. More research is needed to ensure that additives do not alter the properties of human milk or the nutrients available in formulas.

All premature infants should have ongoing nutritional assessments performed. Weight gain should be documented daily, plotted on appropriate growth charts, and weekly average gains noted. Head circumference and body length should be measured and documented weekly. Infants receiving parenteral nutrition require biweekly biochemical testing, which usually includes measurement of sodium, potassium, chloride, carbon dioxide, and blood urea nitrogen levels.

The use of specialized formulas for premature infants after hospital discharge has rapidly increased. Initial evaluations suggest that these formulas provide for more rapid growth and bone mineralization. [8] The optimal length of time these formulas should be used and their long-term benefits remain to be determined. They may be especially useful as a supplement for VLBW infants who, after discharge, primarily receive breast milk. Supplementation with one or two feedings/day of one of these formulas may substantially increase protein and mineral intake.

CYSTIC FIBROSIS

Cystic fibrosis is the most common genetic disease in Caucasians, with an incidence of about 1:2500. It is characterized by recurrent pulmonary infections, pancreatic insufficiency, and fat malabsorption. Poor growth and nutrition are very common in cystic fibrosis. Although in the past children with cystic fibrosis died in childhood, rapid improvements in therapy, especially related to antibiotic therapy, pulmonary management, and nutritional rehabilitation, have led to a marked improvement in the life expectancy for newly diagnosed children with cystic fibrosis.

Pancreatic fibrosis, ductal obstruction, and subsequent pancreatic insufficiency occur in about 85% of children with cystic fibrosis and lead to fat malabsorption. A low-fat diet was considered to be important in management for many years until Corey et al.[23] demonstrated a significant difference in survival between cystic fibrosis centers in Boston (average life expectancy of 21 years) and Toronto (average life expectancy of 30 years). The difference appeared to be related to the use of a diet higher in fat in Toronto, supported by sufficient pancreatic enzyme replacement therapy to prevent fat malabsorption and subsequent gastrointestinal side effects.

The Toronto cohort also had normal heights and weights, disproving the belief that growth retardation was an almost unavoidable aspect of cystic fibrosis.[100]

Nutritional management of cystic fibrosis is now aimed at promoting normal growth and development[74] through the intake of dietary fats and caloric, vitamin, and mineral supplements. It is interesting that the self-selected diets of children with CF are low in fat,[18] possibly due to perceived poor tolerance, if enzyme replacement is suboptimal.

Poor growth and nutrition are associated with poor pulmonary function and increased risk of mortality. Whether nutrition is the cause or the effect of poor pulmonary status is controversial,[16,91,115] although some nutritional interventions have been shown to improve growth and pulmonary status.[102]

SPECIFIC NUTRITIONAL ISSUES IN CF

Energy Metabolism and Intake

Subjects with CF may have increased energy expenditure, decreased energy intakes, and increased energy losses. These combine to make maintaining an adequate energy balance difficult.[84]

Energy expenditure is increased because of the pulmonary component of the disease as well as the frequent respiratory infections that result. Even in relatively healthy subjects with CF, the resting metabolic rate is significantly increased, perhaps involving an underlying gene mutation. Energy expenditure increases curvilinearly with declining pulmonary function.[84] Whether this is due to the increased work of breathing, or to pulmonary inflammation, subclinical pulmonary infections, or medications such as ß-agonists is not clear. Acute pulmonary exacerbations, characteristic of CF, increase the resting metabolic rate by up to 25%, an increase that can persist for up to a week after treatment is completed.

Energy intake is reduced during periods of intercurrent illness. This is due to anorexia, depression, behavioral and psychological problems, as well as other illnesses, such as gastroesophageal reflux and esophagitis, conditions that are common in advanced pulmonary disease. Fat malabsorption leads to abdominal bloating and discomfort. Therefore, patients or their physicians may choose to restrict energy intake to minimize these symptoms.

Pancreatic insufficiency leads to fat malabsorption, which in turn causes increased energy losses in the stool. Unabsorbed fats bind bile salts and prevent their reabsorption, worsening the malabsorption. Diabetes mellitus, which may occur secondary to pancreatic insufficiency in CF, further increases energy losses due to glucosuria. Some additional energy losses also occur in the sputum.

Current recommendations are that subjects with CF receive 120 to 150% of the recommended dietary energy intake for their age. If subjects have difficulty maintaining that intake, numerous interventions are available, including behavioral modifications, energy supplements, additional enteral feeds (overnight, nasogastric feeds or gastrostomy feeds) or parenteral nutrition. All of these appear to be effective in increasing weight gain.[52]

Vitamins and Minerals

Deficiency in fat-soluble vitamins has been reported in CF, secondary to fat malabsorption. Vitamin E deficiency is perhaps the most common, partly because vitamin E is very hydrophobic and most affected by fat malabsorption. Vitamin E-deficient neuropathy (characterized by reduced proprioception and reduced visual acuity, reduced deep tendon reflexes, and tremor and ataxia) is widely reported in CF. Vitamin A deficiency is also a concern because it could theoretically lead to worsening lung function. Vitamin D deficiency may also occur.

Most centers now provide vitamin A, D, and E supplements to subjects with CF. In children, typical doses would be 8000 IU vitamin A, 800 IU vitamin D, and 100 to 200 mg vitamin E. Vitamin K is typically not given, although subclinical vitamin K deficiency is reported in CF, especially if the fecal microflora (which can synthesize vitamin K) is disturbed by antibiotic treatment.

Water-soluble vitamin metabolism is usually normal in CF, although many patients take vitamin C supplements because of its antioxidant properties. Vitamin B_{12} absorption can be low, and supplementation is needed for subjects with extensive ileal resections for meconium ileus.

The self-selected diets of children with CF are low in iron and zinc.[18] However, routine iron supplementation is not advised because it can act as a pro-oxidant and may favor the growth of the pulmonary pathogen *Pseudomonas aeruginosa*. Iron deficiency anemia should, however, receive adequate treatment, especially if severe pulmonary disease coexists. Levels of zinc in the serum may be reduced, often in conjunction with poor vitamin A status. Aminoglycoside antibiotics increase magnesium losses in the urine, and hypomagnesemia has been described in CF. Copper status may also be abnormal.[75,76] Selenium levels are also low, and there has been interest in selenium supplementation because of its antioxidant actions (see the following paragraphs).

A number of studies have examined antioxidant status among subjects with CF. Generally, these show normal levels of glutathione, glutathione peroxidase, catalases, and superoxide dismutase.[57,65] Serum selenium levels may be decreased.[65] Selenium supplementation changes selenium status[80] but does not affect markers of lipid oxidation.[79]

Serum calcium is tightly regulated and therefore remains normal in children with CF.[55] Bone mineral mass is decreased in CF patients with malnutrition (weight less than 90% of ideal body weight) and appears to be inversely related to pulmonary function.[86] In well-nourished children with CF, however, bone mineral mass appears normal,[87] and the previously reported association between CF and osteopenia seems to be due to generalized poor nutrition, not some specific effect of CF on calcium or bone metabolism.[87]

Effects of Nutritional Intervention in CF

Several observational studies have shown that supplementary nasogastric or gastrostomy feedings, oral supplements, behavioral modifications,[52] or short-term supplementary parenteral nutrition[60] can improve growth in CF. Growth hormone treatment improves weight and height velocity[45] but is an invasive procedure.

CONCLUSIONS

The principal nutritional defect in CF is fat malabsorption. The pioneering work of Corey et al. has shown that growth failure is not an unavoidable consequence of CF.[23] Normal growth can be maintained in CF by ensuring an adequate caloric intake (typically 120 to 150% of age-appropriate values) through a normal or high-fat diet. Pancreatic enzyme supplements sufficient to prevent abdominal symptoms are required, bearing in mind that very high intakes of pancreatic lipase increase the risk of fibrosing colonopathy.[16] Fat malabsorption leads to malabsorption of fat-soluble vitamins (especially A and E), and supplementation with these is recommended. Despite many reports of low mineral status in CF, there is little data from well-designed, randomized clinical trials to support the use of zinc, magnesium, selenium, or other mineral supplements.

BRONCHOPULMONARY DYSPLASIA

One of the most common current pediatric problems is the care of infants with chronic pulmonary insufficiency. Although there are numerous causes of this problem, one of the most important is the disorder *bronchopulmonary dysplasia* (BPD). The etiology of BPD is unknown, although it is closely related to extreme prematurity (< 26 weeks) and prolonged ventilatory support. Children with BPD are characterized by severe, but frequently reversible, damage to their lung parenchyma. Therapy includes the use of bronchodilators, oxygen, and inhaled or systemic steroids. It is not uncommon for children with BPD to require oxygen for 1 to 2 years after birth.

ANTHROPOMETRIC ASSESSMENTS OF GROWTH IN INFANTS WITH BPD

Inadequate growth is a well-recognized complication of BPD. Increased energy utilization and decreased energy intake have been reported in infants with BPD.[113] Infants who develop BPD may also experience growth failure early in their hospital course. DeRegneir et al.[26] evaluated early growth in very low-birth-weight infants. They found that between 2 and 4 weeks of age, infants with developing BPD consumed less protein and energy, accreted less arm fat and muscle, and grew more slowly than similarly sized infants who did not develop BPD. After achieving full enteral intakes, the rates of growth of the BPD infants were similar to those of the controls, but catch-up growth did not occur.

Poor growth often continues in infants with BPD even after hospital discharge. Estimates of growth failure range from 30 to 67% during the initial post-discharge period.[53] In one study, growth failure was less in infants with BPD who were receiving oxygen after discharge than those who were not.[21] This finding suggests that adequate tissue oxygenation may ameliorate some growth limitations in infants with BPD.

Medications utilized to treat infants with BPD may also affect growth. The most important of these are corticosteroids. Gibson et al.[35] reported that knee–ankle growth, which is approximately 0.5 mm/d in premature infants, was reduced to zero after 9 d of dexamethasone therapy and did not return to predicted values until 30 d post-treatment.

Leitch et al.[59] have shown that although growth was impaired in infants with BPD during dexamethasone treatment, this difference was not related to differences in energy expenditure or intake. Dexamethasone may therefore alter the tissue composition of weight gain by increasing fat and decreasing protein accretion compared to growth when dexamethasone is not administered.

RELATIONSHIP BETWEEN ENERGY METABOLISM, BPD, AND GROWTH

Evidence of an increased energy requirement in infants with BPD comes from a series of studies using indirect calorimetry in babies with BPD. Increases of 15 to 25 kcal/kg/d are frequently reported.[25] However, significant methodological questions regarding these studies persist, and the exact level of increased energy intake required by infants with BPD remains uncertain.

One approach to nutritional management of BPD involves altering the formula composition so as to increase the fat intake relative to the carbohydrate intake. This has the potential benefit of decreasing carbon dioxide production and the respiratory quotient in infants with chronic lung disease. This approach was effective in a short-term study by Pereira et al.[77] In contrast, however, although Chessex et al.[19] also found higher carbon dioxide production associated with high fat intakes, they found no rise in oxygen consumption, but did find an increase in the transcutaneously measured partial pressure of oxygen in the blood. Longer-term studies of these and other nutritional strategies to treat BPD are ongoing.[8]

NUTRITIONAL MANAGEMENT OF INFANTS WITH BRONCHOPULMONARY DYSPLASIA

Goals should be established for the growth and biochemical monitoring of infants with chronic pulmonary insufficiency. In infants receiving enteral nutrition, it is appropriate to routinely monitor the serum albumin, calcium, phosphorus, and alkaline phosphatase activity at least every 2 weeks. Other tests, including blood urea nitrogen, electrolytes, and prealbumin, may also be evaluated as indicated. A reasonable goal of 15 g/kg/d weight gain for the small infant and 20 g/day for the infant > 35 weeks should be targeted, as well as head circumference gains of 1.0 cm/week.

The need for fluid restriction often leads to premature infants having lower than optimal energy intakes. For example, Wilson et al.[113] showed that actual energy intakes were far below optimal intakes throughout the first 8 weeks of life in small infants who developed BPD. In many cases, balanced supplementation can be achieved by mixing a powdered formula to a higher nutrient concentration. For example, 24 kcal/oz formula is frequently concentrated to 27 kcal/oz. This approach may be preferred over the use of poorly absorbed or tolerated supplements.

Even after hospital discharge, infants with BPD are at risk for ongoing growth failure. Reports have indicated very high rates of growth failure after hospital discharge. This poor growth is probably caused by the infants' ongoing increased energy utilization.[56] In addition to increased energy expenditure, poor oral feeding tolerance and skills and recurrent infections and hospitalizations may also contribute to growth failure. Reliance on high-caloric-density feedings post-discharge may not resolve these issues entirely without close nutritional supervision.[53] Singer et al.[96] reported that post-discharge, infants with BPD spent less time sucking and took in less formula per feeding than infants without BPD, whereas this difference was not observed when comparing other VLBW infants with full-term infants. Of particular interest was their observation that symptoms of maternal depression or anxiety may have caused some mothers to fail in prompting their infants to feed.

CONCLUSION

Nutritional support in children requires the joint efforts of physicians, nurses, and dietetic staff. Tools are readily available to assess the causes of growth failure in otherwise healthy children and in those with chronic illnesses. Treatment of nutritional deficiencies is not always straightforward but should be focused on the specific identified causes. This includes increasing macro- and micronutrient intake, as well as identifying and managing other factors that lead to increased nutrient losses. Benefits to improved nutritional management in children include short-term improvements in health and growth and the very real possibility of decreased long-term morbidity.

ACKNOWLEDGMENTS

This work is a publication of the U.S. Department of Agriculture (USDA)/Agricultural Research Service (ARS) Children's Nutrition Research Center, Department of Pediatrics, Baylor College of Medicine and Texas Children's Hospital, Houston, TX. This project has been funded in part with federal funds from the USDA/ARS under Cooperative Agreement number 58 to 6250–6 to 001. Contents of this publication do not necessarily reflect the views or policies of the USDA, nor does mention of tradenames, commercial products, or organizations imply endorsement by the U.S. government.

We thank Penni Davila Hicks, RD, for carefully reviewing the manuscript and for helpful editorial suggestions.

REFERENCES

1. Adamkin DH. Issues in the nutritional support of the ventilated baby. *Clin. Perinatol.* 25:79–96, 1998.
2. American Academy of Pediatrics, Assessment of nutritional status. In: *Pediatric Nutrition Handbook, 4th ed.*, edited by Ronald E. Kleinman. 165–173, 1998.
3. American Academy of Pediatrics, Committee on Nutrition. Commentary on parenteral nutrition. *Pediatrics.* 71:547–552, 1983.

4. American Academy of Pediatrics, Work Group on Breastfeeding. Breastfeeding and the use of human milk. *Pediatrics.* 100:1035–1039, 1997.

5. American Academy of Pediatrics. Nutritional needs of preterm infants. In: *Pediatric Nutrition Handbook, 4th ed.*, edited by Ronald E. Kleinman. 55–77, 1998.

6. Anderson DM and Kleigman RM. The relationship of neonatal alimentation practices to the occurrence of endemic necrotizing enterocolitis. *Am. J. Perinatol.* 8:62–67, 1991.

7. Arens R and Reichman B. Groove plated palate associated with prolonged use of orogastric feeding tubes in premature infants. *J. Oral. Maxillofac. Surg.* 50:64–65, 1992.

8. Atkinson SA. Special nutritional needs of infants for prevention of and recovery from bronchopulmonary dysplasia. *J. Nutr.* 131:942S-946S, 2001.

9. Aynsley-Green A. Hormones and postnatal adaptation of enteral nutrition. *J. Pediatr. Gastroenterol. Nutr.* 2:418–427, 1983.

10. Beale EF, Nelson, RM, Bucciarelli RL, Donnelly WH, and Eitzman DV. Intrahepatic cholestasis associated with parenteral nutrition in premature infants. *Pediatrics.* 64:342–347, 1979.

11. Beisel WR. Single nutrients and immunity. *Am. J. Clin. Nutr.* 35:417–468, 1982.

12. Bell EF, Warburton D, Stonestreet BS, and Oh W. High-volume fluid intake predisposes premature infants to necrotising enterocolitis. *Lancet.* 2:90, 1979.

13. Benjamin DR. Laboratory tests and nutritional assessment. Protein-energy status. *Pediatr. Clin. North. Am.* 36:139–161, 1989.

14. Berseth CL, Nordyke CK, Valdes MG, Furlow BL, and Go VL. Responses of gastrointestinal peptides and motor activity to milk and water feedings in preterm and term infants. *Pediatr. Res.* 31:587–590, 1992.

15. Berseth CL. Breast-milk-enhanced intestinal and somatic growth in neonatal rats. *Biol. Neonate.* 51:53–59, 1987.

16. Borowitz D. The interrelationship of nutrition and pulmonary function in patients with cystic fibrosis. *Curr. Opin. Pulm. Med.* 2:457–461, 1996.

17. Brans YW, Dutton EB, Andrew DS, Menchaca EM, and West DL. Fat emulsion tolerance in very low birth weight neonates: effect on diffusion of oxygen in the lungs and on blood pH. *Pediatrics.* 78:79–84, 1986.

18. Buchdahl RM, Fulleylove C, Marchant JL, Warner JO, and Brueton MJ. Energy and nutrient intakes in cystic fibrosis. *Arch. Dis. Child.* 64:373–378, 1989.

19. Chessex P, Belanger S, Piedboeuf B, and Pineault M. Influence of energy substrate on respiratory gas exchange during conventional mechanical ventilation of preterm infants. *J. Pediatr.* 126:619–624, 1995.

20. Chwals WJ, Lally KP, Woolley MM, and Mahour GH. Measured energy expenditure in critically ill infants and young children. *J. Surg. Res.* 44:467–472, 1988.

21. Chye JK and Gray PH. Rehospitalization and growth of infants with bronchopulmonary dysplasia: a matched control study. *J. Paediatr. Child. Health.* 31:105–111, 1995.

22. Collins JW, Hoppe M, Brown K, Eddin D, Padbury J, and Ogala ES. A controlled trial of insulin infusion and parenteral nutrition in extremely low birth weight infants with glucose intolerance. *J. Pediatr.* 118:921–927, 1991.

23. Corey M, McLaughlin FJ, Williams M, and Levison H. A comparison of survival, growth, and pulmonary function in patients with cystic fibrosis in Boston and Toronto. *J. Clin. Epidemiol.* 41:583–591, 1988.

24. Dancis J, O'Connell JR, and Holt LE. A grid for recording the weight of premature infants. *J. Pediatr.* 33:570–572, 1948.

25. Denne S. Energy expenditure in infants with pulmonary insufficiency: is there evidence for increased energy needs? *J. Nutr.* 131: 935S-937S, 2001.

26. DeRegneir RA, Guilbert TW, Mills MM, and Georgieff MK. Growth failure and altered body composition are established by one month of age in infants with bronchopulmonary dysplasia. *J. Nutr.* 126:168–175, 1996.

27. Dhanireddy R, Hamosh M, Sivasubramanian KN, Chowdhry P, Scanlon JW, and Hamosh P. Postheparin lipolytical activity and intralipid clearance in very low-birth-weight infants. *J. Pediatr.* 98:617–622, 1981.

28. Duffy B and Pencharz, P. The effects of surgery on the nitrogen metabolism of parenterally fed human neonates. *Pediatr. Res.* 20:32–35, 1986.

29. Eggert LD, Rusho WJ, MacKay MW, and Chan GM. Calcium and phosphorus compatibility in parenteral nutrition solutions for neonates. *Am. J. Hosp. Pharm.* 39:49–53, 1982.

30. Ehrenkranz RA, Younes N, Lemons JA, Fanaroff AA, Donovan EF, Wright LL, Katsikiotis V, Tyson JE, Oh W, Shankaran S, Bauer CR, Korones SB, Stoll BJ, Stevenson DK, and Papile LA. Longitudinal growth of hospitalized very low birth weight infants. *Pediatrics.* 104:280–289, 1999.

31. Ekblad H, Kero P, Takala J, Karveranta H, and Valimaki I. Water, sodium, and acid-base balance in premature infants: therapeutic aspects. *Acta Pediatr. Scand.* 76:47–53, 1987.

32. Fitzgerald KA and Mackay MW. Calcium and phosphate solubility in neonatal parenteral nutrient solutions containing Trophamine. *Am. J. Hosp. Pharm.* 43:88–93, 1986.

33. Friedman Z, Danon A, Stahlman MT, and Oates JA. Rapid onset of essential fatty acid deficiency in the newborn. *Pediatrics.* 58:640–649, 1976.

34. Frisancho AR. New norms of upper limb fat and muscle areas for assessment of nutritional status. *Am. J. Clin. Nutr.* 34:2540–2545, 1981.

35. Gibson AT, Pearse RG, and Wales JK. Growth retardation after dexamethasone administration: assessment by knemometry. *Arch. Dis. Child.* 69:505–509, 1993.

36. Goldberg RN, Chung D, Goldman SL, and Bancalari E. The association of rapid volume expansion and intraventricular hemorrhage in the preterm infant. *J. Pediatr.* 96:1060–1063, 1980.

37. Gray GE and Gray LK. Anthropometric measurements and their interpretation: principles, practices, and problems. *J. Am. Diet. Assoc.* 77:534–539, 1980.

38. Greene HL, Hambidge KM, Schanler R, and Tsang RC. Guidelines for the use of vitamins, trace elements, calcium, magnesium, and phosphorus in infants and children receiving total parenteral nutrition: report of the Subcommittee on Pediatric Parenteral Nutrient Requirements from the Committee on Clinical Practice Issues of the American Society for Clinical Nutrition. *Am. J. Clin. Nutr.* 48:1324–1342, 1988.

39. Greene HL, Phillips BL, Franck L, Fillmore CM, Said HM, Murrell JE, Moore ME, and Briggs R. Persistently low blood retinal levels during and after parenteral feeding of very low birth weight infants: examination of losses into intravenous infusion sets and a method of prevention by adding to a lipid emulsion. *Pediatrics.* 79:894–900, 1987.

40. Greer FR, McCormick A, and Loker J. Changes in fat concentration of human milk during delivery by intermittent bolus and continuous mechanical pump infusion. *J. Pediatr.* 105:745–749, 1984.

41. Griffin IJ, Pang NM, Perrring J, and Cooke RJ. Knee-heel length measurement in healthy preterm infants. *Arch. Dis. Child. Fetal Neonatal. Ed.* 81:F50-F55, 1999.

42. Gryboski JD. Suck and swallow in the premature infant. *Pediatrics.* 43:96–102, 1969.

43. Gurney JM and Jelliffe DB. Arm anthropometry in nutritional assessment: Nomogram for rapid calculation of muscle circumference and cross-sectional muscle and fat areas. *Am. J. Clin. Nutr.* 26:912–915, 1973.

44. Hall RT and Carroll RE. Infant feeding. *Pediatr. Rev.* 21:191–199, 2000.

45. Hardin DS, Stratton R, Kramer JC, Reyes de la Rocha S, Govaerts K, and Wilson DP. Growth hormone improves weight velocity and height velocity in prepubertal children with cystic fibrosis. *Horm. Metab. Res.* 30:636–641, 1998.

46. Haumont D, Deckelbaum RJ, Richelle M, Dahlan W, Coussaert E, Bihain BE, and Carpenter YA. Plasma lipid and plasma lipoprotein concentrations in low birth weight infants given parenteral nutrition with twenty or ten percent lipid emulsion. *J. Pediatr.* 115:787–793, 1989.

47. Hay WW. Nutritional needs of the extremely low birthweight infant. *Semin. Perinatol.* 15(6):482–92, 1991.

48. Hediger ML, Overpeck, MD, Maurer, KR, Kuczmarski RJ, McGlynn A, and Davis WW. Growth of infants and young children born small or large for gestational age: findings from the Third National Health and Nutrition Examination Survey. *Arch. Pediatr. Adolesc. Med.* 152:1225–1231, 1998.

49. Heird WC, Hay W, Helms RA, Storm MC, Kashyap S, and Dell RB. Pediatric parenteral amino acid mixture in low birth weight infants. *Pediatrics.* 81:41–50, 1988.

50. Hey EN and Katz G. Evaporative water loss in the newborn baby. *J. Physiol.* 200:605–619, 1969.

51. Husami T and Abumrad NN. Adverse metabolic consequences of nutritional support: micronutrients. *Surg. Clin. North Am.* 66:1049–1069, 1986.

52. Jelalian E, Stark LJ, Reynolds L, and Seifer R. Nutrition intervention for weight gain in cystic fibrosis: a meta analysis. *J. Pediatr.* 132:486–492, 1998.

53. Johnson DB, Cheney C, and Monsen ER. Nutrition and feeding in infants with bronchopulmonary dysplasia after initial hospital discharge: risk factors for growth failure. *J. Am. Diet. Assoc.* 98:649–656, 1998.

54. Kao LC, Cheng MH, and Warburton D. Triglycerides, free fatty acids, free fatty acids/albumin molar ratio, and cholesterol levels in serum of neonates receiving long-term lipid infusions: controlled trial of continuous and intermittent regimens. *J. Pediatr.* 104:429–435, 1984.

55. Kelleher J, Goode HF, Field HP, Walker BE, Miller MG, and Littlewood JM. Essential element nutritional status in cystic fibrosis. *Hum. Nutr. Appl Nutr.* 40:79–84, 1986.

56. Kurzner SI, Garg M, Bautista DB, Bader D, Merritt RJ, Warburton D, and Keens TG. Growth failure in infants with bronchopulmonary dysplasia: nutrition and elevated resting metabolic expenditure. *Pediatrics.* 81:379–384, 1988.

57. Lands LC, Grey V, Smountas AA, Kramer VG, and McKenna D. Lymphocyte glutathione levels in children with cystic fibrosis. *Chest.* 116:201–205, 1999.

58. Larsson LE, Olegard R, Ljung BM, Niklasson A, Rubensson A, and Cederblad G. Parenteral nutrition in preterm neonates with and without carnitine supplementation. *Acta. Anaesthesiol. Scand.* 34:501–505, 1990.

59. Leitch CA, Ahlrichs J, Karn C, and Denne SC. Energy expenditure and energy intake during dexamethasone therapy for chronic lung disease. *Pediatr. Res.* 46:109–113, 1999.

60. Lester LA, Rothberg RM, Dawson G, Lopez AL, and Corpuz Z. Supplemental parenteral nutrition in cystic fibrosis. *J. Parenter. Enteral Nutr.* 10:289–295, 1986.

61. Lloyd-Still JD, Smith AE, and Wessel HU. Fat intake is low in cystic fibrosis despite unrestricted dietary practices. *J. Parenter. Enteral. Nutr.* 13:296–298, 1989.

62. Lucas A, Morely R, Cole TJ, Lister G, and Leeson-Payne C. Breast milk and subsequent intelligent quotient in children born preterm. *Lancet.* 339:261–264, 1992.

63. Lucas A, Bloom SR, and Aynsley-Green A. Gut hormones and minimal enteral feeding. *Acta Pediatr. Scand.* 75:719–723, 1986.

64. Macdonald PD, Skeoch CH, Carse H, Dryburgh F, Alroomi LG, Galea P, and Gettinby G. Randomised trial of continuous nasogastric, bolus nasogastric, and transpyloric feeding in infants of birth weight under 1400g. *Arch. Dis. Child.* 67:429–431, 1992.

65. Madarasi A, Lugassi A, Greiner E, Holics K, Biro L, and Mozsary E. Antioxidant status in patients with cystic fibrosis. *Ann. Nutr. Metab.* 44:207–211, 2000.

66. Mayne PD and Kovar IZ. Calcium and phosphorus metabolism in the premature infant. *Ann. Clin. Biochem.* 28:131–142, 1991.

67. Meetze WH, Valentine C, Mcguigan JE, Canlon M, Sacks N, and Neu J. Gastrointestinal priming prior to full enteral nutrition in very low birth weight infants. *J. Pediatr. Gastroenterol. Nutr.* 15:163–170, 1992.

68. Merritt RJ and Suskind RM. Nutritional Survey of Hospitalized Pediatric Patients. *Am. J. Clin. Nutr.* 32:1320–1325, 1979.

69. Michaelsen KF, Skov L, Badsberg JH, and Jorgensen M. Short-term measurement of linear growth in preterm infants: validation of a hand-held knemometer. *Pediatr. Res.* 30:464–468, 1991.

70. Mulvihill SJ, Stone MM, Fankalsrud EW, and Debas HT. Trophic effect of amniotic fluid on fetal gastrointestinal development. *J. Surg. Res.* 40:291–296, 1986.

71. Neu J. Functional development of the fetal gastrointestinal tract. *Semin. Perinatol.* 13:224–235, 1989.

72. Neville MC, Keller RP, Seacat J, Casey CE, Allen JC, and Archer P. Studies on human lactation.I. Within-feed and between-breast variation in selected components of human milk. *Am. J. Clin. Nutr.* 40:635–46, 1984.

73. Parker P, Stroop S, and Greene H. A controlled comparison of continuous versus intermittent feeding in the treatment of infants with intestinal disease. *J. Pediatr.* 99:360–364, 1981.

74. Pencharz PB and Durie PR. Nutritional management of cystic fibrosis. *Annu. Rev. Nutr.* 13:111–136, 1993.

75. Percival SS, Bowser E, and Wagner M. Reduced copper enzyme activities in blood cells of children with cystic fibrosis. *Am. J. Clin. Nutr.* 62:633–638, 1995.

76. Percival SS, Kauwell GP, Bowser E, and Wagner M. Altered copper status in adult men with cystic fibrosis. *J. Am. Coll. Nutr.* 18:614–619, 1999.

77. Pereira GR, Baumgart S, Bennett MJ, Stallings VA, Georgieff MK, Hamosh M, and Ellis L. Use of high-fat formula for premature infants with bronchopulmonary dysplasia: Metabolic, pulmonary, and nutritional studies. *J. Pediatr.* 124:605–611, 1994.

78. Piper JM, Xenakis EMJ, McFarland M, Elliott BD, Berkus MD, and Langer O. Do growth-retarded premature infants have different rates of perinatal morbidity and mortality than appropriately grown premature infants? *Obstet. Gynecol.* 87:169–174, 1996.

79. Portal B, Richard MJ, Coudray C, Arnaud J, and Favier A. Effect of double-blind cross-over selenium supplementation on lipid peroxidation markers in cystic fibrosis patients. *Clin. Chim. Acta.* 234:137–146, 1995.

80. Portal B, Richard MJ, Ducros V, Aguilaniu B, Brunel F, Faure H, Gout JP, Bost M, and Favier A. Effect of double-blind crossover selenium supplementation on biological indices of selenium status in cystic fibrosis patients. *Clin. Chem.* 39:1023–1028, 1993.

81. Prestridge LL, Schanler RJ, Shulman RJ, Burns PA, and Laine LL. Effect of parenteral calcium and phosphorus therapy on mineral retention and bone mineral content in very low birth weight infants. *J. Pediatr.* 122:761–768, 1993.

82. Raiha NC. Biochemical basis for nutritional management of preterm infants. *Pediatrics.* 53:147–156, 1974.

83. Rayyis S, Abalavanan N, Wright L, and Carlo W. Randomized trial of "slow" versus "fast" feed advancements on the incidence of necrotizing enterocolitis in very low birth weight infants. *J. Pediatr.* 134:293–297, 1999.

84. Reilly JJ, Ralston JM, Paton JY, Edwards CA, Weaver LT, Wilkinson J, and Evans TJ. Energy balance during acute respiratory exacerbations in children with cystic fibrosis. *Eur. Respir. J.* 13:804–809, 1999.

85. Rivera A, Bell EF, and Bier DM. Effect of intravenous amino acids on protein metabolism of preterm infants during the first three days of life. *Pediatr. Res.* 33:106–111, 1993.

86. Rochat T, Slosman DO, Pichard C, and Belli DC. Body composition analysis by dual-energy x-ray absorptiometry in adults with cystic fibrosis. *Chest.* 106:800–805, 1994.

87. Salamoni F, Roulet M, Gudinchet F, Pilet M, Thiebaud D, and Burckhardt P. Bone mineral content in cystic fibrosis patients: correlation with fat- free mass. *Arch. Dis. Child.* 74:314–318, 1996.

88. Sasanow SR, Georgieff MK, and Pereira GR. Mid-arm circumference and mid-arm/head circumference ratios: standard curves for anthropometric assessment of neonatal nutritional status. *J. Pediatr.* 109:311–315, 1986.

89. Scaglia F and Longo N. Primary and secondary alterations of neonatal carnitine metabolism. *Semin. Perinatol.* 23:152–161, 1999.

90. Schanler RJ, Shulman RJ, and Lau C. Feeding strategies for premature infants: beneficial outcomes of feeding fortified human milk versus preterm formula. *Pediatrics.* 103:1150–1157, 1999.

91. Schoni MH and Casaulta-Aebischer C. Nutrition and lung function in cystic fibrosis patients: review. *Clin. Nutr.* 19:79–85, 2000.

92. Schwarz SM, Gewitz MH, See CC, Berezin S, Glassman MS, Medow CM, Fish BC, and Newman LJ. Enteral nutrition in infants with congenital heart disease and growth failure. *Pediatrics.* 86:368–373, 1990.

93. Shenai JP, Chytil F, and Stahlman MT. Vitamin A status of neonates with bronchopulmonary dysplasia. *Pediatr. Res.* 19:185–188, 1985.

94. Shenai JP, Stahlman M, and Chytil F. Vitamin A delivery from parenteral alimentation solution. *J. Pediatr.* 99:661–663, 1981.

95. Shulman RJ, DeStefano-Laine L, Petitt R, Rahman S, and Reed T. Protein deficiency in premature infants receiving parenteral nutrition. *Am. J. Clin. Nutr.* 44:610–613, 1986.

96. Singer LT, Davillier M, Preuss L, Szekely L, Hawkins S, Yamashita T, and Baley J. Feeding interactions in infants with very low birth weight and bronchopulmonary dysplasia. *J. Dev. Behav. Pediatr.* 17:69–76, 1996.

97. Skuse D, Pickles A, Wolke D, and Reilly S. Postnatal growth and mental development: Evidence for a sensitive period. *J. Child. Psychol. Psychiatry.* 35:521–545, 1994.

98. Solomon SM and Kirby DF. The refeeding syndrome: a review. *J. Parenter. Enteral Nutr.* 14: 90–97, 1990.

99. Soskeno IR and Frank L. Nutritional influences on lung development and protection against chronic lung disease. *Semin. Perinatol.* 15:462–468, 1991.

100. Sproul A and Huang H. Growth patterns in children with cystic fibrosis. *J. Pediatr.* 65:664–676, 1964.

101. Steichen JJ, Gratton TL, and Tsang RC. Osteopenia of prematurity: the cause and possible treatment. *J. Pediatr.* 96:528–534, 1980.

102. Steinkamp G and von der Hardt H. Improvement of nutritional status and lung function after long-term nocturnal gastrostomy feedings in cystic fibrosis. *J. Pediatr.* 124:244–249, 1994.

103. Sturman JA, Gaull G, and Raiha NC. Absence of cystathionase in the human fetal liver: is cysteine essential. *Science.* 169:74–76, 1970.

104. Sunehag A, Gustafsson J, and Ewald U. Very immature (<30wk) respond to glucose infusion with incomplete suppression of glucose production. *Pediatr. Res.* 36:550–555, 1994.

105. Suskind RM and Varma RN. Assessment of nutritional status of children. *Pediatr. Rev.* 5:195–202, 1984.

106. Thureen PJ, Scheer B, Anderson SM, Tooze JA, Young DA, and Hay WW. Effect of hyperinsulinemia on amino acid utilization in the ovine fetus. *Am. J. Physiol. Endocrinol. Metab.* 279:E1294–1304, 2000.

107. Valentine CJ, Hurst NM, and Schanler RJ. Hindmilk improves weight gain in low-birth-weight infants fed human milk. *J. Pediatr. Gastroenterol. Nutr.* 18:474–477, 1994.

108. Valentine CJ. Neonatal conditions. In: *Quick Reference to Clinical Dietetics*, edited by Lysen L. Maryland: Aspen Publication, 1997, 73–88.

109. Verma RP, McCulloch KM, Worrell L, and Vidyasagar D. Vitamin A deficiency and severe bronchopulmonary dysplasia in very low birthweight infants. *Am. J. Perinatol.* 13:389–393, 1996.

110. Waterlow JC. Classification and definition of protein-calorie malnutrition. *Br. Med. J.* 3:566–569, 1972.

111. Werkman SH, Peeples JM, Cooke RJ, Tolley EA, and Carlson SE. Effect of vitamin A supplementation of intravenous lipids on early vitamin A intake and status of premature infants. *Am. J. Clin. Nutr.* 59:586–92, 1994.

112. Widdowson EM. Fetal and neonatal nutrition. *Nutr. Today.* Sept/Oct:16–21, 1987.

113. Wilson DC and McClure G. Energy requirements in sick preterm babies. *Acta Paediatr. Suppl.* 405:60–64, 1994.

114. Zelikovic I, Chesney RW, Friedman AL, and Ahlfors CE. Taurine depletion in very low birth weight infants receiving prolonged total parenteral nutrition: role of renal immaturity. *J. Pediatr.* 116:301–306, 1990.

115. Zemel BS, Jawad AF, FitzSimmons S, and Stallings VA. Longitudinal relationship among growth, nutritional status, and pulmonary function in children with cystic fibrosis: analysis of the Cystic Fibrosis Foundation National CF Patient Registry. *J. Pediatr.* 137:374–380, 2000.

116. Ziegler EE, O'Donnell AM, Nelson SE, and Fomon SJ. Body composition of the reference fetus. *Growth.* 40:329–341, 1976.

117. Zlotkin SH and Buchanan BE. Meeting zinc and copper intake requirements in the parenterally fed preterm and full-term infant. *J. Pediatr.* 103:441–446, 1983.

2 Nutrition and Cardiovascular Health

Theodore A. Kotchen and Jane Morley Kotchen

CONTENTS

Despite progress in prevention, diagnosis, and treatment, cardiovascular disease remains the leading cause of death in industrialized nations. In the U. S., cardiovascular disease is responsible for more years of potential life lost before the age of 75 than any other condition and creates an immense economic burden in health care costs and lost productivity.[15]

CARDIOVASCULAR DISEASE RISK FACTORS

The Framingham Heart Study has played a vital role in defining the contribution of risk factors to coronary heart disease.[31,40] The major recognized risk factors include cigarette smoking, hypertension, high serum total cholesterol and low-density lipoprotein (LDL) cholesterol concentrations, low levels of high-density lipoprotein (HDL) cholesterol, and diabetes mellitus.[1,30] Follow-up data from individuals screened in the Multiple Risk Factor Intervention Trial (MRFIT) indicate that approximately 85% of excess risk for premature coronary heart disease can be explained by these risk factors.[85] Additional risk factors include obesity, physical

0-8493-0945-X/03/$0.00+$1.50
© 2003 by CRC Press LLC

inactivity, family history of premature coronary heart disease, hypertriglyceridemia,[69] small dense LDL particles, increased lipoprotein (a) (Lp[a]), increased serum homocysteine,[98] abnormalities in several clotting factors favoring coagulation, and serum C-reactive protein — a biomarker of inflammation.[30,47,55,69,98]

Risks ensue incrementally over a wide range of both blood pressure and serum cholesterol levels. Even among nonhypertensive persons, blood pressure levels are predictive of morbidity and mortality from stroke, heart disease, and end-stage renal disease.[43,82,85] Based on the relationship between blood pressure level and cardiovascular endpoints, it has been estimated that a reduction in diastolic blood pressure of 2 mmHg in the U.S. population would result in a 15% reduction in risk of stroke and transient ischemic attacks, and a 5% reduction in risk of coronary heart disease.[23] The relationship between total cholesterol and coronary heart disease is also continuous over a broad range of cholesterol levels, extending down to a total cholesterol level of at least 150 mg/dl (Figure 2.1), which corresponds to an LDL cholesterol level of approximately 100 mg/dl. The shape of the line defining this relationship is curvilinear.[29] Reductions of total and LDL levels are associated with reduced coronary disease risk, in patients with or without established heart disease.[1,29] Observational studies indicate that a 10% reduction in serum cholesterol is associated with a 20% reduction in risk for coronary heart disease.[48] In primary prevention, a 20% reduction of cholesterol has produced a 31% reduction in coronary morbidity and a 33% reduction in coronary mortality.[30] Despite the large body of evidence that high HDL cholesterol levels are inversely related to coronary disease risk, it has not been conclusively demonstrated that increases in HDL cholesterol induced by diet and lifestyle modification lead to reduced risk.

Several expert panels have defined *overweight* as a body mass index (BMI) of 25.0 to 29.9 kg/m^2 and *obesity* as a BMI ≥ 30.0 kg/m^2.[67] In addition to being associated with hypertension, diabetes mellitus, and dyslipidemia (high LDL cholesterol, low HDL cholesterol, and high triglyceride levels), overweight and obesity are also independently associated with coronary heart disease and stroke.[1,44,67] Of individuals with type 2 diabetes, 67% have a BMI ≥ 27.0 kg/m,^{2} and 46% have a BMI ≥ 30.0 kg/m^2.[67] Sixty percent of hypertensives are more than 20% overweight.[44] The risk of death from cardiovascular disease increases throughout the range of moderate and severe overweight for both men and women.[21] Compared with lean men, men with BMI of 25 to 29 kg/m^2 reportedly have a 70% greater risk of coronary heart disease, whereas men with BMI of 29 to 33 kg/m^2 have almost a threefold greater risk of coronary heart disease.[74] The cardiovascular risks of obesity are particularly related to increases of abdominal fat, or centripetal obesity.

The U.S. population is a high-risk population for cardiovascular disease. More than 50% of adult Americans are overweight or obese; 25% of U.S. women and 20% of U.S. men are obese.[67] Nearly 25% of children are overweight or obese. The prevalence of obesity has increased more than 50% over the past 10 to 15 years. A majority of adults in the U.S. have a blood cholesterol level high enough to increase the risk of coronary heart disease. Based on data from the second National Health and Nutrition Examination Survey (NHANES II), the mean cholesterol level for the adult U.S. population is approximately 210 to 215 mg/dl. Approximately 30% of adult Americans have cholesterol levels of 200 to 239 mg/dl, and more than 25%

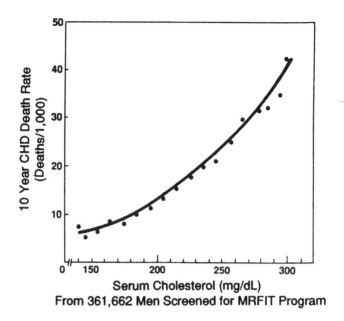

FIGURE 2.1 Relationship between serum cholesterol level and CHD death rate (MRFIT screenees). (From National Cholesterol Education Program. Report of the Expert Panel on Population Strategies for Blood Cholesterol Reduction. NIH publication No. 93–3047, 1993.)

have levels ≥ 240 mg/dl (National Cholesterol Education Program Report). According to the Third National Health and Nutrition Examination Survey (NHANES III) (1988–1991), 24% of the U.S. population was classified as having hypertension.[20] However, between 1971 and 1991, national health examination surveys documented a downward trend in blood pressure levels and the prevalence of hypertension in the U.S.[19] Adoption of a healthier lifestyle may have contributed to this favorable trend. Nevertheless, not all subgroups have benefited equally, particularly African Americans. An estimated 15.6 million people in the U.S., 8% of men and women aged 20 years or older, have diabetes mellitus, and type 2 diabetes accounts for 90 to 95% of these individuals.[67]

Risk factors for cardiovascular disease tend to cluster within individuals. In the Framingham cohort, at baseline, clustering of three or more risk factors occurred at twice the rate predicted by chance.[40] In adolescents and adults, individuals with higher blood pressures tend to have higher serum concentrations of total cholesterol, triglycerides, and apolipoprotein B and lower HDL cholesterol.[64] It has been estimated that 30% to 40% of adults younger than 55 years of age with blood pressure values >140/90 mmHg have serum cholesterol concentrations >240 mg/dl, compared with a prevalence of approximately half that in age-matched normotensive subjects.[66] Likewise, 46% of individuals with blood cholesterol levels >240 mg/dl have blood pressures >140/90 mmHg. The clustering of risk factors may have a genetic basis, and resistance to insulin-stimulated glucose uptake may be the link between hypertension and dyslipidemia.

Epidemiologic observations, including results of the Framingham Study, clearly document the additive risk associated with an increase in the number of risk factors.[40] Observations from autopsy studies also document a strong relation between cardiovascular disease risk factors and atherosclerosis in young people.[89] The severity of asymptomatic coronary and aortic atherosclerosis is related to the number of premortem cardiovascular risk factors.[13] Conversely, according to a recent report of cohort studies conducted in 366,559 young and middle-aged men and women, persons with a low cardiovascular disease risk profile (serum cholesterol <200 mg/dl, blood pressure ≤ 120/80 mmHg, and no cigarette smoking) have a 72 to 85% lower mortality from cardiovascular disease compared with those with one or more of these three risk factors.[83]

IMPACT OF DIET ON RISK FACTORS

Table 2.1 lists dietary factors that may influence blood pressure levels, determined on the basis of animal studies, observational studies, and clinical trials. Observational studies describing the relationship between single nutrients and blood pressure levels are hampered by the difficulty of measuring nutrient consumption and by the multicollinearity among nutrients. Concurrence between results of observational studies and clinical trials provides the most compelling evidence for a nutrient–blood pressure interaction.

BLOOD PRESSURE

Primarily over the past two decades, observational studies and clinical trials have consistently documented a modest effect of dietary NaCl on blood pressure.[45] In cross-sectional, population-based studies, age-related increases of blood pressure are augmented by dietary sodium consumption.[78]

In three meta-analyses of randomized trials, reduced NaCl intakes (mean or median reductions of sodium excretion in the range of 75 to 125 mEq/day) have been shown to result in systolic and diastolic blood pressure reductions of approximately 4 to 6 mmHg and 2 to 4 mmHg, respectively, in hypertensive individuals and lesser reductions in nonhypertensive individuals.[24,27,61] Consistent with results of earlier trials,[38,84] in the Trials of Hypertension Prevention (TOHP), NaCl reduction (sodium intake of 80 mEq/day), alone or combined with weight loss, decreased the incidence of hypertension by 20% over a 3- to 4-year period.[92] In the Trials of Non-Pharmacologic Intervention in the Elderly (TONE), a reduced NaCl intake (alone or combined with weight loss) reduced blood pressure and the need for antihypertensive medications in the elderly.[100]

Individuals vary in their responsiveness to NaCl, and the full expression of NaCl-sensitive hypertension depends on the concomitant intake of both sodium and chloride.[14] In both experimental models and humans, blood pressure is not increased by a high dietary sodium intake provided as non-chloride salts of sodium. In the usual diet, however, most sodium is consumed as NaCl.

There continues to be some lingering controversy about the benefits of NaCl restriction. It has been argued that the blood pressure reduction achieved by a reduced

TABLE 2.1
Dietary Factors that May Influence Blood Pressure

Increase Blood Pressure	Decrease Blood Pressure
Sodium chloride (8.3 g/day = 140 mEq/day)	Potassium (2.7 g/day = 69 mEq/day)
Alcohol (8.8 g/day)	Calcium (767 mg/day)
Cholesterol (298 mg/day)	Magnesium (283 mg/day)
Saturated fat (26.3 g/day)	Protein (79 g/day)
Carbohydrate (254 g/day)	

Note: Numbers in parentheses refer to average daily consumption by US adults, as determined from the NHANES III survey.

NaCl intake is modest and may not be clinically significant. However, on a population basis, this modest blood pressure reduction may have an important impact on the incidence of cardiovascular disease. More recently, it has been argued that blood pressure is only an intermediate or surrogate endpoint and that the true measure of the benefits of reduced NaCl intake is to determine its impact on cardiovascular disease morbidity and mortality. In a prospective study of almost 3000 patients treated for hypertension, Alderman et al. reported an inverse association between sodium excretion and incidence of myocardial infarction.[4] Subsequently, in an analysis based on data from the NHANES I Epidemiologic Follow-up Study, Alderman et al. again reported an inverse association between sodium intake and mortality from both cardiovascular disease and from all causes.[3] Both of these reports have been criticized for serious methodologic flaws. In apparent contrast to the Alderman reports, analyses from the Multiple Risk Factors Intervention Trial (MRFIT) did not find an inverse association between sodium intake and incidence of myocardial infarction.[80] More recently, based on another analysis of 19-year follow-up experience of participants in NHANES I, He et al. reported that a high NaCl intake is strongly and independently associated with an increased risk of cardiovascular disease and all-cause mortality in overweight persons.[34] Similarly, in a prospective cohort study in Finnish men and women, a higher NaCl intake was again associated with increased risk for coronary heart disease, cardiovascular disease, and all-cause mortality.[95] Consistent with these observations, two additional reports have also noted lower cardiovascular mortality rates in people with lower urinary sodium excretion.[77,94] Salt intake may also be a determinant of left ventricular mass,[60] and it has been suggested that sodium sensitivity of blood pressure is an independent risk factor for cardiovascular disease.[62]

Among individuals, there is considerable variability of blood pressure responses to NaCl. Family studies suggest a heritable contribution to salt sensitivity of blood pressure. Recent preliminary evidence suggests that the M235T variant of the angiotensinogen gene may have a modest influence on the blood pressure response to NaCl reduction and to weight loss.[37] It is likely that additional genetic markers of salt sensitivity will be identified. Similar to blood pressure itself, it is also likely that the magnitude of the effect of dietary NaCl on blood pressure will reflect the culmination of a number of genetic variants.

NaCl is not the only dietary constituent that affects blood pressure. Observational studies document inverse associations of blood pressure with dietary potassium, calcium, and magnesium consumption.[45] On the basis of these observations, however, it is difficult to relate blood pressure levels to specific nutrients because of strong correlations among dietary intakes of potassium, magnesium, fiber, and to a lesser extent, calcium. The inverse relationship between calcium intake and blood pressure is more convincing at low levels of calcium consumption, i.e., 300 to 600 mg/day. Low intakes of potassium and calcium may augment the effect of NaCl on blood pressure.

A number of small clinical trials have reported the effects of potassium or calcium supplements on blood pressure, and the effects are modest at best. Based on two reported meta-analyses, the magnitude of the blood pressure lowering effect of potassium supplementation (60 to 120 mEq/day) is greater in hypertensive than in nonhypertensive persons and is more pronounced in persons consuming a high NaCl diet.[22,102] Meta-analyses of controlled clinical trials have shown that calcium supplementation (1000 to 2000 mg/day) results in small but significant reductions of systolic, but not diastolic, blood pressure, and only in hypertensive individuals.[5,18] The effect of supplemental calcium in the diet is at least as great as non-dietary supplementation.[28] Calcium supplementation may preferentially lower blood pressure in patients with NaCl-sensitive hypertension. The overall antihypertensive response to magnesium supplementation among hypertensive individuals is small, and several trials have failed to show a significant effect of magnesium supplementation on blood pressure.[45]

In a number of population surveys, moderate alcohol intake has been associated with reduced cardiovascular events.[1] A relationship between alcohol consumption (≥ 3 drinks/day) and blood pressure level has also been documented in observational studies.[1,44] Further, several small trials have demonstrated that reductions in alcohol intake among heavy drinkers lower blood pressure in both normotensive and hypertensive individuals.[1,44,106]

Limited evidence suggests that other macronutrients may also influence blood pressure. In the MRFIT trial, blood pressure was positively related to dietary cholesterol, saturated fatty acids, and starch; blood pressure was inversely related to dietary protein and the ratio of dietary polyunsaturated to saturated fatty acids.[79] Other observational studies provide inconsistent results about an inverse association between protein intake and blood pressure.[32,68,79,81] In a cross-sectional study based on NHANES III data, a higher protein intake, as well as a higher calcium intake, attenuate the age-related increase of blood pressure.[32] Intervention studies have generally found no significant effect of protein intake on blood pressure.[68] Compared to patients consuming a "usual" protein diet (1.3 g protein/kg/day), dietary protein restriction (0.58 g protein/kg/day) may slow the decline of renal function in patients with moderate renal insufficiency (glomerular filtration rate 25 to 55 ml/min/1.73m^2), but not in patients with severe renal insufficiency (glomerular filtration rate 13 to 24 ml/min/1.73m^2).[46,52] According to two meta-analyses, fish oil supplementation lowers blood pressure in untreated hypertensive individuals.[9,63]

Rather than focusing on any single nutrient, advice about nutrition and blood pressure more appropriately addresses the impact of the overall diet. In a randomized, multicenter study, the Dietary Approaches to Stop Hypertension (DASH) trial evaluated the effects of three dietary patterns on blood pressure: a) a control diet with potassium, calcium, and magnesium levels close to the 25th percentile of U.S. consumption; b) a diet rich in fruits and vegetables; and c) a "combination" diet rich in fruits, vegetables, and fat-free or low-fat dairy products.[10] The study intervention diets were relatively high in potassium, calcium, and magnesium. Compared with the control diet, the study diets were also higher in fiber, protein, carotenoid, and folate, and lower in total fat, saturated fat, and cholesterol. NaCl content was equivalent in all three (7.5 g/day). Both diets lowered blood pressure, and a greater effect was observed with the combined diet. Among nonhypertensive individuals, the combined diet reduced systolic and diastolic blood pressures by 3.5 and 2.1 mmHg, respectively. Corresponding blood pressure reduction in persons with stage 1 hypertension were 11.4 and 5.5 mmHg. African Americans had greater blood pressure reductions than non-African Americans. Although not designed to assess the impact of specific nutrients on blood pressure, the DASH trial convincingly affirms the importance of overall diet for blood pressure control.

The DASH 2 study extended these observations by evaluating the effects of three levels of NaCl intake (target intakes of 150, 100, and 50 mEq/day) on blood pressure in subjects consuming either the control diet or the combination diet in DASH 1.[76] In both normotensive subjects and in subjects with stage 1 hypertension, the DASH diet lowered blood pressure at each of the three levels of NaCl intake. Reduction of NaCl intake below 100 mEq/day augmented the effect of the DASH diet on blood pressure and also lowered blood pressure on the control diet.

The DASH diet also lowered plasma concentrations of homocysteine, another risk factor for coronary artery disease. One of the pathways for homocysteine metabolism is dependent on folic acid. During the 8-week intervention phase, consumption of the combination DASH diet resulted in a significant reduction in serum homocysteine concentrations compared to the control diet.[8] Change in serum homocysteine was inversely associated with change in serum folate.

Serum Cholesterol

The major dietary components that raise LDL cholesterol are saturated fatty acids, trans-unsaturated fatty acids, and to a lesser extent, cholesterol.[1] Dietary factors that lower LDL cholesterol include polyunsaturated fatty acids, monounsaturated fatty acids (when substituted for saturated fatty acids), and, to a lesser extent, soluble fiber, plant sterols/stanols, and soy protein (Table 2.2). In addition, sustained weight reduction can lower LDL levels in some individuals. Average LDL cholesterol levels in the American population have progressively decreased as average saturated fat intake has declined from approximately 20% to approximately 13% of energy intake over the past several decades.[1] However, in part, the percent reduction of dietary saturated fat is related to increased total caloric consumption. On average, further reduction of saturated fat intake to 10% of total energy consumption will lower LDL cholesterol approximately 7 to 9%;[88] reduction of saturated fat to < 7% of energy

TABLE 2.2
Dietary Factors that May Influence Serum
LDL Cholesterol

Increase LDL	Decrease LDL
Saturated fat	Polyunsaturated fatty acids
Trans unsaturated fatty acids	Monosaturated fatty acids
Cholesterol	Soluble fiber
	Plant sterols/stanols
	Soy protein

intake will lower cholesterol 10 to 20%, and more aggressive dietary reductions of fat intake may result in greater reductions of total and LDL cholesterol.[71]

Individuals vary in their responsiveness to dietary interventions designed to lower cholesterol. Results of animal studies suggest a significant genetic component to the impact of dietary interventions on serum lipoproteins. In humans, the contribution of several candidate genes to serum lipid responsiveness to dietary interventions (including APO lipoproteins E, A4, A1, C3, and B, lipoprotein lipase, and cholesterol ester transfer protein) has been evaluated under different experimental conditions.[70] Because of apparently conflicting results, further studies in additional populations will be required.

Dietary consumption of trans fatty acids has a detrimental effect on serum lipid levels and on cardiovascular disease risk.[53] Trans fatty acids are naturally present at low levels in meat and dairy products as a result of bacterial fermentation in ruminant animals. They are also formed in varying amounts during the hydrogenation of oil, a process used in the manufacture of margarines and vegetable shortening. Consumption of products that are low in trans fatty acids and saturated fat has beneficial effects on serum lipoprotein cholesterol levels.[54]

A growing number of metabolic studies have reported total cholesterol reductions of 10 to 15% with diets enriched with fiber from oats, beans, or psyllium, but these diets were also reduced in fat.[1] In a meta-analysis controlling for the impact of a fat-modified diet alone, fiber from two servings of oats enhanced cholesterol reduction by 2 to 3% beyond that achieved by fat modification.[75] Studies of fiber supplements containing psyllium have reported reductions of up to 15% in LDL cholesterol levels.[97] Foods that contain stanol/sterol esters (plant sterols) also decrease plasma cholesterol by decreasing cholesterol absorption from the gut.[1,49,99] Intakes of 2 to 3 g/day have been reported to decrease total and LDL cholesterol levels by 9 to 20%. Plant sterols appear to have little effect on HDL cholesterol or triglycerides.[35] Although consumption of plant sterols and stanols appears to be safe, a potential concern is that they may lower plasma concentrations of carotenoids.[73] Consumption of soy protein in place of animal protein also lowers total cholesterol, LDL cholesterol, and triglycerides, without affecting HDL cholesterol. In double-blind, placebo-controlled trials, in mildly hypercholesterolemic individuals following the National

Cholesterol Education Program (NCEP) Step 1 diets, an intake of 20 to 50 g/day of soy protein has been shown to reduce LDL cholesterol.[1]

BODY WEIGHT

Maintenance of a healthy body weight depends on matching energy intake (calories) with energy requirements. Regular aerobic physical activity, adequate to achieve at least a moderate level of physical fitness, is an integral management strategy for weight reduction, maintenance of weight reduction, and prevention of weight gain.[1,91] Blood pressure can be lowered with moderately intense physical activity (40 to 60% of maximal oxygen consumption), such as 30 to 45 minutes of brisk walking most days of the week. In one prospective, controlled trial, the National Cholesterol Education Program Step 2 diet (< 30% total fat, < 7% saturated fat, and < 200 mg cholesterol/day) failed to reduce elevated LDL cholesterol levels in both men and women. When the diet was combined with aerobic exercise, however, LDL cholesterol was significantly reduced, with no adverse effects on HDL cholesterol.[87] Regular physical activity may also prevent the decrement of HDL cholesterol levels that usually results from a low-fat diet in overweight men and women.[105]

In longitudinal studies, there is a direct correlation between change in weight and change in blood pressure over time, even when dietary salt intake is held constant.[44] In hypertensive individuals, it has been estimated that a mean reduction in body weight of 9.2 kg is associated with a 6.3 mmHg reduction in systolic blood pressure and a 3.1 mmHg reduction in diastolic blood pressure.[59]

Regular aerobic exercise may also lower blood pressure.[101] In overweight subjects with impaired glucose tolerance, the development of type 2 diabetes may be prevented by a combination of lifestyle interventions that include weight loss, reduction of intakes of total fat and unsaturated fat, increasing intake of fiber, and physical activity.[96] The recently terminated Diabetes Prevention Program, funded by the NIH, found that lifestyle intervention (with the goal of reducing weight by 7% through diet and exercise) reduced the risk of type 2 diabetes in people at high risk for diabetes by 58% during an average 3-year follow-up. These lifestyle changes have also been shown to have a favorable impact on blood pressure and serum lipid levels.[25] Weight loss may also increase HDL cholesterol levels in moderately overweight men.[104]

MULTIPLE RISK FACTORS

In a randomized, prospective clinical trial, the impact of a controlled meal plan designed to reduce cardiovascular disease risk was recently evaluated in men and women with essential hypertension, dyslipidemia, type 2 diabetes, or any combination of these diseases.[58] The meal plan met the National Academy of Sciences daily nutritional guidelines for the intakes of sodium, total and saturated fat, cholesterol, refined sugars, fiber, and complex carbohydrates. Over a 10-week follow-up period, compared to participants consuming a self-selected diet, the intervention diet resulted in improvements in multiple cardiovascular risk factors, including hypertension, dyslipidemia, hyperinsulinemia, and excessive body weight.

IMPACT OF DIET AND CALORIC BALANCE ON CARDIOVASCULAR DISEASE

Observational evidence consistently demonstrates a lower incidence of coronary heart disease among groups of people with the highest intake of fruits, vegetables, and grains.[97] The protective constituents of such a dietary pattern are unidentified.[103] Although this dietary pattern is typically lower in total fat, saturated fat, and cholesterol, the protective effects appear to transcend lipid lowering. In addition, a growing body of evidence indicates that foods rich in omega-3 polyunsaturated fatty acids confer cardioprotective effects beyond those that can be ascribed to improvements in blood lipoproteins.[97] Food sources of omega-3 fatty acids include fish, especially fatty fish such as salmon, as well as plant sources such as flaxseed and flaxseed oil, canola oil, soybean oil, and nuts. Observational and clinical studies suggest that an increased intake of fish oil may reduce the risk of coronary heart disease and stroke.[12]

Several prospective studies confirm these observations. In a cohort of 44,875 men, aged 40 to 75 years without diagnosed cardiovascular disease, overall dietary patterns were predictive of the risk of coronary heart disease over 8 years.[36] Based on data obtained from food frequency questionnaires, a diet high in vegetables, fruit, legumes, whole grains, fish, and poultry and low in red meat, processed meat, high-fat dairy products, and refined grains reduced the risk of coronary heart disease in that cohort. Similarly, in a cohort of 42,254 women, with a median follow-up of 5.6 years, participants reporting dietary patterns that included fruits, vegetables, whole grains, low-fat dairy products, and lean meats were found to have a lower risk of all-cause mortality, as well as lower risk of mortality from cancer, coronary heart disease, and stroke.[41]

Regular physical activity (e.g., walking >20 km/wk) decreases stroke incidence and overall mortality in men.[33,50] Results of the Nurses' Health Study further document the beneficial impact of healthy lifestyles, including diet, on prevention of coronary heart disease.[86] In that study, 84,129 middle-aged women were monitored for more than 14 years. Those women who did not smoke cigarettes, were not overweight, maintained a "healthful diet," exercised moderately or vigorously for half an hour a day, and consumed alcohol moderately had an incidence of coronary events that was more than 80% lower than that in the remainder of the population. Specifically, concerning the diet, subjects were considered to be at low risk if they scored in the highest 40% of the cohort on a composite measure based on a diet low in trans fat and glycemic load (which reflects the extent to which diet raises blood glucose levels), high in cereal fiber, marine n-3 fatty acids, and folate, and with a high ratio of polyunsaturated to saturated fat.

Several prospective studies, including the Nurses' Health Study, have reported that low daily intakes of calcium, potassium, and magnesium are each associated with an increased risk of ischemic stroke in both men and women; diets rich in potassium, magnesium, and cereal fiber reduce the risk of stroke.[2,11,39,42,51]

Antioxidants may inhibit multiple proatherogenic and prothrombotic oxidative events in the artery wall. Antioxidants are produced endogenously and are also derived from the diet. Dietary antioxidants include vitamin C (ascorbic acid), vitamin E (alpha-tocopherol), and vitamin A (beta-carotene). Observational and experimental studies suggest that the amount of antioxidants ingested in food is associated with a lower risk of coronary heart disease and atherosclerosis.[93] The data are strongest for vitamin E and carotenoids, whereas the results regarding other antioxidants such as vitamin C are equivocal.[1,90] However, several primary prevention clinical trials have failed to demonstrate an effect of antioxidant supplementation on cardiovascular outcomes.[93] Results of secondary prevention trials evaluating the potential health benefits of antioxidants have been inconsistent.

DIETARY GUIDELINES

Interventions for both the primary and secondary prevention of cardiovascular disease should address overall risk, rather than any single risk factor. Additionally, it would seem prudent to develop strategies to encourage children and adults of all ages to adopt healthy lifestyles, e.g., weight management, a healthful diet, promotion of physical activity, and restriction of smoking.

Expert panels periodically publish guidelines for the prevention, evaluation, and management of obesity, hyperlipidemia, hypertension, and diabetes.[6,26,67,91] The most recent American Heart Association dietary guidelines for reducing cardiovascular disease risk in the general population (including all individuals above 2 years of age) are presented in Table 2.3.[1] More stringent guidelines are recommended for high-risk individuals, including those with preexisting cardiovascular disease, elevated LDL cholesterol, diabetes mellitus, and insulin resistance. For individuals with LDL cholesterol concentrations above target levels (target levels vary, depending on the presence or absence of coronary heart disease and the number of risk factors present), the Third Report of the National Cholesterol Education Program (NCEP) Expert Panel on Detection, Evaluation, and Treatment of High Blood Cholesterol in Adults recommends reducing saturated fats to less than 7% of total calories and dietary cholesterol intake to less than 200 mg/day.[26] The report also recommends consumption of 20 to 30 grams of soluble fiber per day. The NCEP panel allows total fat to range from 25 to 35% of total calories, provided consumption of saturated fats and trans fatty acids is low.[65] The rationale for this is that a higher intake of total fat, mostly in the form of unsaturated fat, may reduce triglycerides and raise HDL cholesterol in persons with the insulin resistance syndrome. In addition, these expert panels provide target goals for plasma concentrations of LDL cholesterol, triglycerides, and blood pressure, as well as recommendations for the pharmacologic treatment of dyslipidemia and hypertension.

Fruits, vegetables, and whole grains are rich in antioxidants, plant sterols, and fiber. Currently, based on available information related to both efficacy and safety, the AHA Guidelines do not recommend additional antioxidant or plant sterol supplementation beyond that in the diet for either the primary or secondary prevention of cardiovascular disease. Fiber from natural dietary sources is also preferred to fiber supplements to avoid a myriad of gastrointestinal disturbances associated with the supplements.

TABLE 2.3
Summary of American Heart Association Dietary Guidelines

1. Maintain a healthy body weight by avoiding excess total energy intake and engaging in a regular pattern of physical activity.
2. Restrict total fat to <30% of total energy consumption, limit dietary saturated fat to <10% of energy, and cholesterol to < 300 mg/day.
3. Consume at least 2 fish servings/week (particularly fatty fish).
4. Consume a diet with a high content of vegetables and fruits (5 or more servings/day), and low-fat dairy products.
5. Limit salt intake (<6 g/day).
6. For those who drink, limit alcohol (no more than 2 drinks/day for men and 1 drink/day for women).
7. Consume a variety of grain products, including whole grains (6 or more servings/day).

Adapted from AHA Dietary Guidelines. Revision 2000: a statement for healthcare professionals from the Nutrition Committee of the American Heart Association. *Circulation.* 102:2284–2299, 2000.

The metabolism of homocysteine is dependent on several vitamins, including folic acid, cyanocobalamin (vitamin B_{12}), and pyridoxine (vitamin B_6). Since January 1, 1998, the U.S. Food and Drug Administration (FDA) has required all "enriched" cereal grains to be fortified with folic acid at a concentration of 1.4 mg/kg to prevent neural tube birth defects. Whether this level of fortification will lower homocysteine concentrations in patients with coronary artery disease remains to be determined. Cereal products fortified with 4 to 5 times the levels of fortification required by the FDA have been reported to lower homocysteine levels in patients with coronary artery disease.[56]

IMPLEMENTATION STRATEGIES

Implementation of these recommendations continues to be a challenge. National nutrition data (Continuing Survey of Food Intakes by Individuals) show that only approximately 30% of Americans are consuming diets that meet dietary fat guidelines.[72] In a recent survey, 40% of respondents indicated that they know they should be eating a healthy diet, but do not do it.[7] Effective strategies will require a multifaceted approach for dealing with the population as a whole, targeted subgroups, and individuals with cardiovascular disease risk factors, or clinically evident cardiovascular disease.

INDIVIDUAL LEVEL

To assist individuals in adhering to dietary guidelines, a practical approach is to focus on appropriate food choices that should be included in an overall dietary program. This includes choosing an overall balanced diet with emphasis on fruits and vegetables (five or more servings/day), a variety of grain products (including whole grains), avoiding high-salt foods, and maintaining an appropriate body weight by balancing total energy intake with energy expenditure. For the primary prevention of cardiovascular disease, it is important that healthy dietary patterns be established at a young age.

In adults, patterns of dietary intake are difficult to alter because they tend to be acquired early in life and are strongly reinforced by cultural and environmental factors. Education alone rarely produces lasting change. However, research related to behavioral change has identified approaches and strategies that can be successfully employed to assist individuals make and maintain difficult lifestyle changes.[16,17] In planning for the implementation of changes, specific, reasonable, and acceptable goals for dietary intake should be specified. Goals should be based on specific guidelines. These can provide a clear indication of expectations and can subsequently be used in determining progress in dietary changes.

Preliminary steps prior to the adoption of dietary changes are key, but often overlooked. These include assessments of the individual's readiness to make dietary changes, current eating patterns and dietary intake, and the extent of family or social support for the recommended dietary changes.[57] Based on research studies focused on the processes involved in behavioral changes, Prochaska and colleagues have suggested a model of the stages through which people pass in making lifestyle changes.[17] These stages include pre-contemplation (no consideration of making changes), contemplation (consideration of changes), preparation (planning for change), action (implementing changes), and maintenance of lifestyle changes. This model has been successfully employed to identify individuals who are interested and ready to make lifestyle changes and to influence those who are not.

An initial diet assessment provides information for recommended changes, and the assessment serves as a baseline comparison to measure change. Diet diaries, 24-hour food recalls, and food frequency lists are often used for dietary assessment and for self-monitoring of dietary changes. Even though these methods are subject to conscious or unconscious reporting bias, they do provide a means for patient feedback on eating patterns and allow the provider to identify problem areas and gain insight into how dietary changes might be implemented. Initial assessment of the social setting and of available social support is also important. Social support for recommended dietary changes is a major predictor of success in dietary change. Assessment of the social context in the environment includes the identification of the individual who shops and prepares food and determination of whether other household members are amenable to dietary changes. If meals are frequently taken away from home, this must be addressed in the plan to change the patient's diet.

Because physicians may not have sufficient time to spend with patients, the role of the physician may be to recommend dietary changes and assign the implementation of the dietary change program to a dietitian. The educational effort must result in skill building: reading food labels, recognizing which food selections are appropriate, learning to prepare foods that meet the recommended guidelines, self-monitoring and problem solving to control stimuli, and dealing appropriately with difficult situations that arise. Finally, it is important that the practitioner track the patient's progress and provide appropriate feedback that indicates which changes in health status may be attributable to changes in the diet.

POPULATION LEVEL

At the national level, it is essential that various expert panels provide consistent and unambiguous messages concerning "healthy" diets and other lifestyles designed to reduce the risk of cardiovascular disease in the general population. Recommendations must be practical and achievable. Dietary guidelines should be presented to the public in terms of overall diet and food choices, rather than as percentages or grams and milligrams. The U.S. Dietary Guidelines and the Food Guide Pyramid of the U.S. Department of Health and Human Services, the U.S. Department of Agriculture, and the American Heart Association are increasing their efforts to do this. Environmental modifications may also facilitate the adoption of healthy lifestyles. For example, simplified food labeling will assist consumers in making informed choices about food selection at the supermarket. City planners and employers might create additional opportunities that will allow individuals to incorporate physical activity into their daily schedules. In addition, it is incumbent on the food industry to develop products that will be of value in the prevention of cardiovascular disease. This may require sensory research and new technologies to create good-tasting products that meet various health needs.

To facilitate change, different and culturally sensitive approaches need to be targeted to special populations, e.g., children, the elderly, and minorities. More intensive interventions may be developed for individuals with cardiovascular disease risk factors or clinically evident cardiovascular disease. Attempts to affect individual health behaviors, including diet and physical activity, generally emphasize knowledge-based approaches to changing behavior. Development of strategies that are based on an understanding of the process of behavior change may further assist in motivating people to make enduring lifestyle changes.

SUMMARY

Cardiovascular disease continues to be a major cause of morbidity and mortality in the U.S., although many of the recognized risk factors for cardiovascular disease are modifiable through alterations of diet and physical activity. Several prospective studies have also documented the beneficial impact of lifestyle interventions on the incidence of cardiovascular disease itself. Nevertheless, the U.S. population remains at high risk for cardiovascular disease. Despite some progress, increased efforts are required to facilitate the widespread adoption of healthier lifestyles for the primary and secondary prevention of cardiovascular disease..Table 2.4 lists several areas for future research that may eventually further improve the ability to reduce cardiovascular disease risk through appropriately designed and targeted dietary interventions.

TABLE 2.4
Research Needs

1. Develop, implement, and evaluate effective strategies for facilitating the adoption of behaviors and lifestyle modifications designed to reduce cardiovascular disease risk.
2. Identify specific components of the "vegetarian" diet that favorably affect cardiovascular disease risk.
3. Determine physiologic mechanisms by which micronutrients affect cardiovascular disease risk.
4. Identify specific genes and genetic variants that affect cardiovascular disease risk directly and indirectly by their interaction with nutrients.

REFERENCES

1. AHA Dietary Guidelines. Revision 2000: a statement for healthcare professionals from the Nutrition Committee of the American Heart Association. *Circulation.* 102:2284–2299, 2000.
2. Abbott RD, Curb D, Rodriguez BL, Sharp DS, Burchfiel CM, and Yano K. Effects of dietary calcium and milk consumption on risk of thromboembolic stroke in older middle-aged men: The Honolulu Heart Program. *Stroke.* 27:813–818, 1996.
3. Alderman MH, Cohen H, and Madhaven S. Dietary sodium intake and mortality: the National Health and Nutrition Examination Survey (NHANES I). *Lancet.* 351:781–785, 1998.
4. Alderman MH, Madhaven S, Cohen H, Sealey JE, and Laragh JH. Low urinary sodium is associated with greater risk of myocardial infarction among treated hypertensive men. *Hypertension.* 25:1144–1152, 1995.
5. Allender PS, Cutler JA, Follmann D, Cappuccio FP, Pryer J, and Elliott P. Dietary calcium and blood pressure: a meta-analysis of randomized clinical trials. *Ann. Intern. Med.* 124:825–831, 1996.
6. American Diabetes Association. Clinical Practice Recommendations, 2000. *Diabetes Care.* 23(Suppl):S1–S116,2000.
7. American Dietetic Association. Nutrition and you: trends 2000. Available at http://www.eatright.org/pr/2000/010300a.html.
8. Appel LJ, Miller ER, Jee SH, Stolzenberg-Solomon R, Lin P-H, Erlinger T, Nadeau MR, and Selhub J. Effect of dietary patterns on serum homocysteine: Results of a randomized, controlled feeding study. *Circulation.* 102:852–857, 2000.
9. Appel LJ, Miller ER, Seidler AJ, and Whelton PK. Does supplementation of diet with "fish oil" reduce blood pressure? A meta-analysis of controlled clinical trials. *Arch. Intern. Med.* 153:1429–1438, 1999.
10. Appel LJ, Moore TJ, Obarzanek E, Vollmer WM, Svetkey LP, Sacks FM, Bray GA, Vogt TM, Cutler JA, Windhauser MM, Lin P-H, and Karanja N. A clinical trial of the effects of dietary patterns on blood pressure. *N. Engl. J. Med.* 336:1117–1124, 1997.
11. Ascherio A, Rimm EB, Hernan MA, Giovannucci EL, Kawachi I, Stampfer MJ, and Willett WC. Intake of potassium, magnesium, calcium, and fiber and risk of stroke among U.S. men. *Circulation.* 98:1198–1204, 1998.

12. Ascherio A, Rimm EB, Stampfer MJ, Giovannucci EL, and Willett WC. Dietary intake of marine n-3 fatty acids, fish intake, and the risk of coronary disease among men. *N. Engl. J. Med.* 338:1650–1656, 1998.

13. Berenson GS, Srinivasan SR, Bao W, Newman WP, Tracy RE, and Wattigney WA. Association between multiple cardiovascular risk factors and atherosclerosis in children and young adults. *N. Engl. J. Med.* 338:1650–1656, 1998.

14. Boegehold M and Kotchen TA. Importance of dietary chloride for salt sensitivity of blood pressure. *Hypertension.* 17(Suppl I):I-158-I-161, 1991.

15. Braunwald E. Shattuck Lecture – Cardiovascular medicine at the turn of the millennium: Triumphs, concerns, and opportunities. *N. Engl. J. Med.* 337:1360–1369, 1997.

16. Brownell KD and Cohen LR. Adherence to dietary regimens 1: an overview of research. *Behavioral Medicine.* 20:149–154, 1955.

17. Brownell KD and Cohen LR. Adherence to dietary regimens 2: components of effective interventions. *Behavioral Medicine.* 20:155–164, 1955.

18. Bucher HC, Cook RJ, Guyatt GH, Lang LD, Cook DJ, Hatala R, and Hunt DL. Effects of dietary calcium supplementation on blood pressure: a meta-analysis of randomized controlled trials. *JAMA.* 275:1016–1022, 1996.

19. Burt VL, Cutler JA, Higgins M, Horan MJ, Labarthe D, Whelton P, Brown C, and Roccella EJ. Trends in the prevalence, awareness, treatment, and control of hypertension in the adult U.S. population: data from the Health Examination Surveys, 1960 to 1991. *Hypertension.* 26:60–69, 1995.

20. Burt VL, Whelton P, Roccella EJ, Brown C, Cutler JA, Higgins M, Horan MJ, and Labarthe D. Prevalence of hypertension in the U.S. adult population: results from the Third National Health and Nutrition Examination Survey, 1988–1991. *Hypertension.* 25:305–313, 1995.

21. Calle EE, Thun MJ, Petrelli JM, Rodriguez C, and Heath CW. Body-mass index and mortality in a prospective cohort of U.S. adults. *N. Engl. J. Med.* 341:1097–1105, 1999.

22. Cappuccio FP and MacGregor GA. Does potassium supplementation lower blood pressure? A meta-analysis of published trials. *J. Hypertens.* 9:465–473, 1991.

23. Cook NR, Cohen J, Hebert P, Taylor JO, and Hennekens CH. Implications of small reduction in diastolic blood pressure for primary prevention. *Arch. Intern. Med.* 155:701–709, 1995.

24. Cutler JA, Follmann D, and Allender PS. Randomized trials of sodium reduction: an overview. *Am. J. Clin. Nutr.* 5(suppl):643S-651S, 1997.

25. Eriksson J, Lindstrom J, Valle T, Aunola S, Hamalainen H, Ilanne-Parikka P, Peinanen-Kiukaanniemi S, Laakso M, Lauhkonin M, Lehto P, Lehtonen A, Louheranta A, Mannelin M, Martikkala V, Rastas M, Sundvall J, Turpeinen A, Viljanen T, Uusitupa M, and Tuomilehto J. Prevention of Type II diabetes in subjects with impaired glucose tolerance: the Diabetes Prevention Study (DPS) in Finland. Study design and 1-year interim report on the feasibility of the lifestyle intervention programme. *Diabetologia.* 42:793–801, 1999.

26. Expert Panel on Detection, Evaluation, and Treatment of High Blood Cholesterol in Adults. Executive summary of the third report of the National Cholesterol Education Program (NCEP). *JAMA.* 285:2486–2497, 2001.

27. Graudal NA, Galloe AM, and Garred P. Effects of sodium restriction on blood pressure, renin, aldosterone, catecholamines, cholesterol, and triglycerides: a meta-analysis. *JAMA.* 279:1383–1391, 1998.

28. Griffith LE, Guyatt GH, Cook RJ, Bucher HC, and Cook DJ. The influence of dietary and nondietary calcium supplementation on blood pressure: an updated meta-analysis of randomized controlled diets. *Am. J. Hypertens.* 12:84–92, 1999.

29. Grundy SM. Cholesterol lowering therapy in secondary prevention: remaining issues. *J. Clin. Endocrinol. Metab.* 85:2092–2098, 2000.

30. Grundy SM. Management of high serum cholesterol and related disorders in patients at risk for coronary heart disease. *Am. J. Med.* 102 (2A):15–22, 1997.

31. Grundy SM, Balady GJ, Criqui MH, Fletcher G, Greenland P, Hiratzka LF, Houston-Miller N, Kris-Etherton P, Krumholz HM, LaRosa J, Ockene IS, Pearson TA, Reed J, and Washington R. Primary prevention of coronary heart disease: guidance from Framingham. A statement for healthcare professionals from the AHA task force on risk reduction. *Circulation.* 97:1876–1887, 1998.

32. Hajjar IM, Grim CE, George V, and Kotchen TA. Impact of diet on blood pressure and age-related changes in blood pressure in the U.S. population. *Arch. Intern. Med.* 161:589–593, 2001.

33. Hakim AA, Pertovitch H, Burchfiel CM, Ross GW, Rodriguez BL, White LR, Yano K, Curb JD, and Abbott RD. Effects of walking on mortality among nonsmoking retired men. *N. Engl. J. Med.* 338:94–99, 1998.

34. He J, Ogden LG, Vapputuri S, Bazzano LA, Loria C, and Whelton PK. Dietary sodium intake and subsequent risk of cardiovascular disease in overweight adults. *JAMA.* 282:2027–2034, 1999.

35. Howard BV and Kritchevsky D. Phytochemicals and cardiovascular disease: a statement for healthcare professionals from the American Heart Association. *Circulation.* 95:2591–2593, 1997.

36. Hu FB, Rimm EB, Stampfer MJ, Ascherio A, Spiegelman D, and Willett WC. Prospective study of major dietary patterns and risk or coronary heart disease in men. *Am. J. Clin. Nutr.* 72:912–921, 2000.

37. Hunt SC, Cook NR, Oberman A, Cutler JA, Hennekens CH, Allender PS, Walker WG, Whelton PK, and Williams RR. Angiotensinogen genotype, sodium reduction, weight loss, and prevention of hypertension: Trials of Hypertension Prevention, Phase II. *Hypertension.* 32:393–401, 1998.

38. Hypertension Prevention Trial Research Group. The Hypertension Prevention Trial: three-year effects of dietary changes on blood pressure. *Arch. Intern. Med.* 150:153–162, 1990.

39. Iso H, Stampfer MJ, Manson JE, Rexrode K, Hennekens CH, Colditz GA, Speizer FE, and Willett WC. Prospective study of calcium, potassium, and magnesium intake and risk of stroke in women. *Stroke.* 30:1772–1779, 1999.

40. Kannel WB. Contributions of the Framingham Study to the conquest of coronary artery disease. *Am. J. Cardiol.* 62:1109–1112, 1988.

41. Kant AK, Schatzkin A, Graubard BI, and Schairer C. A prospective study of diet quality and mortality in women. *JAMA.* 283:2109–2115, 2000.

42. Khaw K-T and Barrett-Connor E. Dietary potassium and stroke-associated mortality: a 12-year prospective population study. *N. Engl. J. Med.* 316:235–240, 1987.

43. Klag MJ, Whelton PK, Randall BL, Neaton JD, Brancati FL, Ford CE, Shulman NB, and Stamler J. Blood pressure and end-stage renal disease in men. *N. Engl. J. Med.* 334:13–18, 1996.

44. Kotchen TA and Kotchen JM. Nutrition, diet, and hypertension in *Modern Nutrition in Health and Disease, 8th ed.* ME Shils, JA Olson, M Shike, Eds. Lea & Febiger, 1994. Chapter 71, pp. 1287–1297.

45. Kotchen TA and McCarron DA. Dietary electrolytes and blood pressure: a statement for healthcare professionals from the American Heart Association Nutrition Committee. *Circulation.* 98:613–617, 1998.

46. Klahr S, Levey AS, Beck GJ, Caggiula AW, Hunsicker LG Kusek JW, and Striker G. The effects of dietary protein restriction and blood-pressure control on the progression of chronic renal disease. *N. Engl. J. Med.* 330:877–884, 1994.

47. Lagrand WK, Visser CA, Hermens WT, Niessen HWM, Verheugt FWA, Wolbink GJ, and Hack CE. C-reactive protein as a cardiovascular risk factor: more than a epiphenomenon? *Circulation.* 1000:96–102, 1999.

48. LaRosa JC, Hunningshake D, Bush D, Criqui MH, Getz GS, Gotto AM Jr, Grundy SM, Rakita L, Robertson RM, and Weisfeldt ML. The cholesterol facts. A summary of the evidence relating dietary fats, serum cholesterol, and coronary heart disease. A joint statement by the American Heart Association and the National Heart, Lung, and Blood Institute. *Circulation.* 81:1721–1733, 1990

49. Law, M. Plant sterol and stanol margarines and health. *BMJ.* 320:861–864, 2000.

50. Lee IM and Paffenbarger RS. Physical activity and stroke incidence. The Harvard Alumni Health Study. *Stroke.* 29:2049–2054, 1998.

51. Lee CN, Reed DM, MacLean CJ, Yano K, and Chiu D. Dietary potassium and stoke. *N. Engl. J. Med.* 318:995–996, 1988.

52. Levey AS, Greene T, Beck GJ, Caggiula AW, Kusek JW, Hunsicker LG, and Klahr S. Dietary protein restriction and the progression of chronic renal disease: what have all the results of the MDRD study shown? *J. Am. Soc. Nephrol.* 10:2426–2439, 1999.

53. Lichtenstein AH. Trans fatty acids, plasma lipid levels, and risk of developing cardiovascular disease: a statement for healthcare professionals from the American Heart Association. *Circulation.* 95:2588–2590, 1997.

54. Lichtenstein AH, Ausman LM, Jalbert SM, and Schaefer EJ. Effects of different forms of dietary hydrogenated fats on serum lipoprotein cholesterol levels. *N. Engl. J. Med.* 340:1933–1940, 1999.

55. Malinow MR, Bostom AG, and Krauss RM. Homocyst(e)ine, diet, and cardiovascular diseases. A statement for healthcare professionals from the Nutrition Committee, American Heart Association. *Circulation.* 99:178–182, 1999.

56. Malinow MR, Duell PB, Hess DL, Anderson PH, Kruger WD, Phillipson BE, Gluckman RA, Block PC, and Upson BM. Reduction of plasma homocyst(e)ine levels by breakfast cereal fortified with folic acid in patients with coronary heart disease. *N. Engl. J. Med.* 338:1009–1015, 1998.

57. McCann BS and Bovbjerg VE. Promoting dietary change, in *The Handbook of Health Behavior Change, 2nd ed.* SA Shumaker, EB Schron, JK Ockene, WL McBee, Eds. Springer Publishing Co, 1998. Chapter 7, pp. 166–188.

58. McCarron DA, Oparil S, Chait A, Haynes B, Kris-Etherton P, Stern JS, Resnick LM, Clark S, Morris CD, Hatton DC, Metz JA, McMahon M, Holcomb S, Snyder GW, and Pi-Sunyer FX. Nutritional management of cardiovascular risk factors — a randomized clinical trial. *Arch. Intern. Med.* 157:169–177, 1997.

59. MacMahon SW, Cutler J, Brettan E, and Higgins M. Obesity and hypertension: epidemiological and clinical issues. *Eur. Heart J.* 8(Suppl B):57–70, 1987.

60. Messerli FH, Schmieder RE, and Weir MR. Salt: a perpetrator of hypertensive target organ disease? *Arch. Intern. Med.* 157:2449–2452, 1997.

61. Midgley JP, Matthew AG, Greenwood CM, and Logan AG. Effect of reduced dietary sodium on blood pressure: a meta-analysis of randomized controlled trials. *JAMA.* 275:1590–1597, 1996.

62. Morimoto A, Uzu T, Fujii T, Nishimura M, Kuroda S, Nakamura S, Inenaga T, and Kimura G. Sodium sensitivity and cardiovascular events in patients with essential hypertension. *Lancet.* 350:1734–1737, 1997.

63. Morris MC, Sacks F, and Rosner B. Does fish oil lower blood pressure? A meta-analysis of controlled trials. *Circulation.* 88:523–533, 1993.

64. Nadig VS and Kotchen TA. Insulin sensitivity, blood pressure, and cardiovascular disease. *Cardiology in Review.* 5:213–219, 1997.

65. National Cholesterol Education Program. Report of the Expert Panel on Population Strategies for Blood Cholesterol Reduction. NIH publication No. 93–3047, 1993.

66. National High Blood Pressure Education Program and National Cholesterol Education Program: Working Group report on management of patients with hypertension and high blood cholesterol. NIH publication No. 90–2361, 1990.

67. National Task Force on the Prevention and Treatment of Obesity. Overweight, obesity, and health risk. *Arch. Intern. Med.* 160;898–904, 2000.

68. Obarzanek E, Velletri PA, and Cutler JA. Dietary protein and blood pressure. *JAMA.* 275:1598–1603, 1996.

69. Oberman A and Kreisberg RA. Hypertriglyceridemia and coronary heart disease. *J. Clin. Endocrinol. Metab.* 85:2098–2105, 2000.

70. Ordovas JM. The genetics of serum lipid responsiveness to dietary intervention. *Proceedings of the Nutrition Society.* 58:171–187, 1999.

71. Ornish DM, Brown SE, Scherwitz LW, Billings JH, Armstrong WT, Ports TA, McLanahan SM, Sirkeeide RL, Brand RJ, and Gould KL. Can lifestyle changes reverse coronary heart disease? The Lifestyle Heart Trial. *Lancet.* 336:129–133, 1990.

72. Petersen S, Sigman-Grant M, Eissenstat B, and Kris-Etherton P. Impact of adopting lower-fat food choices on energy and nutrient intakes of American adults. *J. Am. Diet. Assoc.* 99:177–183, 1999.

73. Plat J, Kerckhoffs DAJM, and Mensink R. Therapeutic potential of plant sterols and stanols. *Curr. Opin. Lipidol.* 11:571–576, 2000.

74. Rimm EB, Stampfer MJ, Ciovannuci E, Ascherio A, Spiegelman D, Colditz GA, and Willett WC. Body size and fat distribution as predictors of coronary heart disease among middle-aged and older U.S. men. *Am. J. Epidemiol.* 141:1117–1127, 1995.

75. Ripsin CM, Keenan JM, Jacobs DR, Elmer PJ, Welch RR, Van Horn L, Liu K, Turnbull WH, Thye FW, Kestin M, Hegsted M, Davidson DM, Davidson MH, Dugan LD, Demark-Wahnefried W, and Beling S. Oat products and lipid lowering: a meta-analysis. *JAMA.* 267:3317–3325, 1992.

76. Sacks FM, Svetkey LP, Vollmer WM, Appel LJ, Bray GA, Harsha D, Obarzanek E, Conlin PR, Miller ER, Simons-Morton DG, Karanja N, and Lin PH. Effects on blood pressure of reduced dietary sodium and the dietary approaches to stop hypertension (DASH) diet. *N. Engl. J. Med.* 344:3–10, 2001.

77. Sasaki S, Zhang X-H, and Kesteloot H. Dietary sodium, potassium, saturated fat, alcohol and stroke mortality. *Stroke.* 26:783–789, 1995.

78. Stamler J. The INTERSALT Study: background, methods, findings, and implications. *Am. J. Clin. Nutr.* 65(Suppl)626S-642S, 1997.

79. Stamler J, Caggiula A, Grandits GA, Kjelsberg M, and Cutler JA. Relationship to blood pressure of combinations of dietary macronutrients: findings of the Multiple Risk Factor Intervention Trial (MRFIT). *Circulation.* 94:2417–2423, 1996.

80. Stamler J, Cohen J, Cutler JA, Grandits G, Kjelsberg M, Kuller L, Neaton J, and Okene J. Sodium intake and mortality from myocardial infarction: Multiple Risk Factors Intervention Trial. *Can. J. Cardiol.* 13(Suppl B):272B, 1997.

81. Stamler J, Elliott P, Kesteloot H, Nichols R, Claeys G, Dyer AR, and Stamler R. Inverse relation of dietary protein markers with blood pressure: findings for 10 020 men and women in the INTERSALT study. *Circulation.* 94:1629–1634, 1996.

82. Stamler J, Stamler R, and Neaton JD. Blood pressure, systolic and diastolic, and cardiovascular risks. *Arch. Intern. Med.* 153:598–615, 1993.

83. Stamler J, Stamler R, Neaton JD, Wentworth D, Daviglus ML, Garside D, Dyer AR, Liu K, and Greenland P. Low risk-factor profile and long-term cardiovascular and noncardiovascular mortality and life expectancy: findings for 5 large cohorts of young adult and middle-aged men and women. *JAMA.* 282:2012–2018, 1999.

84. Stamler R, Stamler J, Gosch FC, Civinelli J, Fishman J, McKeever P, McDonald A, and Dyer AR. Primary prevention of hypertension by nutritional-hygienic means: final report of a randomized, controlled trial. *JAMA.* 262:1801–1807, 1989.

85. Stamler J, Wentworth D, and Neaton JD. Is the relationship between serum cholesterol and risk of premature death from coronary heart disease continuous and graded? Findings in 356,222 primary screenees of the Multiple Risk Factor Intervention Trial (MRFIT). *JAMA.* 256:2823–2828, 1986.

86. Stampfer MJ, Hu FB, Manson JE, Rimm EB, and Willett WC. Primary prevention of coronary heart disease in women through diet and lifestyle. *N. Engl. J. Med.* 343:16–22, 2000.

87. Stefanick M, Mackey S, Sheehan M, Ellsworth N, Haskell WL, and Wood PD. Effects of diet and exercise in men and postmenopausal women with low levels of HDL cholesterol and high levels of LDL cholesterol. *N. Engl. J. Med.* 339:12–20, 1998.

88. Stone NJ, Nicolosi RJ, Kris-Etherton P, Ernst ND, Krauss RM, and Winston M. Summary of the scientific conference on the efficacy of hypocholesterolemic dietary interventions. *Circulation.* 94:3388–3391, 1996.

89. Strong JP, Malcom GT, Oalmann MC, and Wissler RW. The PDAY study: natural history, risk factors, and pathobiology. Pathobiological Determinants in Atherosclerosis in Youth. *Ann. N.Y. Acad. Sci.* 811:226–235, 1997.

90. The Heart Outcomes Prevention Evaluation Study Investigators. Vitamin E supplementation and cardiovascular events in high-risk patients. *N. Engl. J. Med.* 342:154–160, 2000.

91. The Sixth Report of the Joint National Committee on Prevention, Detection, Evaluation, and Treatment of High Blood Pressure. *Arch. Intern. Med.* 157:2413–2446, 1997.

92. The Trials of Hypertension Prevention Collaborative Research Group. Effects of weight loss and sodium reduction intervention on blood pressure and hypertension incidence in overweight people with high-normal blood pressure: the Trials of Hypertension Prevention, Phase II. *Arch. Intern. Med.* 157:657–667, 1997.

93. Tribble DL. Antioxidant consumption and risk of coronary heart disease: emphasis on vitamin C, vitamin E, and beta-carotene. A statement for healthcare professionals from the American Heart Association. *Circulation.* 99:591–595, 1999.

94. Tunstall-Pedoe H, Woodward M, Tavendale R, A'Brook R, and McCluskey MK. Comparison of the prediction by 27 different factors of coronary heart disease and death in men and women of the Scottish Heart Health Study: cohort study. *BMJ.* 315:722–729, 1997.

95. Tuomilehto J, Jousilahti P, Rastenyte D, Moltchanov V, Tanskanen A, Pietinen P, and Nissinen A. Urinary sodium excretion and cardiovascular mortality in Finland: a prospective study. *Lancet.* 357:848–851, 2001.

96. Tuomilehto J, Lindstrom J, Eriksson JG, Valle TT, Hamalainen H, Ilanne-Parikka P, Keinanen-Kiukaanniemi S, Laasko M, Louheranta A, Rastas M, Salminen V, and Uusitupa M. Prevention of type 2 diabetes mellitus by changes in lifestyle among subjects with impaired glucose tolerance. *N. Engl. J. Med.* 344:1343–1350, 2001.

97. Van Horn L. Fiber, lipids, and coronary heart disease. *Circulation.* 95:2701–2704, 1997.

98. Welch GN and Loscalzo J. Homocysteine and atherothrombosis. *N. Engl. J. Med.* 338:1042–1950, 1998.

99. Westrate JA and Meijer GW. Plant sterol-enriched margarines and reduction of plasma total- and LDL-cholesterol concentration in normocholesterolaemic and mildly hyper-cholesterolaemic subjects. *Eur. J. Clin. Nutr.* 52:334–343,1988.

100. Whelton PK, Appel LJ, Espeland MA, Applegate WB, Ettinger WH, Kostis JB, Kumanyika S, Lacy CR, Johnson KC, Folmar S, and Cutler JA. Sodium reduction and weight loss in the treatment of hypertension in older persons: a randomized controlled trial of nonpharmacologic interventions in the elderly (TONE). *JAMA.* 279:839–846,1998.

101. Whelton SP, Chin A, and He J. Effect of aerobic exercise on blood pressure: a meta-analysis of randomized controlled trials (abstract). *Circulation.* 103:1369-C, 2001.

102. Whelton PK, He J, Cutler JA, Brancati FL, Appel LJ, Follmann D, and Klag MJ. Effects of oral potassium on blood pressure: meta-analysis of randomized controlled clinical trials. *JAMA.* 277:1624–1632, 1997.

103. Willett WC. Diet and health: What should we eat? *Science.* 264:532–537, 1994.

104. Wood PD, Stefanick ML, Direon DM, Frey-Hewitt B, Garay SC, Williams PT, Superko HR, Fortmann SP, Albers JJ, Vranizan KM, Ellsworth NM, Terry RB, and Haskell WL. Changes in plasma lipids and lipoproteins in overweight men during weight loss and through dieting as compared with exercise. *N. Engl. J. Med.* 319:1173-179, 1998.

105. Wood PD, Stefanick ML, Williams PT, and Haskell WL. The effects on plasma lipoproteins of a prudent weight-reducing diet, with or without exercise, in overweight men and women. *N. Engl. J. Med.* 325:461–466, 1991.

106. Xin X, Frontini MG, Ogden LG, Motsamai OI, and He J. Effects of alcohol reduction on blood pressure: a meta-analysis of randomized controlled trials (abstract). *Circulation.* 102:II-844, 2000.

3 Nutritional Support in Chronic Diseases of the Gastrointestinal Tract and the Liver

Khursheed P. Navder and Charles S. Lieber

CONTENTS

INTRODUCTION

The gastrointestinal (GI) tract involves a complex system of organs that are responsible for converting the foods we eat into nutrients required by the body. It has a remarkable capacity to distinguish nutrients from substances for which the body has no use. All disturbances related to food intake, digestion, absorption, and elimination affect the GI tract, and usually require special diets. Such diets were among the very first ever used in the treatment of diseases. Unfortunately, many of these diets have not changed much since they were first used, even though research has shown that some of them are ineffective in treating diseases and incompatible with the clinical conditions of the patients.

Contrary to the popular belief, diet plays a relatively minor role in the *onset* of GI diseases. Diseases once thought to be *caused* by dietary or emotional factors have been found to be the result of bacteria, or of viruses interacting with the body's immune system, or disturbances in the motility patterns of the digestive organs. Although diet has little to do with causing GI disorders, the diseases and their treatment often have serious nutritional consequences.

DISORDERS OF THE ESOPHAGUS

GASTROESOPHAGEAL REFLUX DISEASE (GERD)

Definition, etiology, prevalence: GERD is a common disorder in which the cardiac sphincter (also known as lower esophageal sphincter (LES) or gastroesophageal sphincter) fails to close tightly, causing a backflow of the highly acidic gastric contents into the esophagus, leading to its inflammation. When the LES does not maintain a pressure that is higher than the pressure in the stomach, the contents of the stomach back up into the esophagus, resulting in gastroesophageal reflux. GERD frequently develops as a consequence of aging.[2] The most common cause is hiatal hernia (discussed in the next section). In addition, the acid reflux may result from obesity, pregnancy, pernicious vomiting, or nasogastric tubes. The true incidence of GERD is difficult to establish because many cases go undetected, and it is estimated that only a minority of chronic sufferers are seen clinically.

Symptoms: The most common symptom is pyrosis (heartburn), severe pain, and pressure sensation. The symptoms are aggravated by lying down or by any increase of abdominal pressure, such as that caused by tight clothing. Respiratory symptoms such as nocturnal coughing, wheezing, and hoarseness can also be caused by GERD. When esophagitis becomes chronic, regurgitation of the acidic gastric contents (hydrochloric acid and pepsin) may cause esophageal stricture, with the inner diameter of the esophagus narrowed by inflammation and scarring. Dysphagia and its potential complications can result.

Treatment: The main objective of therapy is to reduce the backflow of gastric acid into the esophagus.[12] Since obesity is often an associated or precipitating factor, weight reduction is essential. Some spicy foods may have to be avoided. Beneficial lifestyle changes, including diet, are listed in Table 3.1. In addition, antacids, or other drugs that neutralize gastric acidity, or antisecretory agents may be useful. If medical management fails, however, surgery may be indicated.

TABLE 3.1
Lifestyle Changes Beneficial in Gastroesophageal Reflux Disease

- Eat small, frequent meals and drink liquids one hour before or after meals to avoid distending the stomach.
- Refrain from lying down for at least 3–4 hours after meals.
- Refrain from wearing tight-fitting clothing, avoid constipation, and refrain from prolonged activities that increase abdominal pressure such as repeated bending.
- Lose weight, if overweight.
- Limit consumption of foods that lower esophageal sphincter pressure including fatty foods, coffee (with or without caffeine), chocolate, peppermint, caffeine, and alcohol.
- Limit consumption of acidic foods that irritate the lining of the esophagus or stimulate gastric acid production including citrus fruits and juice, tomato-based products, carbonated beverages, coffee and alcohol.
- Refrain from smoking cigarettes.
- Avoid certain medications (such as anticholinergic drugs, calcium channel blockers, and other smooth muscle relaxants).
- Sleep with head of bed elevated.

HIATAL HERNIA

Definition, etiology, prevalence: Normally, the stomach lies completely below the diaphragm. The esophagus passes through an opening in the diaphragm known as the esophageal hiatus. A *sliding* hiatal hernia occurs when a portion of the upper part of the stomach protrudes through the hiatus.[15] A majority of people older than age 60 have a small hiatal hernia, and in most cases, it does not cause problems. In some, however, hiatal hernia is accompanied by weakness of the lower esophageal sphincter, causing stomach contents to regurgitate into the esophagus and resulting in esophagitis.

Symptoms: The hiatal hernia can be associated with reflux of acidic gastric contents and cause symptoms similar to those already described above in GERD. Additionally, gastritis may occur in the herniated portion of the stomach and cause bleeding and anemia.

Treatment: Relieving the discomfort and pain associated with this condition is important, and treatment is the same as that described above for GERD. Large hiatal hernias may require surgical repair.

DISORDERS OF THE STOMACH AND DUODENUM

PEPTIC ULCER DISEASE

Definition, etiology, prevalence: A peptic ulcer is an erosion or open sore in the upper gastrointestinal tract, usually near the pylorus, the opening between the stomach and the small intestine. An ulcer on the stomach side of the pylorus is called a gastric ulcer; one on the duodenal side is called a duodenal ulcer. The erosion leaves the underlying layers of cells exposed to gastric juices. The erosion may proceed until the gastric juices reach the nerves and capillaries that feed the area, leading to bleeding and pain.

Peptic ulcer disease (PUD) affects as many as 25 million Americans, with about 75% of people having a recurrence of the disorder within 1 to 2 years. Gastric ulcers are less common; most ulcers occur in the duodenal bulb, where gastric contents empty through the pyloric valve. When excessive gastric acid and pepsin secretion overpower the protective action of the mucus lining, an ulcer develops. Exactly why this occurs is unknown, but a combination of factors is apparently involved. Cigarette smoking, the use of nonsteroidal anti-inflammatory drugs (NSAIDS), e.g., aspirin and ibuprofen (Advil,® Motrin®), and genetic factors (relating to protective functions of prostaglandins, which regulate mucus production, bicarbonate secretion, mucosal blood flow, and acid secretion by the parietal cells) appear to play a role in the formation of peptic ulcers by lowering the resistance of the mucosal lining. The bacterium *Helicobacter pylori* has been identified as a significant factor in weakening the body's mucosal defense mechanism.[3] This rod-shaped bacteria inhabits the gastrointestinal area around the pyloric valve and is a major determinant of chronic active gastritis.[16]

Symptoms: The usual symptom is a burning pain that occurs when the stomach is empty (between, before, and 1 to 3 hours after meals). The amount and concentration of hydrochloric acid is increased with duodenal ulcer, but may be normal with gastric ulcer. Hemorrhage may be the first sign in some patients. If the ulcer erodes a small blood vessel, the blood may seep out slowly, eventually causing anemia, weakness, tiredness, and dizziness. If the damaged blood vessel is large, the bleeding is more rapid and may cause fainting, vomiting of blood, or sudden collapse, requiring immediate medical attention and possibly surgery.

General treatment: The treatment aims at relieving pain, healing the ulcer, minimizing the likelihood of recurrence, and preventing complications. General therapeutic measures, including adequate rest, relaxation, and sleep, have long been a foundation for general care of peptic ulcer disease to enhance the body's natural healing processes. Habits that contribute to ulcer development, such as smoking and alcohol use, should be eliminated. Common drugs such as aspirin and other NSAIDS should be avoided and substituted by nonirritating analgesic drugs such as acetaminophen (Tylenol®).

Medical management: Drug therapy plays the primary role in treatment; the specific type of drug depends on the cause of the ulcer. Antibiotics are used to eradicate *H. pylori*. Acid neutralizing drugs (antacids) are rarely, if ever, used as the primary therapeutic agent but instead are often taken by patients for symptomatic relief of dyspepsia. H_2 receptor antagonists reduce gastric acid secretion and are currently the class of drug most often used for treatment of active ulcers in combination with antibiotics. Proton pump inhibitors are acid inhibitory agents and potently inhibit all phases of gastric acid secretion. Cytoprotective agents protect the stomach and duodenal walls from the eroding effect of gastric acid.

Nutritional care: Diet therapy once played a major role in ulcer treatment, but current practice is simply to prescribe medication and eliminate food that routinely causes indigestion or pain.[5] Avoid foods that are strong stimulants of gastric acid: coffee (both regular and decaffeinated), tea, chocolate, caffeinated soft drinks, and alcohol-containing beverages. Other than these restrictions, patients who are taking medication should be able to tolerate all foods without discomfort.

TABLE 3.2
Dietary Principles for Peptic Ulcer Treatment

- Eat three regular meals a day, without frequent snacks, especially at bedtime. Any food intake stimulates increased acid output.
- Avoid stomach distention with large quantities of food at a meal.
- Avoid drinking milk frequently. It stimulates significant gastric acid secretion, has only a transient buffering effect.
- Seasonings should be used as tolerated. However, some spices like hot chili peppers, black pepper, and chili powder are best avoided.
- Intake of dietary fiber, especially soluble fiber, does not have to be restricted.
- Citric fruits and juices may need to be restricted since they are acidic and induce gastric reflux in some people.
- Avoid regular and decaffeinated coffee, since it stimulates acid secretion and may cause dyspepsia. Tea and caffeinated beverages may also stimulate acid secretion.
- Avoid alcohol and alcoholic beverages, since they are stimulants of gastric acid secretion.
- Avoid smoking, since it not only increases gastric acid secretion but also influences the effectiveness of drug therapy.
- Avoid aspirin and other nonsteroidal anti-inflammatory drugs (NSAIDs).

Sometimes the dietary approach ordered by the physician depends on the patient's needs. Some patients expect dietary restrictions and feel more confident with them. Although there is no evidence that fatty, acidic, or spicy foods will harm people with ulcers, some find that these foods cause discomfort. Most gastroenterologists, both academic and clinical, now ascribe no therapeutic value to the old restrictive milk-based "ulcer diets." The American Dietetic Association has long stated the ineffectiveness and potential harm of the past generation of bland diet routines.

Table 3.2 lists a summary of the principles of nutrition and diet for support of medical management of this disease. It is important to remember, however, that the course of the disease is conditioned by the patient's unique makeup and life situation, and dietary modifications must respect individual responses or tolerances to specific foods.

DISORDERS OF THE SMALL INTESTINE

CELIAC DISEASE

Definition, etiology, prevalence: Celiac disease (celiac sprue, nontropical sprue) is a hereditary disorder of the small intestine in which there is sensitivity to *gliadin*, a protein fraction found in wheat, rye, barley, and possibly oats. Gliadin (the damaging fraction) and glutenin protein of grains combine to form a protein complex called *gluten.* Thus, celiac disease is also referred to as gluten-induced enteropathy. The exact cause of the disease is unknown, but interactions between immunologic and genetic factors have been suggested.[25] It has an incidence of about 1 in every 2000 to 3000 births, and typically appears in infancy when cereals are added to the diet. Sometimes the child has a remission of the disease in adolescence, and then the disease recurs. It is a lifelong condition. It may also occur as a transitory condition secondary to intestinal damage in other disorders. It is felt that the detection rate of celiac disease greatly underestimates its prevalence, due to a lack of awareness of the many manifestations of the condition and the requirement for at least one biopsy of the small intestine for diagnosis.[17]

The basis for the toxicity of gluten in wheat, rye, barley, and oat grains is unknown but might be related to the content of amide or bound glutamine and proline. The toxic glutens, for example, have a higher percentage of amide nitrogen than the well-tolerated corn and rice. Wheat gliadin contains a high proportion of glutamine and proline. When the affected person eats foods containing gliadin, it acts as a toxic substance, causing the intestinal villi to atrophy, become partially or totally flattened and thereby seriously reduces the absorptive surface of the intestinal tract. Protein, fat, and carbohydrates are malabsorbed, as are fat-soluble vitamins.

Symptoms: The patient with celiac disease often experiences distention, flatulence, steatorrhea, diarrhea, weight loss, and malnutrition. The malabsorption of fat results in steatorrhea and malabsorption of carbohydrates, which leads to production of gas and foul-smelling stool due to fermentation of carbohydrate by bacteria. Microcytic or macrocytic anemia may occur as a result of iron, folate, or vitamin B12 deficiencies. Because protein is malabsorbed, serum proteins can decline dramatically, resulting in edema. A vitamin K deficiency may precipitate clotting abnormalities, and the person may bleed easily. Furthermore, deficiencies in fat-soluble vitamins can lead to osteomalacia, rickets, tetany, bone pain, and night blindness.

Nutritional therapy: The primary treatment is the strict avoidance of gluten in the diet. Foods containing wheat, rye, barley, and oats, must be carefully eliminated.[24] Once gluten is removed from the diet, the intestinal changes reverse almost completely. Generally, improvement occurs within a few weeks of strict adherence to the diet. However, lactase deficiency and lactose intolerance may be permanent. If the person fails to follow the diet, symptoms will return.

The gluten-restricted diet is easier to prescribe than to follow, however, because wheat, rye, barley, and oats are common in many foods (Table 3.3). People with celiac disease and their caregivers need help to understand what foods they can eat and what foods they must avoid. With an increasing number of processed foods being marketed, as well as increasing use of ethnic foods, it is difficult to detect all foods containing gluten; therefore, a home test kit for gluten has been developed.[22] In addition, food labels may be confusing. Newer products may be labeled "gluten free" and may contain wheat *starch* as a thickening agent, which is also not tolerated.[4] Dietitians often suggest the use of corn, potato, rice, and soybean flour as substitutes for wheat flour in recipes. They can also offer tips about books, recipes, and special food products that can help patients manage their diet restrictions. More information is available at internet site, www.csaceliacs.org/celiacdisease.html.

INFLAMMATORY BOWEL DISEASE

The general term, *inflammatory bowel disease* (IBD), applies to two chronic disorders having similar symptoms but which are different clinical entities: Crohn's

TABLE 3.3
Gluten-Restricted Diet

Grains and Starches
Permitted: Baked products made from cornmeal, soybean flour, rice flour, potato flour, and gluten-free starch; hominy; tapioca; cornmeal; cornflakes; popcorn; rice; cream of rice; puffed rice; rice flakes
Not Permitted: Bread, cereal or dessert products made from wheat, rye, barley or oats; commercially prepared mixes for cakes, cookies, biscuits, cornbread, muffins, pancakes, waffles; pasta, macaroni, and noodles; wheat germ; pretzels, doughnuts, ice cream cones; matzo.

Milk and Milk Products
Patients without lactose intolerance may consume most milk products but not ovaltine, commercial chocolate milk with a cereal additive, pudding thickened with wheat flour, ice creams and sherbets containing gluten stabilizers.

Meat Products, Fruits and Vegetables
All allowed except those that are breaded, prepared with bread crumbs, or creamed (thickened with wheat flour or starch).

Miscellaneous Items
Not Permitted: Commercial salad dressings that contain gluten stabilizers; creamed soups containing wheat flour as thickener; cereal beverages (Postum); root beer; beer; ale; certain whiskeys.

disease and ulcerative colitis.[14] IBD has severe, often devastating tissue effects and nutritional consequences, but the major difference between the two appears to be the extent of damage to the gastrointestinal tract.

Ulcerative colitis is an inflammation confined to the colon. The areas of ulceration may involve only part of the rectum or colon or the entire large intestine. Additionally, the tissue changes are usually acute, lasting for a brief period and limited to the outermost mucosal and submucosal tissue layers of the intestinal wall.

The inflammation in *Crohn's disease* (also called ileitis or regional enteritis) usually involves the lower part (ileum) of the small intestine. However, it may also affect other portions of the digestive tract including the mouth, esophagus, stomach, and large intestine. In Crohn's disease, the inflammation appears to be more penetrating, leading to stricture and to fistula formation.

Incidence: The incidence of these diseases, especially Crohn's disease, has increased worldwide. Crohn's disease and ulcerative colitis have similar clinical and pathologic features even though they differ in incidence according to ethnic populations, climates, dietary habits, and customs. Crohn's disease is particularly prevalent in industrialized areas of the world. It also appears among otherwise low-risk people who move from rural to urban areas. These factors suggest a role for pathogenic agents in the environment. The incidence of both diseases is highest among teenagers, with a secondary peak at ages 55 to 60.

Etiology: Although epidemiologic and clinical studies continue to increase knowledge of the disease processes involved, the causes of these two inflammatory bowel diseases remain unknown.[9,19] A number of observations indicate that genetic factors may predispose people to the development of IBD: 1) its increased incidence in children whose parents have had IBD; 2) its incidence in certain close-knit population groups; and 3) the high rate of IBD among identical (monozygotic) twins.

Symptoms: The common clinical symptoms during the acute phase are abdominal pain and bloody diarrhea. Other symptoms include fever, nausea, vomiting, fatigue, and weight loss. Sometimes problems outside the digestive tract — skin lesions, joint pain, and inflamed eyes — develop, which could possibly be due to abnormal immunologic responses. In colitis, the patient may have 15 to 20 stools a day, causing profound effects on the individual's nutritional status. Nutrients are poorly absorbed, fluids and electrolytes are lost in diarrhea, and protein and iron are lost in bleeding. Low serum albumin is common due to malnutrition and GI losses. The individual may be seriously underweight, and children show slowed growth and sexual maturation. In Crohn's disease, fistulas can penetrate through the bowel wall into surrounding tissues and organs, and are associated with pockets of infection or abscesses.

General treatment: At present there is no cure for IBD. In the case of ulcerative colitis, there may be remission for months or even years. For most patients, though, the symptoms eventually return. Recent studies suggest that marine fish-oil supplements, which are high in omega-3 fatty acids, may reduce the inflammation associated with ulcerative colitis.[20]

Medical management: The drug therapy's aim for IBD is to control the inflammatory process, relieve pain, and promote tissue healing.[26] Corticosteroids, anti-inflammatory agents, and other drugs that affect the immune system are commonly used in therapy. Infliximab (Remicade®), a monoclonal antibody that neutralizes tumor necrosis factor-alpha (TNF-alpha), has shown promising results. Effects of these drugs become important aspects of planning supportive nutritional care. Most people who develop fistulas or obstructions require surgery to remove the affected segment.[10]

Nutritional therapy: This centers on supporting the healing process and avoiding nutritional deficiency states. When the disease is active, eating aggravates abdominal pain. Furthermore, the accompanying drug therapy can also impair the nutritional status. Multiple nutrient deficiencies threaten immune function and may reduce the effectiveness of drug therapy. If the person requires surgery or develops an infection, nutrient needs become even greater. Because surgery removes a portion of the bowel, the procedure itself contributes to malabsorption.

Restoring and maintaining nutrition status is a challenging task. Protein and energy malnutrition, as well as deficiencies of calcium, magnesium, zinc, iron, vitamin B12, folate, vitamin C, fat-soluble vitamins, fluids and electrolytes, are common problems.

Enteral nutrition may be the feeding route of choice in serious conditions, with hydrolyzed formulas offering some advantages over intact formulas. The upper small intestine easily absorbs hydrolyzed formulas, with the large bowel minimally stimulated. Hydrolyzed formulas, however, are unpalatable to many people. If people cannot accept hydrolyzed formulas orally, tube feedings are frequently employed. Tube feedings offer an advantage in that the tube can be placed in an area of the intestine that bypasses a fistula or a partial obstruction. In some cases, however, enteral nutrition is contraindicated and total parenteral nutrition (TPN) is used. TPN is chosen when enteral nutrition significantly aggravates diarrhea, when the bowel is obstructed, when complete bowel rest might help, or when enteral nutrition cannot meet nutrient requirements.

Once the acute stage has resolved, the person with IBD can gradually progress to an oral diet. A high-caloric diet (2500 to 3000 kcal/d) may be needed to restore nutritional deficits from daily losses in stools. Also the negative nitrogen balance is overcome only if sufficient calories are present to support and protect the main anabolic functions of protein. A high-protein diet is also needed to counteract the large protein losses due to exudation and bleeding from intestinal mucosal tissue, as well as deficits associated with impaired intestinal absorption. The total diet needs to supply about 100 g/d. Fat content may need to be restricted for people with fat malabsorption. To help secure additional calories, medium-chain triglycerides (MCT) may be used instead of regular fats. Low-fiber diets may be recommended for individuals with partial obstruction, but once healing is established, a regular diet including normal use of grains, fruits, and vegetables is recommended.

Vitamin and mineral supplements are routinely prescribed. When anemia is present, iron supplements may be useful. There is a need for extra B-vitamins associated with the healing process and the increase in metabolism due to the increased intake of calories and protein. Trace minerals such as zinc, which participates in tissue restoration, along with vitamin E, which contributes to tissue integrity, are also important. Potassium supplementation may be indicated if losses from diarrhea and tissue destruction have led to hypokalemia.

Because of the distressing symptoms that follow meals, people with IBD may severely restrict what they eat; yet adequate nutritional intake is imperative. These patients require much patience, understanding, and encouragement from the health care team to ensure that their therapeutic needs are met.

DISORDERS OF THE LARGE INTESTINE

Irritable Bowel Syndrome

Definition, etiology, prevalence: Irritable bowel syndrome (IBS) is not a disease but a syndrome characterized by abdominal pain and spasm, bloating, and abnormal bowel movements. It is also known as *spastic constipation.* It is a complex motility disorder influenced by physical factors and emotional stress. The term *irritable* refers to the unusual sensitivity of the nerve endings lining the colon and the hyperactivity of the nerves that control the muscles of the intestinal tract. Some people experience constipation, others experience diarrhea, and some experience diarrhea alternating with constipation. Women appear to have more symptoms during menstrual periods. Currently, gastroenterologists estimate about 20% of the general U.S. population (rates are higher for women than men) has the symptoms of IBS, about 5% of whom seek medical help.[7] Although IBS is one of the most common gastrointestinal disorders, its precise nature and cause continue to be puzzling.[8]

Symptoms: The two major symptoms of IBS are: 1) pain, chronic and recurrent, occurring in any area of the abdominal region, and 2) bowel dysfunction, varying from constipation or diarrhea to an intermittent combination of both. Patients often also appear tense and anxious.

Medical management: Medical evaluation of stress level is important in treating this disorder. Antispasmodics, antidepressants, and antianxiety medications can be prescribed. Other drugs that may be used include stool bulk formers and antidiarrheal agents.

Nutritional management: IBS, unlike IBD, is not life threatening and does not result in maldigestion or malabsorption of nutrients. Dietary practices, however, are very important in controlling symptoms. Because the nature of IBS varies among individual patients, a highly individualized and personal approach to nutritional care is essential. A high-fiber diet is recommended, because it increases and softens the fecal output, regulates transit time of the final food mass through the colon, and lowers segmental pressure, especially in the sigmoid area. A regular diet with optimal energy-providing nutrients and fiber sources — such as whole grains, fruits, and vegetables — should constitute the basic therapy. Moderate supplemental dietary fiber may come from controlled additions of bran, or of more soluble forms such as psyllium. An excess intake of fiber should be avoided because it may trigger symptoms in people with diarrhea-dominant IBS. Accordingly, it is germane to reduce total fat intake, avoid large meals, and identify and reduce gas formers (legumes, cabbage, brussels sprouts, broccoli, turnips, radishes, etc.). Adequate rest, exercise, mental health counseling, and relaxation training are also helpful elements of treatment.

DIVERTICULAR DISEASE

Definition, etiology, prevalence: Diverticular disease is characterized by outpouchings, called *diverticula*, of the wall of the large intestine. Outpouchings, about the size of large peas, usually protrude at points of weakened musculature in the bowel walls. It has been suggested that diverticular disease is caused by a lack of fiber, or roughage, in the diet, a situation that leads to small, hard, and dry fecal material.[1] The pressure and straining required when passing such stools, pushes the lining through weak spots in the muscle covering the colon, thereby forming pouches. Formation and presence of these small diverticula in the colon is called diverticulosis. This common disorder occurs in 10% of people older than 40, and in nearly 60% of people older than 60. The condition does not present any problems unless these pouches become infected and inflamed from fecal irritation and colon bacteria, a state called *diverticulitis*. The commonly used collective term covering diverticulosis and diverticulitis is *diverticular disease*.

Symptoms: Diverticulosis may be asymptomatic, or the person may have pain and constipation sometimes alternating with diarrhea. Diverticulitis involves inflammation with fever and pain. The pain and tenderness are usually localized in the lower left side of the abdomen and are accompanied by nausea, vomiting, distention, and intestinal spasms.

Nutritional therapy: Diet has become an important aspect of the long-term management of chronic diverticular disease. Studies and clinical practice have demonstrated that the condition is better managed when the amount of dietary fiber is increased, as opposed to the former practice of restricting fiber intake.[6] A high-fiber diet increases the bulk of the material reaching the colon. This enlarges the diameter of the lumen of the colon, minimizes the need for excessive segmentation, and reduces symptoms. Between 10 and 15 g of bran and 150 to 200 g of fruits and vegetables are recommended to increase stool weight to at least 150 g/d.

In diverticulitis, antibiotics may be required. Sometimes analgesics or tranquilizers are prescribed if the patient is in pain. A liquid diet followed by a low-fiber diet may be used for a short time. Colonic smooth muscle contractions, which intensify after a high-fat meal, may contribute to the discomfort felt by some.[23] Therefore, a low-fat diet may be useful for these individuals. When acute symptoms subside, the high-fiber diet should be reinstituted. Continued intestinal obstruction or perforation may necessitate surgical intervention.[6]

DISEASES OF THE LIVER

HEPATITIS

Etiology: Hepatitis, or inflammation of the liver, is usually caused by a virus (see the following sections). Hepatitis can also be caused by a drug or a chemical toxin, such as chloroform, carbon tetrachloride, or alcohol. In most cases, hepatitis is completely reversible; however, sometimes the condition deteriorates.

Viral inflammation of the liver is a major public health problem throughout the world, affecting millions of people. It is a cause of considerable illness and often death from the acute infection or its effects, which may include chronic hepatitis, cirrhosis, and primary liver cancer. Several human hepatitis viruses have been identified, and their number is increasing.

Hepatitis A virus (HAV): This virus enters the body through the intestinal tract and is usually spread when food has been contaminated by fecal content (fecal–oral route). Infected persons who do not wash their hands properly and then handle food are probably the most common route of transmission. Sometimes shellfish become contaminated with the virus (for example, if they are taken from a bay contaminated with sewage). This disease can attain epidemic proportions when people eat the contaminated shellfish. The infection is often minor in children and young adults, but mortality increases with the age of the patient. Although HAV infection does not lead to chronic hepatitis and cirrhosis, it is a serious illness. Travelers to countries where the virus is endemic and where food and water may be contaminated can be protected by vaccination.

Hepatitis B virus (HBV): This virus is spread by direct blood or sexual contact, or by sharing needles when injecting drugs. A screening test that can detect contaminated blood has, to a large degree, eliminated blood transfusions as a source of this hepatitis. The hepatitis B virus has been implicated worldwide as a major etiologic factor in chronic liver disease and associated liver cancer; it is endemic in Asia. An improved vaccine is now being used with high-risk groups, such as health care personnel. In some endemic areas, all newborn babies are routinely vaccinated.

Hepatitis C virus (HCV): This virus is also known as non-A, non-B hepatitis and can be transmitted through blood contact. It is found in a variety of high-risk populations, such as intravenous drug users, hemophiliacs receiving blood transfusions, and hemodialysis patients. The only antiviral treatment of some value at this time is interferon-alpha, a protein normally made by cells as a defense against viral infections. The therapeutic effect of interferon-alpha is enhanced by the antiviral drug ribavarin. Hepatitis C is becoming an increasing cause of cirrhosis and hepatocellular carcinoma.

Hepatitis D virus (HDV): This is a defective virus that can replicate only in the presence of HBV. It is endemic in Italy, the Middle East, and parts of Africa and South America, but is relatively rare elsewhere. HDV appears to be directly toxic for functional liver cells. Interferon–alpha has been shown to have only a transient effect on HDV. Control of HDV infections is by HBV vaccination.

Hepatitis E virus (HEV): The source of this infection is the oral–fecal route. It can severely affect pregnant women. Sporadic cases originating from epidemics in various countries of Asia, Africa, and Central America have been reported in the U.S. Contaminated water appears to be the source of infection, which usually afflicts people living in crowded and unsanitary conditions.

Symptoms: The viral agents of hepatitis produce diffuse injury to liver cells, especially the parenchymal cells. In milder cases, the liver injury is largely reversible, but with increasing severity, more extensive necrosis occurs, which can lead to liver scarring (cirrhosis), liver failure, and death. An obvious symptom is jaundice. An important complication is anorexia, which contributes to the risk of malnutrition. Other general symptoms include malaise, weakness, nausea, vomiting, diarrhea, headache, fever, enlarged and tender liver, and enlarged spleen. Convalescence can require from 3 weeks to 3 months.

General treatment: Optimal nutrition is the major therapy. It provides essential foundations for overall return of strength and recovery of the injured liver cells.

Nutritional therapy: The symptoms during the early stage of hepatitis make it difficult for the patient to consume adequate nutrients. Food may need to be given in liquid form at first, using concentrated formulas, and as the patient improves, healthy and appetizing solid foods can be introduced. If excessive nausea and vomiting persist after small amounts of food have been eaten, tube feeding or even intravenous feeding may be necessary. The goal in either case is to return the patient to normal eating. Once oral intake is resumed, several small meals are usually better tolerated than three large ones.

High protein: Protein catabolism is increased in liver disease and may be exacerbated by inadequate protein in the diet. Protein is needed not only to promote liver tissue repair, but also to provide lipotropic agents such as methionine and choline for the conversion of fats to lipoproteins for removal from the liver, thus preventing fatty infiltration. Unless there is encephalopathy (*vide infra*), the diet should provide high-quality protein in the amount of 1.5 to 2 g/kg body weight (100 to 150 g/d). This amount should achieve a positive nitrogen balance (Table 3.4).

Adequate carbohydrate: Adequate carbohydrate intake ensures the use of protein for vital tissue regeneration, the so-called protein-sparing action. The diet should supply about 45 to 50% of total calories as carbohydrate, or about 300 to 400 g of carbohydrate daily.

Moderate fat: An adequate amount of fat makes foods more palatable, and therefore encourages the anorectic patient to eat. If steatorrhea is present, the more easily absorbed medium-chain triglycerides, such as MCT oil, may be used for a

TABLE 3.4
Comparison of Diet Recommendations for Liver Disease

Nutrients	Hepatitis	Cirrhosis	Encephalopathy
Kilocalories	2500–3000 (150% of REE)	2000–2500 (125% of REE)	1800–2000 (115% of REE)
Carbohydrate	(45–50% of Cals) 300–400 g	(55–60% of Cals) 300–400 g	(65–70% of Cals) 300–400 g
Protein	High: 1.5–2 g/kg (20–22% of Cals) (100–150 g)	Moderate: 1–1.2 g/kg (15–20% of Cals) (70–100 g)	Low: 0.5 g/kg (0–5% of Cals depending on severity) (20–30 g)
Fat	30% of Cals (80–100 g)	25–30% of Cals as tolerated (60–75 g)	Increase as tolerated up to 20–25% of Cals
Other	Vit. Supplement including folic acid	Vit. Supplement especially thiamin, folic acid, B_6 and B_{12}. Possible Na and fluid restrictions REE-resting energy expenditure (see Table 3.5)	Vit. Supplement

brief time, and then regular vegetable oil can be resumed to ensure adequate amounts of essential linoleic acid. The diet should supply about 30% of calories as fat, or about 85 to 90 g/d.

Increased energy: For malnourished hospitalized patients, the energy expenditure should be 150% of the calculated resting energy expenditure (REE), estimated by the classic Harris–Benedict equation (Table 3.5) value. Ambulatory patients require more energy intake to cover added, though limited, physical activity, and need about 2500 to 3000 kcal/day.

Micronutrients: Water-soluble vitamins should be provided in amounts twice the normal RDA levels. Fat-soluble vitamins should also be provided in water-soluble forms.

FATTY LIVER

Fatty liver is a condition in which excess fat deposits in the liver cells. Excessive alcohol consumption can induce the condition (see Chapter 13, *Nutrition and Alcoholism*). It can also be brought about by severe protein malnutrition known as kwashiorkor (in infants), or by diabetes and obesity. Some drugs, including the antibiotic tetracycline, can also cause fatty liver.

Nutritional management: Fatty liver treatment consists of dealing with the underlying condition. Kwashiorkor is treated with a high-protein diet. The overall diet principles outlined for hepatitis also apply to fatty liver. If a toxic drug or alcohol

TABLE 3.5
Methods for Predicting Resting Energy Expenditure (REE)

Harris–Benedict Equation
  Women REE (kcal):
  655 + {9.56 × weight (kg)} + {1.85 × height (cm)} – {4.68 × age (yr)}
  Men REE (kcal):
  66.5 + {13.75 × weight (kg)} + {5.0 × height (cm)} – {6.8 × age (yr)}

Simplified Equation for Persons of Normal Height and Weight
  Women REE (kcal): weight (kg) × 0.95 kcal/kg × 24 h
  Men REE (kcal): weight (kg) × 1 kcal/kg × 24 h

Example
  20-year-old woman, 165 cm tall and weighing 55 kg

  REE by Harris–Benedict Equation:
  655 + {9.56(55)} + {1.85 × (165)} – {4.68 × (20yr)} = 1392.5 kcal

  REE by Simplified Equation:
  55 × 0.95 × 24 = 1254 kcal

---

is responsible, it must be discontinued or decreased. In fatty liver related to diabetes or obesity, a low-sugar, high-protein diet and maintenance of desirable body weight constitute treatment.

CIRRHOSIS

Cirrhosis is the general term used for scarring of the liver, whatever the initial cause of the disease may be. The cirrhotic liver is a firm, fibrous, brown-yellowish mass with nodules projecting from its surface.

Etiology: Cirrhosis may be caused by viral hepatitis, metabolic disorders such as Wilson's disease or hemochromatosis, obstruction of biliary drainage or cholestasis, excess use of hepatotoxic drugs, or self-prescribed mega doses of vitamin A (25,000 to 100,000 IU taken for 2 to 5 years). Alcoholic cirrhosis, historically referred to as *Laennec's cirrhosis,* is the most common type encountered in the Western world.

Even after many liver cells have died, the organ's remarkable recuperative power permits it to regenerate healthy new tissue. Nevertheless, the disease is serious and irreversible. Cirrhosis caused by viral hepatitis is most progressive. Alcoholic cirrhosis has better prognosis if all alcohol consumption ceases.

Symptoms: Early signs include GI disturbances such as nausea, vomiting, anorexia, distention, and epigastric pain. In time, jaundice may appear, with increasing weakness and edema. Since blood cannot flow easily through the fibrous liver tissue, it backs up into the portal vein, causing portal hypertension. Increased portal pressure causes enlargement and dilatation of the collateral circulatory blood vessels, including esophageal varices. Increased intrahepatic pressure also results in accumulation of large quantities of fluid in the abdomen, or ascites. Folic acid deficiency, macrocytic anemia, and low plasma protein levels are also observed.

Steatorrhea is another common symptom. Ammonia from deamination reactions fails to be converted to urea by the sick liver, and ammonia levels build up in the blood and can cause hepatic coma.

Medical management: Primary treatment involves removing the causative agent (for example, alcohol and drugs). Medical management of esophageal varices includes endoscopic sclerotherapy — a procedure whereby a hardening agent is injected into, or adjacent to, the varices to prevent their rupture and resulting bleeding. The patient is also constantly monitored for early signs of hepatic coma: personality changes, forgetfulness, and other symptoms that progress to intellectual deterioration, confusion, and stupor.

Nutritional management: Diet therapy is extremely important in the cirrhotic patient because the use of drugs is limited by the reduced ability of the liver to metabolize them. Because the patient is likely to have less appetite, frequent small meals are preferable. If esophageal varices develop, it is necessary to give soft, smooth foods to prevent variceal rupture.

General nutrition support of the patient in the absence of hepatic encephalopathy consists of a high-calorie diet (125% of the calculated REE value) to minimize endogenous protein catabolism. The protein intake should be sufficient to attain positive nitrogen balance with daily amounts of 1 to 1.2 g/kg body weight (Table 3.4). However, if signs of hepatic encephalopathy appear, proteins are decreased to individual tolerance. If the patient has ascites, the calorie and protein content should be based on an estimate of dry body weight.

Fat may be provided at 25 to 30% of total calories. If fat malabsorption is present, part of the fat may be given as MCT.

Determining carbohydrate needs is often challenging, because of the liver's primary role in carbohydrate metabolism. Glucose intolerance occurs in almost two-thirds of the patients with cirrhosis, and 10 to 35% of patients develop overt diabetes. Glucose intolerance in liver disease is due to insulin resistance in peripheral tissues. Fasting hypoglycemia can also occur because of decreased availability of glucose from glycogen in addition to the failing gluconeogenic capacity. Patients with hypoglycemia should eat frequently to avoid periods of fasting. The diet should provide 55 to 60% of calories from carbohydrate (300 to 400 g/d) to spare protein. Cirrhotic patients are prone to develop diabetes, but insulin requirements can be reduced by giving complex carbohydrates.

A patient weighing 75 kg, for example, may receive a diet containing 340 g of carbohydrate (mainly complex), 83 g of protein, 63 g of fat, and 2250 Cal/d. Since anorexia and nausea may be present, the patient should be fed four to six small meals. In addition, the patient may also be given liquid supplements.

The remaining overall diet principles outlined for hepatitis also apply to cirrhosis. Increased vitamins, especially B-complex vitamins and folic acid, are supplied. Sodium is typically restricted to 250 to 500 mg/d (11 to 22 mEq; see Table 3.6 for conversion) to help reduce fluid retention if ascites or edema is present. The sodium restriction is severe because diuretic drugs often cannot be used in the cirrhotic patient. If any diuretic is used, it should be a potassium-sparing one. Alcohol should be avoided.

TABLE 3.6
Sodium Measurement Equivalents

1 tsp of salt = ~ 6 g NaCl (40% Na and 60% Cl)

6000 mg of NaCl = 2400 mg of Na

To convert mg of Na to milliequivalents (mEq), divide by the atomic weight of 23

2400 mg Na/23 = 104 mEq Na

HEPATIC ENCEPHALOPATHY

The term *encephalopathy* refers to any disease or disorder that affects the brain, especially chronic degenerative conditions. Hepatic encephalopathy is a major serious complication of end-stage liver disease brought on by an accumulation of toxic substances in the blood as a result of liver failure and characterized by impaired consciousness, memory loss, personality changes, tremors, seizures, stupor, and coma.

Etiology: The precise cause of hepatic encephalopathy cannot always be attributed to a single agent or mechanism. However, approaches are based on the general concept that the impaired liver cells no longer detoxify substances such as ammonia, which must be converted by the liver for its excretion by the kidney. Also, as portal blood circulation decreases and collateral circulation develops, increasing amounts of blood bypass the liver and carry potentially toxic substances, such as ammonia, to the brain. Hepatic coma is also related to abnormal neurotransmitters in the brain. Cirrhotic patients have an abnormal amino acid profile: an increase in aromatic amino acids (phenylalanine, tyrosine, tryptophan) and a decrease in branched-chain amino acids (BCAA), such as isoleucine, leucine, and valine. The accumulation of the aromatic amino acids may encourage formation of false neurotransmitters, which then cause the altered neurological behavior. The resulting encephalopathy can range from trivial disorientation to full-blown coma depending on the degree of disease.

Symptoms: The patient's typical response involves disorders of consciousness and alterations in motor function. There is apathy, confusion, and drowsiness, progressing to coma. Facial expression is described as an absent stare. Speech may be slurred and monotonous. A typical motor system change is the coarse, flapping tremor known as *asterixis* (this occurs as a result of sustained contraction in a group of muscles). The breath may have a fecal odor, called "fetor hepaticus."

General treatment: A fundamental principle of therapy is to eliminate excess toxic ammonia. Two main approaches have been successful: 1) administer the non-absorbable synthetic disaccharide lactulose, which ferments in the colon, acidifies the intestinal contents, and impedes intestinal ammonia production. 2) decrease intestinal ammonia production by bacterial conversion of urea to ammonia by oral administration of a "nonabsorbable" antibiotic such as neomycin. By reducing the number of colonic bacteria, more dietary protein can be given. The use of these drugs has decreased the need for very low-protein diets.

Nutritional therapy: The metabolism of all nutrients in hepatic failure is abnormal, but intolerance to protein makes nutrition support particularly different. In addition to removing the sources of excess ammonia (*vida supra*), the following are guidelines for the nutritional support of the patients:

Low protein: Protein in the diet is restricted to 5 to 20 g or even less, depending on the severity of the condition (Table 3.4). As the mental condition improves, proteins are then reintroduced progressively, 0.25 g/kg at a time. Liquid supplements of BCAAs have been used, but the therapeutic effectiveness of these expensive supplements is not convincing. Two such products, Hepatic-Aid® and Travasorb Hepatic®, are available. Hepatic-Aid comes also in a pudding form. It has been suggested that vegetable protein may be preferable to animal protein. This difference results from the amino acid patterns of the various proteins: vegetable proteins contain less methionine and fewer aromatic amino acids than animal proteins. In addition, given the proposed role of ammonia in the pathogenesis of hepatic encephalopathy, foods that contain ammonia are omitted. Among these are various cheeses, salami, bacon, and ham.[21] The practice of protein restriction has been challenged by some[13] who feel that unnecessary protein restriction may only worsen body protein losses, and therefore must be avoided if possible.

Calories, vitamins, and minerals: Complex carbohydrate is the primary energy source along with some fat, as tolerated. Complex carbohydrates may also be advantageous in patients with hepatic encephalopathy because the nonabsorbable fiber found in such foods acts the same way as lactulose. About 2000 Cal/d is needed to prevent tissue catabolism. Vitamin and mineral supplements are usually given.

Fluid intake: Fluid balance is carefully controlled. If water excretion is defective, intake may be restricted to a volume equivalent to the urine output of the previous day. Sodium may be limited if edema is present. Potassium deficiency may develop with diuretic therapy, and the plasma potassium level must be monitored and corrected if necessary.

PRIMARY BILIARY CIRRHOSIS

Definition, etiology: Primary biliary cirrhosis (PBC) is a liver disease of unknown etiology, but with prominent immunological features such as lymphoid aggregates in the liver and circulating antimitochondrial antibodies (AMA). Women of middle age are affected predominantly, 51 years being the mean age when first diagnosed. Men represent only 10% of those affected. The disease is characterized by chronic cholangitis, involving especially the middle-sized ductules, with progressive destruction of bile ducts, cholestasis, bile ductular proliferation, fibrosis, cirrhosis, and liver failure.

Symptoms: The disease typically appears in females between the ages of 40 and 60 years; the patient may be asymptomatic and have only laboratory findings, or may present with pruritus, sometimes accompanied by mild, right, upper quadrant discomfort, dyspepsia, or fatigue. The progression of the disease is quite variable. Cholestasis leads to diminished bile salts in the intestine, failure of micelle-mediated absorption of fat and fat-soluble vitamins, steatorrhea, fat-soluble vitamin deficiency, osteoporosis, and compression fractures of vertebral bodies. Other aspects of the disease are hyperpigmentation, xanthomas, and salivary and lachrymal gland pathology (Sjogren's syndrome). Hepatosplenomegaly, jaundice, portal hypertension, and liver failure eventually ensue. Liver transplantation frequently becomes necessary.

Nutritional management: Although an increase in stool fat may be documented by quantitative analysis, it is not advisable to restrict dietary fat unless the patient has symptoms attributable to steatorrhea. If the patient has bloating, postprandial discomfort, and diarrhea secondary to fat malabsorption, dietary fat may be limited to no more than 40 g/d, approximately 360 calories, which amounts to 15 to 20% of the usual daily energy intake. Triglycerides comprising fatty acids of medium chain length (MCT) may be used to increase energy intake for patients losing weight, since these fatty acids are absorbed without the need for bile salts to promote micelle formation. MCT oil may be used in cooking and baking and to dress salads or be mixed with fruit juices. A tablespoon contains 14 g MCT and provides 115 calories. Commercial powders afford convenience of handling, and they can be reconstituted to aqueous beverages containing high proportions of MCT. Patients are sometimes intolerant of dietary MCT and report mid-abdominal and epigastric distress; gradual introduction of MCT may be necessary.

Deficiencies of fat-soluble vitamins A, D, E, and K have been diagnosed in patients with PBC. The availability of water-soluble preparations of these vitamins has been of great practical importance. It has been suggested that when cholestasis is severe, 100,000 units of parenteral vitamin A should be administered every 4 weeks, as well as 100,000 units of vitamin D and 10 mg of vitamin K.

INHERITED LIVER DISEASES

PHENYLKETONURIA (PKU)

PKU results when the enzyme phenylalanine hydroxylase, which metabolizes phenylalanine to tyrosine, is absent. In that situation, phenylalanine is not adequately converted to tyrosine and phenylalanine blood concentrations are elevated to toxic levels. Disease occurs in 100 cases per 1 million live births. The disorder gives rise to concomitant problems in the brain, because hydroxylation of L-tryptophan and L-tyrosine to 5-hydroxytryptophan and L-dopa, respectively, is interfered with. The latter two compounds are precursors of serotonin and catecholamines. High phenylalanine levels are probably the major cause of toxicity and result in low cognitive function. The normal plasma levels of phenylalanine are 58 ± 15 micromoles/liter.

Nutritional management: Plasma phenylalanine can be kept near normal, i.e., below toxic levels, by restricting phenylalanine intake to 200 to 500 mg/d in infants. In older children, the restriction has to be somewhat greater. This necessitates a diet in which phenylalanine content is approximately 25% of the normal diet. Several semisynthetic diets are commercially available. With sound dietary management, loss of IQ is minimized. Patients should restrict ingestion of phenylalanine forever or until the safety of returning to a normal diet has been demonstrated. The artificial sweetener aspartame, N-aspartyl-phenylalanine methyl ester, may provide 250 to 300 mg/quart of beverage. This is half the desired daily intake and this sweetener, therefore, should be avoided.

WILSON'S DISEASE

Definition, etiology and prevalence: Wilson's disease is an autosomally recessive inherited disorder of copper metabolism. In Wilson's disease, the ability of the liver to excrete copper is greatly impaired. This leads to copper overload and organ damage. In addition, there is poor incorporation of copper into ceruloplasmin. The copper incorporation defect provides an elegant, if not generally available, diagnostic test. However, the pathogenetic significance of the failure to incorporate copper efficiently is not clear.

The disease occurs in every ethnic and geographic population, with a worldwide incidence of about 1 in 30,000 (influenced by extent of intermarriage) and a heterozygous frequency of about 1 in 90.

Symptoms: Patients may present with liver disease, usually chronic and progressive in nature, with fatigue and eventually jaundice and signs of cirrhosis, including spider nevi, portal hypertension, ascites, peripheral edema, and esophageal varices. Some may occasionally exhibit acute hepatic failure. Copper-mediated damage to the basal ganglia of the nervous system causes symptoms such as tremors, rigidity, slurring of speech, and clumsiness of gait. Characteristic deposits of copper in Descemet's membrane of the cornea appear initially as brown crescents at the periphery, superiorly and inferiorly, near the border with the iris, and eventually take the form of the Kayser–Fleischer ring. Another, somewhat less common manifestation of Wilson's disease, hemolytic anemia, is probably related to a sudden elevation of nonceruloplasmin-bound copper. Proximal renal tubular damage with aminoaciduria, glycosuria, phosphaturia, and renal tubular acidosis is also ascribed to copper overload. Osteomalacia and osteoporosis are associated with Wilson's disease, as is hypoparathyroidism. Typically, Wilson's disease appears in childhood as hepatic illness; after the teen age, neuropsychiatric symptoms are more likely, and both may be present at the same time. A diagnosis of Wilson's disease in middle age is rare.

Medical management: The mainstay of treatment is reduction of body copper by penicillamine treatment. Penicillamine is a chelating agent which binds copper and promotes renal excretion of about 1000 to 3000 micrograms of the metal per day. Because D-penicillamine is an antimetabolite of pyridoxine, a daily supplement of 12.5 to 25 mg of pyridoxine should be provided. For patients who are intolerant to penicillamine, oral zinc acetate can promote the fecal loss of copper. The proposed mechanism is the induction of small bowel mucosal cell metallothionein by oral zinc. Metallothionein has a greater affinity for copper than for zinc, and thus fecal loss results as mucosal cells slough. The source of copper is probably salivary and gastric secretions into the gut, estimated at 1.5 mg/d, since biliary copper secretion is much diminished and is usually poorly absorbed.

Dietary management: This involves the avoidance of foods high in copper. The estimated safe and adequate daily dietary intake (ESADDI) of copper is 0.4 to 0.7 mg for infants under 1 year of age, 0.7 increasing to 2 mg for ages 1 to 10, and 1.5 to 3 mg for adults. Examples of foods high in copper are: fried beef liver, 2.4 mg per 3-ounce portion; dry roasted cashews, 0.8 mg per 1/4 cup; black-eyed peas, dried and cooked, 0.7 mg per 1/2 cup; chocolate chips, semisweet, 0.5 mg per 1/4 cup; V-8 Juice®, 0.5 mg per cup. Shellfish and mushrooms should be avoided.

HEMOCHROMATOSIS

Definition, etiology: Disorders of iron overload are characterized by retention of excessive amounts of iron in the body. Patients with hereditary hemochromatosis may store 20 to 40 g iron as compared to 0.3 to 0.8 g in normal individuals. Hepatomegaly, ascites, abnormal skin pigmentation, glucose intolerance, hypogonadism, and hepatocellular carcinoma may develop.

Symptoms: The full-blown clinical syndrome can comprise fatigue and other symptoms of liver involvement, bronze diabetes, cardiac arrhythmia or failure, diminished sexual function, loss of body hair, and chondrocalcinosis radiologically. Clinical symptoms of iron overload may appear earlier, but it most commonly occurs after 40 years of age in men and later in women. Death from progressive organ damage, hepatic or cardiac, will ensue if iron is allowed to continue to accumulate, but normal life expectancy can be approximated when accumulation is prevented or reversed. After cirrhosis occurs, there is an increased incidence of hepatoma despite iron removal.

Medical treatment: The goal of treatment is to prevent accumulation of iron to toxic levels and to reduce body iron levels to normal when they are too high. In a case of hereditary hemochromatosis discovered early in life, periodic phlebotomy will prevent iron overload because each unit of blood removed (500 to 600 ml) accounts for 200 to 250 mg of elemental iron. Symptomatic cases may already have body iron stores that are expanded from the normal adult male value of 3.6 to 4.0 g, up to 50 g. Phlebotomy of a unit or more per week for several years may be necessary to achieve normal body iron content. Patients with iron overload secondary to hemolytic and other anemias, complicated by transfusion siderosis and variable inappropriately high iron absorption, can be managed with blood transfusion to maintain desirable serum hemoglobin levels and with iron chelation (parenteral desferoxamine) to remove excess iron.

Nutritional therapy: This is not utilized by most practitioners because low-iron diets are not practical for the prevention of iron accumulation and cannot reverse established iron accumulation. Given the high gene frequency for hyperabsorption of iron, however, especially in the Caucasian population, which often eats a diet rich in efficiently absorbed iron, and given the several other mechanisms for iron overload discussed above, there have been calls for a change of current public policy. Present policy fosters acceleration of iron overload by allowing indiscriminate intake of medicinal iron (and vitamin C) and promoting iron supplementation of many foods.

DISEASES OF THE GALLBLADDER

CHOLECYSTITIS AND CHOLELITHIASIS

Definition, etiology, prevalence: The most frequently seen gallbladder diseases are cholecystitis (inflammation of the gallbladder) and cholelithiasis (the presence of gallstones).

Cholecystitis: Inflammation of the gallbladder may occur alone, but in 90 to 95% of patients, acute cholecystitis is associated with obstruction of the cystic duct by gallstones.

Cholelithiasis: An estimated 20 million Americans have gallstones, but most of these people have no symptoms.[11] Before age 60, women are twice as likely as men to develop gallstones. Obesity is a major risk factor, and people who go on very low-calorie diets and lose weight rapidly are also at higher risk. Gallstones occur when substances in the bile, primarily cholesterol and bile pigments, form hard, crystal-like particles that can range in size from a small grain to the size of a golf ball. In the U.S. and most Western countries, more than 75% of gallstones are cholesterol stones. The exact cause of cholesterol stones is unknown; however, genetics, obesity, and high-fat, low-fiber intake have been implicated. Pigmented stones are usually associated with disorders such as cirrhosis, infections, and chronic hemolysis in conditions such as sickle cell anemia. Gallstones can become lodged in the cystic duct leading from the gallbladder or in the common bile duct shared by the gallbladder and liver. If gallstones block the flow of digestive fluids from the pancreas, pancreatitis results.

Symptoms: When inflammation, stones, or both are present in the gallbladder, contraction causes severe pain in the upper abdomen. The pain may radiate to the right shoulder or to the back between the shoulder blades. There is fullness and abdominal distention with eating (particularly fatty foods). Nausea and vomiting may also be present.

Medical management: Cholecystectomy, removal of the gallbladder, is the most common method of treatment. Obese patients may be advised to lose weight before surgery. Laparoscopic cholecystectomy used today is much less invasive than the major abdominal surgery of the past and requires only an overnight stay in the hospital, followed by several days of recovery at home. Alternative treatments for people who cannot tolerate surgery include use of oral bile acids to dissolve small stones (litholysis), or extracorporeal (ultrasonic or laser) shock wave lithotripsy, a noninvasive procedure that shatters stones.

Nutritional therapy: Before surgery, diet therapy may require fat reduction to 25 to 40 g/d to reduce contractions of the gallbladder. Calories for energy needs should come principally from carbohydrate foods, especially during the acute phase. After surgery, bile flows directly from the liver to the small intestine. Clients may feel more comfortable limiting fat, but after a month or two, they may resume a diet with normal fat content. In time, a portion of the bile duct may enlarge, providing for temporary storage of bile.

DISEASES OF THE PANCREAS: PANCREATITIS

Definition, etiology: Pancreatitis, or inflammation of the pancreas, may be associated with gallstones, alcoholism, infection, or other diseases, or its origin may be unknown. Eighty percent of all cases of acute pancreatitis are caused by biliary sludge, which blocks normal outflow of the pancreatic juices. Alcohol abuse stimulates release of secretin, which increases secretion of pancreatic enzymes and, when the duct becomes blocked, there is backup of powerful activated proteolytic enzymes such as trypsin, which can digest the tissues of the organ itself.

Symptoms: When the pancreas is inflamed, stimulation of pancreatic secretion causes excruciating pain. A principal sign of progressive pancreatitis is malabsorption caused by a deficiency of pancreatic enzymes, resulting in steatorrhea. Weight loss, general malnutrition, and frequent nausea and vomiting are common. Diabetes mellitus may develop because of insufficient insulin production.

Treatment: In the acute stages, one must avoid stimulating pancreatic secretions and adding to the severe pain. Thus, nothing is given by mouth, and electrolytes are supplied intravenously. Clear, fat-free, high-carbohydrate liquids or chemically defined formulas may be tolerated as the patient's condition improves. If the disease state does not improve and the patient remains unable to consume sufficient food, total parenteral nutrition (TPN) may be required. Subsequently, frequent small meals of easily digested carbohydrate and protein may be given.

Nutritional concerns for chronic pancreatitis focus on preventing malnutrition and treating the malabsorption caused by pancreatic insufficiency and the lack of digestive enzymes. Malabsorption is treated by pancreatic enzyme replacement. This usually improves diarrhea and steatorrhea, although the steatorrhea may not be completely corrected. The major problem is delivering enough active enzyme into the duodenum. Steatorrhea could be abolished if 10% of the normal amount of lipase were available in the duodenum at the proper time. This concentration of lipase cannot be achieved with the current enzyme preparations, even if given in large doses. This is because the lipases are inactivated by gastric acid, and also because food empties from the stomach faster than the enzyme tablets dissolve.

Delivery has been enhanced by enteric coating, which keeps lipase from becoming denatured in the stomach. Incorporation of enzymes into microspheres speeds the emptying of enzymes from the stomach. Higher potency preparations, requiring fewer capsules per meal, contribute to improving compliance. For the usual patient, two or three enteric-coated capsules or eight conventional non-coated tablets of a potent enzyme preparation should be administered with meals. Some patients using conventional tablets may require adjuvant therapy to improve enzyme replacement treatment. H_2 receptor antagonists, sodium bicarbonate, and proton pump inhibitors are effective adjuvants.

Steatorrhea can also be managed by lowering the fat content of the diet to 20 to 25%. Restriction of long-chain triglyceride intake can help patients who do not respond satisfactorily to pancreatic enzyme therapy. Medium-chain triglycerides may be included to provide additional calories. These are absorbed directly into the portal system without lipolysis, but the hyperosmolarity of these preparations may result in cramps and diarrhea.

Malabsorption of protein (azotorrhea) and carbohydrate may also be present but is less overt. Azotorrhea can result in hypoalbuminemia, and carbohydrate malabsorption increases overall energy wastage. Since fecal losses of protein and carbohydrate do not have the negative symptomatic consequences of excess stool fat, increasing the intake of dietary protein and carbohydrate is well tolerated. Thus, a diet high in protein (24%) and carbohydrate (6 g/kg/d or ~ 50 to 55% — unless the patient is diabetic) and low in fat is generally recommended, with vitamin and mineral supplementation to replace losses. Abstinence from alcohol is important.

Diabetes is common in patients with pancreatic insufficiency and its management presents unique challenges given their concurrent difficulties in carbohydrate digestion. Unpredictable insulin requirements further complicate the situation. At times, even small doses of insulin may predispose patients to hypoglycemia, a phenomenon attributed to coexisting glucagon deficiency. Ideally, carbohydrate intake should be limited in these individuals; if possible, oral hypoglycemics should be employed rather than insulin to minimize the risks of hypoglycemia.

ACKNOWLEDGMENTS

Skillful typing of the manuscript by Ms. Y. Rodriguez, as well as the editing assistance of Ms. F. DeMara, is gratefully acknowledged.

REFERENCES

1. Aldoori WH, Giovannucci EL, Rimm EB, Wing AL, Trichopoulos DV, and Willett WC. A prospective study of diet and the risk of symptomatic diverticular disease in men. *Am. J. Clin. Nutr.* 60:757–764, 1994.
2. Barkun AN and Mayrand S. The treatment of peptic esophageal strictures. *Can. J. Gastroenterol.* 11:94B-97B, 1997.
3. Blaser MJ. The bacteria behind ulcers. *Sci. Am.* 274:104–107, 1996.
4. Chartrand LJ, Russo PA, Duhaime AG, and Seidman EG. Wheat starch intolerance in patients with celiac disease. *J. Am. Dietet. Assoc.* 97:612–618, 1997.
5. Damianos AJ and McGarrity TJ. Treatment strategies for *Helicobacter pylori* infection. *Am. Fam. Physician.* 55:2765–2774, 1997.
6. Deckmann R and Cheskin L. Diverticular disease in the elderly. *J. Am. Geriatr. Soc.* 40:986–993, 1993.
7. Drossman DA, Whitehead WE, and Camilleri M. Irritable bowel syndrome: a technical review for practice guideline development. *Gastroenterology.* 112:2120–2137, 1997.
8. Friedman G. Treatment of the irritable bowel syndrome. *Gastroenterol. Clin. North Am.* 20:325–333, 1991.
9. Hanauer SB. Inflammatory bowel disease. *N. Engl. J. Med.* 334:841–848, 1996.
10. Hyams JS. Crohn's disease in children. *Pediatr. Clin. North Am.* 43:255–277, 1996.
11. Johnston D and Kaplan M. Pathogenesis and treatment of gallstones. *N. Engl. J. Med.* 328:412–421, 1993.
12. Kitchin LI and Castell DO. Rationale and efficiency of conservative therapy for gastroesophageal reflux disease. *Arch. Intern. Med.* 151:448–454, 1991.

13. Kondrup J and Muller MJ. Energy and protein requirements of patients with chronic liver disease. *J. Hepatol.* 27:239–247, 1997.
14. Miner PB. Factors influencing the relapse of patients with inflammatory bowel disease. *Am. J. Gastroenterol.* 92:1S-4S, 1997.
15. Mittal RK and Balaban DH. The esophagogastric junction. *N. Engl. J. Med.* 336:924–932, 1997.
16. Munnangi S and Sonnenberg A. Time trends of physician visits and treatment patterns of peptic ulcer disease in the United States. *Arch. Intern. Med.* 157:1489–1494, 1997.
17. Murray JA. Serodiagnosis of celiac disease. *Clin. Lab. Med.* 17:445–464, 1997.
18. Plauth M, Merli M, Kondrup J, Weimann A, Ferenci P, and Muller MJ. ESPEN guidelines for nutrition in liver disease and transplantation. *Clin. Nutr.* 16:43–45, 1997.
19. Podolsky DK. Inflammatory bowel disease. (First of 2 pts). *N. Engl. J. Med.* 325:928–937, 1991.
20. Ross E. The role of marine fish oils in the treatment of ulcerative colitis. *Nutr. Rev.* 51:47–49, 1993.
21. Rudnan D, Smith RB, Salam AA, Warren WD, Galambos JT, and Wenger J. Ammonia content of food. *Am. J. Clin. Nutr.* 26:487–490, 1973.
22. Skerritt JH and Hill AS. Self management of dietary compliance in celiac disease by means of ELISA "home test" to detect *gluten. Lancet.* 337:379–382, 1991.
23. Snape W. Nutrition and colonic diverticular disease. *Nutrition and the MD.* 20:1, 1994.
24. Thompson T. Do oats belong in a gluten-free diet? *J. Am. Dietet. Assoc.* 97:1413–1416, 1997.
25. Trier JS. Celiac sprue. *N. Engl. J. Med.* 325:1709–1719, 1991.
26. Van Hogezand RA and Verspaget HW. Selective immunodulation in patients with inflammatory bowel disease — Future therapy on reality? *Neth. J. Med.* 48:64–7, 1996.

4 Nutritional Therapy of Impaired Glucose Tolerance and Diabetes Mellitus

Harry G. Preuss and Debasis Bagchi

CONTENTS

0-8493-0945-X/03/$0.00+$1.50
© 2003 by CRC Press LLC

69

INTRODUCTION

Possibly caused or potentiated by the lifestyle of the modern industrialized world, various metabolic perturbations have become more common.[5,61] Among these proliferating chronic disorders, which are especially prominent during aging, are disturbances in glucose–insulin metabolism. The worst form of these glucose–insulin perturbations is diabetes mellitus, which itself is not a single disease. Rather, diabetes mellitus is generally considered to be a group of disorders with varying etiologies and pathogenetic mechanisms.[61] Although the exact magnitude of the problem is unknown, it is generally recognized that more individuals than ever have diabetes mellitus or impaired glucose tolerance, a condition which falls short of a clinical diagnosis of diabetes mellitus. Whether the state of impaired glucose tolerance, largely ignored today, influences the general health over many years is uncertain.[43]

Diagnosis of glucose–insulin perturbations is frequently difficult. They are generally characterized by a) elevated blood glucose concentrations; b) real insulin deficiency or relative insulin deficiency secondary to decreased insulin action; c) abnormalities of glucose, lipid, and protein metabolism; and d) the development of both acute and long-term complications (Table 4.1). When hyperglycemia causes an osmotic diuresis, the common initial complaints of patients subsequently diagnosed with diabetes mellitus are frequent urination, coupled with thirst, increased appetite, and fatigue.

CLASSIFICATION OF GLUCOSE–INSULIN
PERTURBATIONS

Table 4.2 lists the categories of glucose–insulin perturbations. These range from simple impaired glucose utilization that does not fit the strict diagnostic criteria for diabetes to severe disturbances that are due to insulin-dependent diabetes (IDDM), also referred to as type I, and non-insulin-dependent diabetes (NIDDM), referred to as type II.

Normal glucose tolerance is characterized by a fasting plasma glucose level of 115 mg/dL or less and by a maximum glucose level of 139 mg/dL or less after 2 hours in response to a 75 g glucose challenge. Diabetes mellitus is defined by a fasting plasma glucose concentration greater than 140 mg/dL (on two occasions), by a random plasma glucose concentration greater than 200 mg/dL in the presence of diabetic symptomatology, or by a plasma glucose level greater than 200 mg/dL

TABLE 4.1
Common Complications of Diabetes Mellitus

Acute Complications

    Hyperglycemia

    Polyuria

    Increased thirst

    Fatigue

    Dehydration

    Weight loss

    Blurred vision

    Hyperosmotic nonketotic coma

    Ketoacidosis

    Prone to infections

    Poor wound healing

Long-term Complications

    Microvascular abnormalities

        Nephropathy

        Retinopathy

    Peripheral neuropathy

    Premature cataract formation

    Accelerated macrovascular disease

        Coronary artery disease

        Cerebrovascular disease

        Peripheral vascular disease

From Preuss HG, Talpur N, Fan AY, Bagchi M, and Bagchi D. Treating impaired glucose tolerance and diabetes mellitus via nutrition and dietary supplement. *The Original Internist.* 8:4, 17–23, 42–44, December 2001. With permission.

TABLE 4.2
Classification of Glucose/Insulin Perturbations

1. Insulin resistance
2. Insulin-dependent diabetes mellitus (IDDM, type I, juvenile onset)
3. Non-insulin dependent diabetes mellitus (NIDDM, type II, adult onset)
4. Gestational diabetes mellitus (GDM)
5. Pancreatic disease with diabetes mellitus

at two time points after a 75 g glucose challenge. The criteria for diagnosing impaired glucose tolerance lie between the normal and diabetic values, i.e., a fasting plasma glucose between 116 and 139 mg/dL and a 2-hour glucose between 140 and 200 mg/dL following an oral challenge with 75 g glucose.

How common is diabetes mellitus? Probably one half of diabetics are still undiagnosed, because symptoms tend to be absent or nonspecific, but estimates suggest that more than 16 million people in the U.S. have diabetes. This represents approximately 6% of the population.[37] Prevalence of adult diabetes in the world is thought to be also between 5 and 6%. By the year 2025, some 300 million adults worldwide may have become diabetic.[34] In addition, a large number of individuals are unaware that they have some degree of impaired glucose tolerance, which, although not severe enough to be classified as diabetes mellitus, might still be harmful over the long run.[12,43,46,59]

Because diabetes mellitus is a serious and costly health problem, it is important to have effective education, care, and support strategies. Also necessary is the promotion of effective strategies for prevention — primarily avoiding, delaying, or ameliorating impaired glucose tolerance and type II diabetes mellitus. Early treatment of glucose–insulin perturbations is preferred, because proper early treatment can lead to a decrease or slowing of long-term complications.[12,43,46,59] Also, measures that improve lifestyle in a healthful fashion, i.e., beneficial changes in nutritional habits and exercise patterns, can prevent or ameliorate many chronic perturbations that are associated with insulin resistance — dyslipidemias, hypertension, and obesity. Even in genetically susceptible individuals, insulin resistance and diabetes mellitus can benefit by appropriate changes in diet, exercise, and other lifestyle habits.

INSULIN-DEPENDENT DIABETES MELLITUS (IDDM)

IDDM (type I) is associated with severe insulinopenia that may develop over months or years. The basic pathogenesis is thought to be an inadequate insulin supply, an inadequacy that has resulted from gradual destruction of the pancreatic beta cells via autoimmune processes. Genetic and infectious processes may contribute to the development of type I diabetes mellitus. By the time of overt glucose intolerance, uptake of peripheral glucose is impaired, and the production of glucose via the liver is markedly increased.

Patients eventually depend on exogenous insulin to prevent ketoacidosis, coma, and death because of profound loss of functioning beta cells. IDDM is most prevalent in the white population. Approximately 5 to 10% of those diagnosed with diabetes mellitus have type I. Although IDDM may occur at any age, it characteristically starts during childhood. In contrast to type II diabetes, the disorder often develops in lean individuals who have undergone a period of weight loss prior to diagnosis.

NON-INSULIN-DEPENDENT DIABETES MELLITUS (NIDDM)

In 1922 at age 20, Frederick Banting with colleagues Best, Macleod, and Collip isolated insulin, establishing its important role in the pathogenesis of diabetes mellitus. This discovery by the Toronto-based group revolutionized the treatment of diabetes mellitus. However, it later became obvious that there is another form of diabetes mellitus in which the initial pathogenic mechanism was not lack of insulin but a poor tissue response to the hormone. When peripheral insulin resistance, along with related phenomena such as failure of the beta cells to produce and release adequate amounts of insulin, and increased hepatic glucose production are severe enough to create hyperglycemia, the entity is called type II diabetes to distinguish it from the originally described type I.

Type II accounts for 90% of diabetes worldwide and is becoming even more prevalent due to an ever-increasing, widespread sedentary lifestyle and associated obesity. In contrast to type I, it is more prevalent in the elderly, in non-whites, and in the obese. Also, NIDDM patients are not prone to ketoacidosis except under severe stress. Similar to type 1, there appears to be a genetic predisposition. In overweight individuals showing insulin resistance, weight reduction brought about by alterations in nutrition and physical exercise can be quite beneficial in prevention, amelioration, and cure of NIDDM.

GESTATIONAL DIABETES MELLITUS

When glucose intolerance is first discovered during pregnancy, i.e., without a prior history of diabetes, the patient is classified as having gestational diabetes. However, many believe that pregnancy, with all its hormonal changes, is just unmasking already existing diabetes mellitus, or at least a predisposition to diabetes — usually type II. Approximately 2% of pregnant females develop this disorder during the second and third trimesters, with most returning to normal after parturition. However, women with this condition must be followed closely, because many will eventually develop diabetes mellitus.

PANCREATIC DISEASE WITH DIABETES MELLITUS

Some common situations for diabetes secondary to pancreatic destruction are pancreatitis, surgical removal of the pancreas, hematochromatosis, exposure to toxins, etc.

IMPAIRED GLUCOSE TOLERANCE

The number of individuals whose glucose tolerance is impaired is probably twice as large as the number of individuals who have NIDDM.[25,38] The condition is particularly prevalent in the elderly, the obese, minority populations, and in persons with a family history of diabetes. The condition is thought to be due to receptor or post-receptor disturbances. Also, the sensitivity of pancreatic beta cells to glucose is diminished. Many recent reports indicate that over the long run impaired glucose tolerance with hyperinsulinemia and/or hyperglycemia contributes to or causes chronic disorders such as obesity, hypertension, lipid abnormalities, and athero-sclerosis.[43] Although impaired glucose intolerance is often only a transient phenomenon, it is not uncommon to have it progress to NIDDM.[24] Information from the NHANES II study indicates that the characteristics associated with impaired glucose tolerance — older age, obesity, family history of diabetes, inadequate physical activity — fall in intensity between subjects with normal glucose tolerance and those with NIDDM.[49]

How can disturbances in glucose–insulin metabolism lead to the many chronic disorders listed in Table 4.1? In individuals with diabetes mellitus or impaired glucose tolerance diabetes, the increase in circulating glucose and other reducing sugars secondary to age-induced insulin resistance can react nonenzymatically with proteins and nucleic acids to form products that affect function and diminish tissue elasticity.[8] Perturbations in glucose–insulin metabolism are also associated with enhanced lipid peroxidation secondary to greater free radical formation.[20] Free oxygen radicals cause tissue damage and are associated with many aspects of aging, including inflammatory diseases, cataracts, diabetes, and cardiovascular diseases.[3] Accordingly, glycosylation and free radical damage may contribute, at least to some extent, to the development of chronic disorders seen in diabetics or even elderly individuals having impaired glucose tolerance.

CHRONIC DISORDERS ASSOCIATED WITH GLUCOSE–INSULIN PERTURBATIONS

For over four decades, Yudkin in England carried out pioneer work examining the role of disturbed insulin homeostasis in various disease states.[59] He began by linking the consumption of excess sucrose with such disease entities as coronary thrombosis, atherosclerosis, obesity, and platelet disturbances. Yudkin believed the initial perturbation behind these common chronic disorders to be hyperinsulinemia, a common sign of insulin resistance.[59] It was subsequently shown in well-controlled rat[45] and human[47] studies that heavy sugar ingestion caused significant insulin resistance with a relative hyperinsulinemia. Reaven further expanded on the potential role of insulin resistance to human disease and named a collection of chronic metabolic disorders Syndrome X (glucose intolerance, circulating lipid disturbances, obesity, and hypertension).[46] The list of perturbations associated with Syndrome X has expanded with time.[55] In addition, the state of insulin resistance and hyperinsulinemia have also been proposed to play a role in tumor formation and aging.

GOALS FOR THE MANAGEMENT OF GLUCOSE–INSULIN PERTURBATIONS

The primary goals of any therapeutic regimen are to achieve normal (or close to normal) metabolism of carbohydrates, lipids, and proteins and, by so doing, maintain normal circulating concentrations of glucose and insulin (Table 4.3). Elevated concentrations of each have been associated with deleterious consequences and conditions (Tables 4.4 and 4.5). Additional therapeutic goals are to avoid acute complications, such as hypoglycemia, hyperglycemia, and ketoacidosis, and chronic disorders, such as microvascular disease of the eyes and kidneys, neuropathy, and large vessel problems such as cardiac, cerebral, and peripheral vascular insufficiency (Table 4.1). Good glycemic control has the potential to reduce the incidence of blindness, dialysis and kidney transplantation, nontraumatic amputation, coronary artery and myocardial disease associated with diabetes mellitus.[53] Nutrition plays a large role in this endeavor. In general, dietary treatment of IDDM, NIDDM, and impaired glucose tolerance is similar, but with more emphasis on maintaining ideal body weight in the latter two categories. For people with type I diabetes, nutrition and physical activity recommendations are given along with blood glucose monitoring and a pharmaceutical regimen which includes insulin use. Despite the need for an overall approach to therapeutics, we will concentrate on the nutritional management of glucose–insulin perturbations.[27,28,58]

TABLE 4.3
Goals of Nutrition Therapy for Diabetes Mellitus

1. Maintain blood glucose levels as near to normal as possible through appropriate food choices, as well as balance food intake with insulin and oral glucose-lowering medications and physical activity
2. Achieve optimal serum lipid levels
3. Improve quality of life and overall health
4. Provide adequate calories

   Maintain or achieve reasonable weight for adults and normal rates of growth and development in children and adolescents

   Maintain optimal nutrition during pregnancy and lactation and during recovery from catabolic illnesses
5. Prevention and treatment of acute complications

   Hypo and hyperglycemia
6. Prevention and treatment of chronic complications

   Renal disease

   Autonomic neuropathy

   Hypertension

   Hyperlipidemia

   Cardiovascular disease

TABLE 4.4
Pathogenic Consequences of Hyperglycemia

1. Microvascular complications, i.e.,
2. Nephropathy

   Neuropathy

   Retinopathy
3. Basement membrane thickening
4. Microvascular disease
5. Protein glycosylation
6. Impaired cellular immunity
7. Cell cycle abnormalities
8. Glucose toxicity: carbohydrate metabolism

   Impaired insulin secretion

   Impaired peripheral sensitivity to insulin

TABLE 4.5
Conditions Associated with Impaired Glucose
Tolerance or Hyperinsulinemia

1. Obesity
2. Impaired glucose tolerance
3. Diabetes type II
4. Non-diabetic offspring of type II diabetic parents
5. Diabetes type I as consequence of treatment
6. Essential hypertension
7. Uremia
8. Pregnancy
9. Acromegaly
10. Cushing's syndrome
11. Alcohol abuse
12. Drugs: estrogens, glucocorticoids, anabolic steroids

NUTRITION IN DIABETES MELLITUS — AN OVERVIEW

Since chronic perturbations such as diabetes mellitus are many times more prevalent in industrialized, Western countries than in underdeveloped countries, many believe nutritional factors play a significant role in the pathogenesis of these conditions.[59] This possibility is firmly supported by the fact that the Western diet is high in refined carbohydrates (sugars) and saturated/trans fatty acids and low in fiber — nutritional settings that are often associated with impaired glucose tolerance (insulin resistance) and hyperinsulinemia.[29] In turn, insulin resistance has been postulated by many to be a prominent factor in the development of often-accompanying obesity, hypertension, and dyslipidemias.[12] Thus, nutrition is one of the most challenging components of successful diabetes management.

Diabetes can be difficult for physicians to treat, because successful management relies heavily on lifestyle and behavioral changes. In a sense, diabetes is a self-management disease. Effective nutritional self-management requires an individualized approach (monitor glucose, glycated hemoglobin, lipids, blood pressure, renal status). Nevertheless, a team approach is often useful to provide guidance and support at many levels. The primary care physician can manage the overall process to aid in behavioral change. Obviously, a nutritionist is a valuable member of the team.

In the 1950s, the U.S. Public Health Service, the American Diabetes Association (ADA), and the American Dietetic Association developed a consensus on dietary recommendations.[7] These were referred to as "ADA diets," because standard diet handout sheets based on calorie level and distribution of macronutrients were used. After many years, these strict dietary guidelines were phased out. Between 1971 and 1994, nutritional recommendations for diabetics were revised many times by the American Diabetes Association. In 1994, the recommendations showed a shift toward individualization.[58] To facilitate adherence, sensitivity to cultural, ethnic, and financial aspects assumes prime importance, and recommended percentages of calories from carbohydrates, fats, and proteins are now based more on individualization.

Lifestyle recommendations given to diabetics currently are not unlike what a nutritionist might recommend to any person to maintain good health over the long term, but one might expect the recommendations to undergo modifications in the future. Perusal of the literature on nutritional treatment of diabetes shows varying outcomes and indicates there is a long way to go before all aspects of a correct therapeutic regimen are fully understood. Accordingly, most nutritional health providers practice the art of medicine more than the science of medicine in the field of diabetes. General recommendations regarding nutrition in diabetes can be given, but special individualized advice should be left to the nutritionist on the therapeutic team.

BODY WEIGHT REDUCTION
(CALORIES AND EXERCISE)

Impaired glucose tolerance and type II diabetes mellitus often respond favorably to weight reduction when obesity or the overweight state is a problem. Weight reduction associated with decreased total body and intra-abdominal fat mass often results in a marked improvement in insulin resistance and in improved blood glucose and lipid profiles. Diet, exercise, and behavioral modification form the cornerstone of treatment, because even small weight loss ameliorates insulin resistance and benefits the various components of the often-accompanying metabolic disturbances (Syndrome X). Low caloric intake frequently improves blood glucose levels and insulin sensitivity and, along with other dietary practices, can help prevent type II diabetes in overweight individuals with impaired glucose tolerance.[56] A moderate energy deficit of about 250 to 500 kcal per day or about 70% of maintenance energy balance achieved by decreasing caloric intake and increasing physical activity is a reasonable goal for most overweight diabetic patients.[32] Unfortunately, loss of fat-free-mass (muscle) occurs during most weight loss programs which, in turn, decreases resting energy expenditure with a tendency to promote weight regain. The obvious goal is to lose fat not muscle.

While most individuals eventually regain lost weight, the addition of exercise and behavioral modification helps some patients to maintain much of the weight loss, at least for a couple of years. Regular exercise (three to five times/week) enhances weight loss in individuals undergoing caloric restriction and helps in weight maintenance. The most significant effect of adding exercise to caloric restriction occurs during the weight maintenance or long-term weight loss period.[17] Exercise enhances glucose uptake, improves insulin sensitivity, and increases lean body mass, which in turn increase overall energy expenditure. Exercise to induce weight loss is more effective in men than women. Severe caloric restriction (<800 kcal/day), pharmaceutical regimens, and surgery should be reserved for the most difficult weight loss situations.

PROTEINS

Available data are too limited to establish firm nutritional recommendations for daily protein consumption that differ from the current average intake of the general population.[26] Accordingly, the daily protein intake in diabetics under most circumstances is similar to that of the general public, i.e., 10 to 20% of total daily energy requirements.[17] This amount is sufficient to ensure normal growth, development, and maintenance of protein stores in the general population. Many believe the average individual in the U.S., with or without diabetes, consumes more protein than necessary.[17] However, additional protein to maintain nitrogen balance is needed when obesity is being treated with a very low energy diet that accelerates protein breakdown, when individuals undergo high levels of physical activity, in pregnant women, in growing children, and in patients with catabolic diseases.[17] It has long been accepted that lowering consumption of proteins can slow the progression of diabetic nephropathy, but some studies have not demonstrated a protective effect in the absence of established renal failure.[26] Nevertheless, it seems prudent to limit protein consumption to 0.8 g/kg body weight in patients showing some evidence of renal failure until new evidence is developed. The average ratio of proteins in the diet that originate from vegetable sources compared to animal sources should be about 2:1. Some evidence, as yet not firmly established, favors a higher proportion of vegetable protein.

FATS AND OILS

Keeping calories derived from proteins at 10 to 20% of the total leaves 80 to 90% of calories to be divided between fats and carbohydrates. The correct distribution between these two macronutrients is not always certain; therefore, the distribution can be individualized depending on a patient's dietary preferences and blood chemical parameters.[15]

Compared to normals, type II diabetics have a two to three times higher incidence of dyslipidemias. The most frequent lipid abnormalities are hypertriglyceridemia, increased LDL cholesterol, and reduced HDL cholesterol. Many dyslipidemias may be related to hyperinsulinemia, insulin resistance, and an increased mass of intra-abdominal fat.[46] In type I diabetics, sufficient insulin replacement often improves the levels of several circulating lipids.

Despite the uncertainty surrounding the recommendations for fat and carbohydrate consumption, it seems prudent to recommend reducing the proportion of trans and saturated fat intake to 10% or less of total energy intake. High saturated fat intake seems to be associated with insulin resistance.[57] Following the suggestions of the National Cholesterol Education Program (NECP), total fat consumption should be limited to less than 30% of total caloric intake. The NECP also suggests that saturated fats be limited to <7% and cholesterol to <200 mg/day if circulating LDL cholesterol values are high. Polyunsaturated fats of the omega-3 series, found in high concentrations in fish and other seafoods, need not be curtailed in people with diabetes but should constitute less than 10% of calories. Change from a polyunsaturated to a monounsaturated fat diet in type II diabetes has been shown to reduce insulin resistance and to restore endothelium-dependent vasodilation; this may be the explanation for many of the cardiovascular benefits of the Mediterranean-type diet.[48]

With the above proportion of ingested proteins and specific fats (saturated, trans, and PUFA) in the daily intake, the remaining 60 to 70% of calories must derive from various combinations of monounsaturated fats and carbohydrates. Most nutritionists have preferred carbohydrates to fats in the diabetic diet. However, diets high in total carbohydrates have been associated with hypertriglyceridemia, reduced HDL cholesterol, increased postprandial glycemia, and insulinemia in some type II diabetes.[21] A current remedy favored for this form of dyslipidemia is to replace a portion of the carbohydrates with monounsaturated fats.

REFINED AND COMPLEX CARBOHYDRATES

The type and quantity of carbohydrates recommended for diabetics can cause controversy. As indicated above, recommended diets for diabetics are relatively high in carbohydrate content, averaging 50 to 60% of the energy content. This high carbohydrate proportion is the result of the attempt to keep total fat content below 30% of energy intake, thereby avoiding cardiovascular risks. However, carbohydrate intake, like fat intake, can also influence circulating lipid levels through alterations in glucose–insulin metabolism.[18,47] In support, insulin replacement can ameliorate dyslipidemias.

In cases of NIDDM where there is a two- to threefold increase in the prevalence of dyslipidemia compared to similar nondiabetic control subjects, it has been reported that diets high in carbohydrates protect against glucose intolerance and diabetes, probably owing to their high fiber content. Fruits and vegetables, in addition to being rich in fiber, also contain a high content of antioxidants, which may be beneficial in diabetic therapy. The benefit of antioxidants in diabetic management will be discussed in more detail below. Unfortunately, some diabetics experience increased VLDL production, hypertriglyceridemia, and low HDL cholesterol levels, which have been attributed to the high carbohydrate content of the diets.[21] If hypertriglyceridemia and increased VLDL cholesterol are significant problems in diabetics, weight loss via decreased caloric intake should be tried initially. Then, dietary carbohydrates can be restricted by increasing the energy intake from monounsaturated fats up to 20%, with no more than 10% of intake of energy derived from saturated and 10% from polyunsaturated fats each.[21]

Most health professionals classify carbohydrates broadly into three types: 1) simple (refined, sugars); 2) complex (starches), and 3) fibers (soluble, insoluble). Different types of carbohydrates vary in their effects on glucose homeostasis and insulin sensitivity and on Syndrome X.[18,47,60] Many animal studies corroborate clinical studies by clearly showing that these types of carbohydrates have different effects on glucose–insulin metabolism, fat metabolism, and various facets of Syndrome X.[44]

According to some investigators, complex carbohydrates should receive preference over simple carbohydrates in diabetic diets.[18,60] Theoretically, digestion and absorption of complex carbohydrates offer less bolus effect on postprandial hyperglycemia than simple sugars. However, gastric emptying and absorption of sugars vary considerably in the presence of foods that contain fiber, protein, and fat. Health professionals in the nutrition field often refer to a "glycemic index."[30] Expressed as a percentage, this is the increment in blood glucose as a function of time measured after ingestion of the test food divided by the corresponding area after ingestion of a portion of white bread that contains an equal amount of carbohydrate. Foods with relatively high glycemic index show fast glucose absorption, whereas a low glycemic index connotes relatively slower absorption.

Development of chronic hyperglycemia following high sucrose consumption is not always evident. This lack of consistent findings has led many, including the American Diabetic Association panel, to conclude that dietary restrictions of sucrose in diabetics are not necessary.[17,28,58] Other studies find contrasting results.[30,60] It would seem prudent at the moment to generally recommend carbohydrates with a low glycemic index and avoid those with a high glycemic index. In addition, consumption of foods that help lower the glycemic index, such as those containing fiber, should also be recommended.

FIBERS

Dietary fibers are nondigestible cell wall components and are classified as soluble and insoluble. Most Americans eat less than the amount of fiber suggested for good health. Diets high in soluble fibers can improve insulin status and various aspects of Syndrome X. Soluble fibers may be beneficial through their ability to satisfy appetite and limit caloric intake, affect various gastrointestinal hormones, form small carbon chain compounds in the gastrointestinal tract which benefit general health, and slow the absorption of carbohydrates which, in turn, lowers the glycemic index.[22] Although many cereals themselves have a relatively high glycemic index, others containing beta glucans (oats) increase the viscosity of a meal and delay absorption — actually decreasing the glycemic index.

In 1992, a prospective, randomized, double-blind, placebo-controlled study of dietary fiber supplementation in mildly hypertensive patients showed a significantly decreased fasting serum insulin in conjunction with a decreased diastolic blood pressure.[35] In a double-blind, placebo-controlled, crossover study of 25 healthy, non-obese, middle-aged men, 10 g of guar gum given three times daily for 3 weeks decreased fasting blood glucose, cholesterol, triglycerides, and systolic and diastolic blood pressure when compared to placebo.[16] This decrease coincided with increased insulin sensitivity. Seeking other associations, high-fiber diets increase peripheral insulin sensitivity in healthy young and old adults.[19] Accordingly, the need on the part of most Americans to increase dietary fiber intake is generally recognized by the majority of nutritionists.

MINERALS (NA, MG, ZN, CR, VD)

SODIUM

It is generally recognized that the average sodium intake in the U.S. of 4 to 6 g/d is well above the amount needed for normal body function.[42] Since some diabetics, like some hypertensives, are sensitive to salt-induced blood pressure elevations, it seems prudent now to limit their sodium intake to less than 3 g/d. For diabetics with already developed hypertension, the amount should be <2.4 g/d. In a study performed on type II diabetics, limitation of sodium intake caused an approximate 20 mmHg decrease in systolic pressure, while diastolic pressure remained essentially unchanged.[14]

MAGNESIUM

There is some speculation that magnesium deficiency could augment insulin resistance, carbohydrate intolerance, hypertension, and cardiovascular complications in type II diabetes mellitus.[31] In one study, 3 months of oral magnesium supplementation in insulin-requiring patients with type II diabetes mellitus increased thin plasma concentrations and urinary excretion of magnesium but essentially had no effect on glycemic control or plasma lipid levels.[13] The current recommendation is to initiate magnesium supplementation only if hypomagnesemia can be documented using a reliable measurement or in those diabetic patients at high risk of hypomagnesemia. However, in the former case, clinical estimations of magnesium status are not as reliable as desired. It seems wise to recommend that diabetics without renal insufficiency ingest some magnesium in the form of a multivitamin–mineral preparation.

ZINC

There appears to be an interrelationship between zinc status and both type I and type II diabetes mellitus.[9] However, the role of zinc in the clinical management of diabetes and prevention of its complications is unclear. Although a specific role for zinc in therapy cannot yet be assigned, a short period of controlled intake is not unreasonable to see if any benefits ensue.

CHROMIUM

Trivalent chromium may be ideal to test the role of insulin resistance in many chronic disorders and even the aging process.[1,36] The known physiological effects of chromium virtually relate only to the insulin system. Therefore, favorable effects of chromium on glucose and lipid concentrations are attributed to this mechanism.[1] While soluble fibers prolong sugar absorption in the gastrointestinal tract, chromium acts more directly on the insulin system.[1] More than one mechanism may be involved in this adaptation. Mertz postulated that chromium increases insulin sensitivity by forming links between sulfhydryl groups on insulin and membrane proteins.[36] Various chromium complexes also bind directly to insulin, which then becomes more effective than unbound insulin. Chromium supplementation can increase the insulin receptor number in hypoglycemic patients. Potter et al. reported enhanced beta cell sensitivity in older patients taking chromium.[41] Recently, Sun et al.[52] described a model of action whereby four chromium molecules bind to a low-molecular-weight chromium binding substance (LMWCr) which attains a configuration that binds to the active tyrosine kinase site of the insulin receptor.

Chromium is not used therapeutically as a free element but in combination with other ligands, such as nicotinic and amino acids.[1] Attached to various ligands, chromium compounds differ in their bioavailabilities.[1,36] For example, chromium chloride is reported to be less effective than chromium polynicotinate and chromium picolinate in influencing certain physiological parameters. Trivalent forms of chromium are amazingly nontoxic, even at markedly high doses. Doses of chromium calculated to be 300 times above the currently recommended daily dietary intake have been found safe in animals.[36] Reports of toxicity are rare, and those few reported may relate more to the ligand, such as picolinate, than the chromium itself.[51]

In a magnificent effort, Davies et al.[11] examined the chromium content of 51,665 hair, sweat, and serum samples obtained from 40,872 patients, classified according to age and gender. There were good correlations between chromium levels assayed by graphite furnace atomic absorption spectrometry in hair, sweat, and serum (r =0.536 to 0.729, p<.0001 for all correlations). This indicates that hair and sweat chromium levels are as valid as serum levels in estimating chromium status. However, it is uncertain whether serum analysis is sufficient to evaluate the chromium status accurately. Therefore, at present, the only acceptable diagnostic procedure for the need for chromium is to determine whether or not individuals with perturbed glucose and insulin metabolism respond to chromium supplementation. There is a statistically significant age-related decrease in chromium levels in hair, sweat, and serum for both males and females.[11]

The best evidence for the risk/benefits of chromium supplementation is due to a recent clinical study performed in China,[2] which tested the hypothesis of whether an elevated intake of supplemental chromium can control type II diabetes. Patients with type II diabetes, 180 men and women, were divided randomly into three groups — group 1 was given a placebo, group 2 ingested 100 µg chromium twice a day, and group 3 consumed 500 µg chromium in the form of picolinate twice a day. The higher dose of chromium generally produced faster and better results than the lower dose. Measurement of the nonenzymatic glycosylation of hemoglobin in the form of HbA1C is generally used to estimate glucose status — the higher the values, the higher the circulating glucose concentrations over the long term. HbA1C values improved significantly in the higher chromium group after 2 months and were significantly lower in both chromium groups after 4 months, indicating better status of glucose metabolism. Fasting and 2-hour insulin values decreased significantly in both groups after 2 and 4 months. Plasma total cholesterol decreased after 4 months in the groups receiving the higher amounts of chromium. No signs of toxicity occurred at these doses of chromium, considered to be well above the limits of the estimated safe and adequate daily dietary intake (ESADDI). The data demonstrated a significant benefit of supplemental chromium on HbA1C, glucose, insulin, and cholesterol in type II diabetes. A single case report in the U.S. found that 600 µg of chromium given daily returned HbA1C levels toward the normal range 3 months after initiation. It seems prudent to attempt a therapeutic regimen that includes a satisfactory daily dose of chromium over a reasonable time period in some diabetics.

VANADIUM

In the 1980s, it was reported that vanadium is an insulin mimetic agent.[40] For therapeutic considerations, the tetravalent and pentavalent states are the most important forms of vanadium. Most vanadium preparations are poorly absorbed in the gastrointestinal tract. Therefore, a question arises whether oral vanadium would provide therapeutic benefits for those with impaired glucose tolerance and diabetes mellitus.[33] Cam et al. believe that vanadium has some role and hypothesized that vanadium acts selectively by enhancing, rather than mimicking, the effects of insulin *in vivo*.[6] Goldfine et al.[23] examined the effects of 2 weeks of sodium orthovanadate administration to both type I and II diabetic patients. Insulin requirements were lowered, and two of five type I diabetic patients showed improved glucose utilization. Type II diabetic patients receiving vanadium exhibited improved insulin sensitivity that was attributed to an enhanced glucose disposal rate. Cohen et al. gave vanadyl sulfate (100 mg/d) for 3 weeks to six type II diabetic subjects[10] and found reduced fasting plasma glucose and HbA1C without concurrent changes in plasma insulin levels. The major adverse events associated with vanadium are gastrointestinal perturbations such as anorexia and diarrhea.[40] Although new forms of vanadium with greater biological activity and less toxicity are constantly being sought, the usefulness of vanadium to treat diabetes needs further assessment.

ANTIOXIDANTS AND VITAMINS

Theoretically, much of the damage caused by diabetes mellitus may be due to the increased presence of free radicals, which are known to be more prevalent in this condition.[20,43] Two separate clinical studies have shown ascorbic acid and alpha-lipoic acid to have potentially beneficial effects on glycemic control. Antioxidants in fruits, vegetables, and red wine may be responsible for their reported beneficial effects in diabetic patients. Vitamin E and/or C supplementation may provide an additional benefit toward reducing the risks of diabetic retinopathy or nephropathy. Again, even though the ability of antioxidants to ameliorate diabetes and its consequences is as yet uncertain, the risk/benefit ratios would favor reasonable supplementation with a variety of antioxidants until more definitive information is available. Improved glucose parameters in type II diabetic patients on dialysis after they received biotin suggests the need for future studies to discern this vitamin's exact role in treatment.[33]

AMINO ACIDS

A handful of reports have suggested that ℓ-carnitine can improve insulin sensitivity.[33] Animal investigations and incidental reports also suggest the potential benefit from supplementing with taurine, ℓ-arginine and glutathione.[33]

BOTANICAL SUPPLEMENTS

Some botanicals influence glucose/insulin metabolism.[39] However, the research on their therapeutic benefits is, for the most part, incomplete. Nevertheless, reasons to consider them in this brief overview are: they might help in overcoming the great financial and health costs of diabetes; they provide a favorable risk/potential benefit ratio; and, as natural supplements, they lend themselves to self management under the general purview of a physician. Because many natural products have been used as foods for a long time, many would appear to be safe for consistent consumption. It is possible that early intake of these botanical supplements may prevent or at least ameliorate diabetes mellitus, especially type II and impaired glucose tolerance. It is also possible that chronic perturbations, which characteristically develop during aging, may be substantially lessened or delayed. Some of the most examined botanicals with reported hypoglycemic activity are included in Table 4.6. More information on each can be found in References 33, 39, and 54.

GYMNEMA SYLVESTRE (GURMAR)

Gymnema is reported to increase glucose uptake and utilization, to improve function of beta cells, and to decrease glucose absorption of carbohydrates in the gastrointestinal tract. Clinical studies suggest that use of gymnema may have therapeutic utility in type I and type II diabetes mellitus.[4,50] The typical dose of 400 to 600 mg/d has virtually no toxicity when used over years. However, gymnema does have the potential to cause hypoglycemia, because it can enhance the blood glucose lowering effects of insulin and various hypoglycemic drugs.[4,50] A constituent of the leaf, gymnemic acid, inhibits the ability to taste sweet without affecting the ability to taste sour, astringent, or pungent flavors.[54]

FENUGREEK (*Trigonella foenum-gracum*)

Oral fenugreek is used to lower blood glucose in people with diabetes. Fenugreek is a spice used in cooking and has "Generally Recognized as Safe" (GRAS) status in the U.S. Fenugreek affects gastrointestinal transit, slowing glucose absorption probably through the presence of mucilaginous fiber, which comprises roughly one half the seed.[39] Also, glucose uptake and utilization improve in peripheral tissue subsequent to ingestion of fenugreek. Like many other agents that improve glucose–insulin metabolism, fenugreek has the potential also to improve dyslipidemias. In clinical studies on type I and type II diabetes, fenugreek has shown promise, but more extensive larger studies are needed. The more common side effects are cramping, diarrhea, flatulence, and hypersensitivity.

GINSENG (*Panax quinquefolius*)

Studies on type II diabetes reported improved fasting blood glucose levels, postprandial glucose levels, and HbA1C concentrations with use of ginseng.[39] Various reports based principally on animal studies have suggested that American and Asiatic ginseng alter gastrointestinal absorption, increase glucose transporter numbers, increase glucose uptake, and increase insulin release. Major side effects include excitation, insomnia, nervousness, diarrhea, chest pain, headaches, hypertension, and estrogenic effects such as vaginal bleeding. In addition, ginsengs may interact with various drugs: MAO inhibitors, anticoagulants, diuretics, stimulants, and hyperglycemic medications. In the case of the latter, combination of ginsengs with antidiabetic drugs could result in hypoglycemia.

TABLE 4.6
Major Natural Products Affecting
Glucose–Insulin Metabolism

Fibers

   Guar

   Psyllium

   Beta glucan (oats, maitake mushroom)

   Fenugreek

   Nopal cactus

   Aloe

Trace Elements, Minerals, Vitamins

Chromium	Glutathione
Magnesium	Coenzyme Q10
Zinc	Biotin
Vanadium	Vitamin B3
Alpha lipoic acid	Vitamins C and E

Amino Acids

   L-Carnitine

   Leucine

   Taurine

   L-Arginine

Botanicals with Hypoglycemic Activity

   Gymnema sylvestre (gurmar)

   Fenugreek (*Trigonella foenum-gracum*)[a]

   Ginseng (*Panax quinquefolius*)

   Nopal cactus (*Opuntia streptacanthia Lemaire*)[a]

   Garlic

   Maitake mushroom

   Bitter melon (*Momordica charantia*)

   Aloe (Aloe barbadenesis)[a]

   Bilberry (*Vaccinium mytillus*)

   Cinnamon

   Banaba leaf

[a] Fenugreek, Nopal cactus, and aloe are botanicals in which soluble fiber constituents may play a major additional role in their antidiabetic effects. Accordingly, these natural supplements have been listed under both categories.

From Preuss, HG et al. Treating impaired glucose tolerance and diabetes mellitus via nutrition and dietary supplement. *The Original Internist* . 8:4, 17–23, 42–44, December 2001. With permission.

Nopal Cactus (*Opuntia streptacantha* Lemaire)

The hypoglycemic activity of the plant is localized in the leaves and stems of the prickly pear cactus. In addition to diabetes, these extracts have been used to treat the overweight state and high cholesterol levels. Although the mechanisms that lead to hypoglycemia are not understood, the plant is a good source of fiber. Thus, glucose reabsorption may be decreased. Other investigations suggest it may enhance peripheral insulin sensitivity. Very limited studies on small patient populations indicate that the plant extracts can reduce blood glucose 41 to 46% in individuals with diabetes. There are no known serious side effects.

Garlic (*Allium sativum*)

Garlic, commonly used to treat hypercholesterolemia and hypertension, has mild hypoglycemic activity. The active constituents are probably volatile, sulfur-containing compounds. Attempts to understand the clinical therapeutic value of garlic(s) are thwarted by the disparate results from numerous studies that used different types, different doses, and different extracts. Obviously, comparison from study to study is difficult even though many meta-analyses have been reported in the literature. Since the circulating insulin concentrations tend to rise after usage, it is believed that garlic decreases the rate at which insulin is degraded. However, increased beta cell production and release of insulin are alternative possibilities.

Maitake Mushroom (*Grifola frondosa*)

Animal studies suggest that whole maitake powder, a water-soluble fraction, and an ether-soluble fraction can lower fasting and postprandial glucose and insulin levels. The simultaneous decrease in circulating glucose and insulin concentrations suggests that enhanced peripheral insulin sensitivity is a major mechanism behind the hypoglycemic effect. Maitake is rich in beta glucans, which may also slow glucose absorption from the gastrointestinal tract. Another fraction of the mushroom (Fraction D), as well as whole mushroom powder rich in a specific form of beta glucan, have the potential to enhance various functions of the immune system and favorably affect treatment of various cancers. There are no recognized serious side effects.

Bitter Melon (*Momordica charantia*)

The fruit yields the major antihypoglycemic material from the plant, although the leaves and stems are also used. Clinical studies indicate that the liquid extract from the fruit may be more powerful than a powder extract. Both pancreatic and extra-pancreatic mechanisms may underlie the antidiabetic effects. Research studies, mainly on animals, suggest a decreased hepatic glucose output, increased glucose uptake and utilization in the periphery, decreased intestinal absorption, and increased muscle glycogen synthesis. Although clinical studies are limited in certain respects (too few patients and too short a follow-up), they have shown a decrease in HbA1C and postprandial glucose of 15 to 42%. They also showed that there were both responders and nonresponders among type II diabetics. Use of this common Indian food appears to be reasonably safe.

ALOE (*Aloe barbadenesis*)

Aloe gel, which contains the polysaccharide glucomannan, has been used orally to treat type II diabetes and hyperlipidemias. Oral use decreases fasting blood glucose and HBA1C levels in the few clinical studies undertaken. Adverse reactions include multiple gastrointestinal disturbances. Interactions with cardiac glycosides, antiarrhythmic drugs, diuretics, and corticosteroids are possible.

BILBERRY (*Vaccinium mytillus*)

Bilberry extracts from leaves decrease blood glucose and blood triglycerides levels. The berry of the plant has been considered as a natural treatment of diabetic retinopathy. Myritillin, an anthocyanoside, may be the hypoglycemic component. The leaf is a rich source of chromium.

CINNAMON

The reported favorable effects of this spice on glucose–insulin metabolism may be due to the presence of polyphenolic compounds such as chalcones.

BANABA LEAF

Animal studies indicate that banaba leaf produces hypoglycemic effects by enhancing peripheral glucose utilization.

CARBOHYDRATE BLOCKERS

Slowing glucose absorption via the use of viscous fibers is a well-accepted means to improve glucose–insulin metabolism (see section on fibers). The effects on absorption correlate strongly with the viscosity of the fiber. However, a number of natural products can influence glucose absorption by other mechanisms. For instance, ℓ-arabinose has been reported to block the activity of the gastrointestinal enzyme, sucrase. If sucrose cannot be broken down to glucose and fructose, the disaccharide sucrose is not readily absorbed. Dry bean seed extract (*Phaseolus vulgaris*), hibiscus flower extract (*Hibiscus sabdariffa*), green tea leaf extract (*Camellia sinensis*), and apple fruit extract (*Malus silvestris*) have been reported to impede the gastrointestinal absorption of starch.

SUMMARY

Dietary recommendations for a diabetic are not different from those given to the general public to maintain good health. While nutritional advice to individuals may vary, general advice is consistent and must be given on the basis of today's information with the awareness that changes will occur over time as new information is gathered. At the moment, we believe:

a. Calories must be watched closely and, in many cases, limited
b. The type of fat intake is important — more emphasis on polyunsaturated fat intake and less on trans and saturated fats
c. A proportional substitution of monounsaturated fats for carbohydrates may be necessary in some individuals
d. A reasonable protein-to-carbohydrate ratio must be maintained
e. Among carbohydrates ingested, those with a low glycemic index are preferable
f. Intakes of fibers, especially soluble fibers, in excess of 25 g/d can lower the glycemic index of carbohydrates and improve overall health
g. Considering good health, consumption of a variety and reasonable amount of antioxidants provides a low risk/high benefit potential for good health
h. A good supply of minerals is necessary and that a trial of chromium usage among diabetics and individuals with impaired glucose tolerance is reasonable
i. Some botanical supplements may be useful adjuncts in the therapeutic regimens of diabetics

REFERENCES

1. Anderson RA. Recent advances in the clinical and biochemical effects of chromium deficiency, in *Essential and Toxic Trace Elements in Human Health and Disease.* Prasad AS (Ed.). New York:Wiley Liss, 1993, pp. 221–234.
2. Anderson RA, Cheng N, Bryden NA, Polansky MM, Cheng N, Chi J, and Feng J. Elevated intakes of supplemental chromium improve glucose and insulin variables in individuals with type 2 diabetes. *Diabetes.* 46:1786–1791, 1997.
3. Bagchi D, Bagchi M, Stohs SJ, Das DK, Ray SD, Kuszynski A, Joshi SJ, and Preuss HG. Free radicals and grape seed proanthocyanidin extract: importance in human health and disease prevention. *Toxicology.* 148:187–197, 2000.
4. Baskaran K, Ahamath BK, ShanmugasundaramKR, and Shanmugasundaram ER. Antidiabetic effect of a leaf extract from Gymnema sylvestre in non-insulin-dependent diabetes mellitus patients. *J. Ethnopharmacol.* 30:295–300, 1990.
5. Breslow L. Prevention and control of noncommunicable diseases. *World Health Forum.* 3:429–431, 1982.
6. Cam MC, Brownsey RW, and McNeill JH. Mechanisms of vanadium action: insulin-mimetic or insulin enhancing agent? *Can. J. Physiol. Pharmacol.* 78:829–847, 2000.
7. Caso EK. Calculation of diabetic diets. *J. Am. Diet. Assoc.* 26:575–583, 1950.
8. Cerami A, Vlassare H, and Brownlee M. Glucose and aging. *Sci. Am.* 256:90–96, 1987.
9. Chausmer AB. Zinc, insulin, and diabetes. *J. Am. Coll. Nutr.* 17:109–115, 1998.
10. Cohen N, Halberstam M, Shlimovich P, Chang CJ, Shamoon H, and Rossetti L. Oral vanadyl sulfate improves hepatic and peripheral insulin sensitivity in patients with non-insulin dependent diabetes mellitus. *J. Clin. Invest.* 95:2501–2509, 1995.

11. Davies S, Howard JM, Hunnisett A, and Howard M. Age-related decreases in chromium levels in 51,665 hair, sweat, and serum samples from 40,872 patients — implications for the prevention of cardiovascular disease and type II diabetes mellitus. *Metabolism. Clinical and Experimental.* 46:469–473, 1997.

12. DeFronzo RA and Ferinimmi E. Insulin resistance: a multifaceted syndrome responsible for NIDDM, obesity, hypertension, dyslipidemia, and atherosclerotic cardiovascular disease. *Diabetes Care.* 14:173–194, 1991.

13. de Valk HW, Verkaaik R, van Rijn HJ, Geerdink RA, and Struyvenberg A. Oral magnesium supplementation in insulin-requiring type 2 diabetes patients. *Diabet. Med.* 15:503–507, 1998.

14. Dodson PM, Beevers M, Hallworth R, Webberley MJ, Fletcher RF, and Taylor KG. Sodium restriction and blood pressure in hypertensive type II diabetics: randomized blind controlled and crossover studies of moderate sodium restriction and sodium supplementation. *Br. Med. J.* 298:226–230, 1989.

15. Dunn FL. Plasma lipid and lipoprotein disorders in IDDM. *Diabetes.* 41(Suppl 2):102–106, 1992.

16. Eliasson K, Ryttig KR, Hylander B, and Rossner S. A dietary fibre supplement in the treatment of mild hypertension. A randomized, double-blind, placebo-controlled trial. *J. Hypertens.* 10:195–199, 1992.

17. Franz MJ, Horton ES, Bantle JP, Beebe CA, Brunzell JD, Coulston AM, Henry RR, Hoogwerf BJ, and Stacpoole PW. Nutrition principles for the management of diabetes and related complications. *Diabetes Care.* 17:490–518, 1994.

18. Frost G, Keogh B, Smith D, Akinsanya K, and Leeds A. The effect of low-glycemic carbohydrate on insulin and glucose response *in vivo* and *in vitro* in patients with coronary heart disease. *Metabolism.* 45:669–672, 1996.

19. Fukagawa NK, Anderson JW, Hageman G, Young VR, and Ninaker KL. High-carbohydrate, high-fiber diets increase peripheral insulin sensitivity in healthy young and old adults. *Am. J. Clin. Nutr.* 52:524–528, 1990.

20. Gallaher DD, Csallany AS, Shoeman DW, and Olson JM. Diabetes increases excretion of urinary malonaldehyde conjugates in rats. *Lipids.* 28:663–666, 1993.

21. Garg A, Bonanome A, Grundy SM, Zhang AJ, and Unger RH. Comparison of a high-carbohydrate diet with high-monosaturated fat diet in patients with non-insulin dependent diabetes mellitus. *N. Engl. J. Med.* 391:829–834, 1988.

22. Giacco R, Parillo M, Rivellese AA, Lasorella G, Giacco A, D'Episcopo L, and Riccardi G. Long-term dietary treatment with increased amounts of fiber-rich low-glycemic index natural foods improves blood glucose control and reduces the number of hypoglycemic events in type I diabetic patients. *Diabetes Care.* 23:1461–1466, 2000.

23. Goldfine AB, Patti ME, Zuberi L, Goldstein BJ, LeBlanc R, Landaker EJ, Jiang Z, Willsky GR, and Kahn CR. Metabolic effects of vanadyl sulfate in humans with non-insulin-dependent diabetes mellitus: *in vivo* and *in vitro* studies. *Metabolism.* 49:400–410, 2000.

24. Harris MI. Impaired glucose tolerance in the US population. *Diabetes Care.* 14:628–638, 1991.

25. Harris MI, Hadden WC, Knowler WC, and Bennett PH. Prevalence of diabetes and impaired glucose tolerance and plasma glucose levels in the U.S. population aged 20–74. *Diabetes.* 36:523–534, 1987.

26. Henry RR. Protein content of the diabetic diet. *Diabetes Care.* 17:1502–1517, 1994.

27. Holler HJ and Pastors JG. *Diabetes Medical Nutrition Therapy.* Chicago, American Dietetic Association, 1997, pp. 99–123.

28. Horton ES and Napoli R. Diabetes mellitus, in *Present Knowledge in Nutrition, 7th ed.* EE Ziegler and LJ Filer, Jr. (Eds.) ILSI Press, Washington, D.C., 1996, pp. 445–455.

29. Hu FB, van Dam RM, and Liu S. Diet and risk of type II diabetes: the role of types of fat and carbohydrate. *Diabetologia.* 44:805–817, 2001.

30. Jenkins DJA, Wolever TMS, and Jenkins AL. Starchy foods and glycemic index. *Diabetes Care.* 11:149–159, 1988.

31. Kao WH, Folsom AR, Nieto FJ, Mo JP, Watson RL, and Brancati FL. Serum and dietary magnesium and the risk for type 2 diabetes mellitus: the Atherosclerosis Risk in Communities Study. *Arch. Intern. Med.* 159:2119–2120, 1999.

32. Katsilambros NL. Nutrition in diabetes mellitus. *Exp. Clin. Endocrinol. Diabetes.* 109:S250-S258, 2001.

33. Kelly GS. Insulin resistance: lifestyle and nutritional interventions. *Altern. Med. Rev.* 5:109–132, 2000.

34. King H, Aubert RE, and Herman WH. Global burden of diabetes, 1995–2025: prevalence, numerical estimates, and projections. *Diabetes Care.* 21:1414–1431, 1998.

35. Landin K, Holm G, Tengborn L, and Smith U. Guar guar improves insulin sensitivity, blood lipids, blood pressure, and fibrinolysis in healthy men. *Am. J. Clin. Nutr.* 56:1061–1065, 1992.

36. Mertz W. Chromium in human nutrition: a review. *J. Nutr.* 123:626–633, 1993.

37. Mokdad AH, Ford ES, Bowman BA, Nelson DE, Engelgay M, Vinicor F, and Marks JS. Diabetes trends in the US: 1990–1998. *Diabetes Care.* 23:1278–1283, 2000.

38. National Diabetes Data Group: classification and diagnosis of diabetes mellitus and other categories of glucose intolerance. *Diabetes.* 28:1039–1057, 1979.

39. O'Connell B. The diabetes connection: botanicals with hypoglycemic activity. *Nutraceuticals World.* May: 28–38, 2000.

40. Orvig C, Thompson KH, Batrtell M, and McNeill JH. The essentiality and metabolism of vanadium, in *Metal Ions in Biological Systems, Vol 31, Vanadium and Its Role in Life.* Sigel H and Sigel A (Eds.) Marcel Dekker, New York, 1995, pp. 575–594.

41. Potter JF, Levin P, Anderson RA, Freiberg JM, Andres R, and Elahi D. Glucose metabolism in glucose-intolerant older people during chromium supplementation. *Metabolism.* 34:199–204, 1985.

42. Preuss HG. Sodium, chloride and potassium, in *Present Knowledge in Nutrition.* BA Bowman and RM Russell (Eds.) ILSI Press: Washington, D.C., 2001, pp. 302–310.

43. Preuss HG, Bagchi D, and Clouatre D. Insulin resistance; a factor in aging, in *The Advanced Guide to Longevity Medicine.* Ghen MJ et al. (Eds.) Ghen, Landrum SC, 2001, pp. 239–250.

44. Preuss HG, Gondal JA, Bustos E, Bushehri N, Lieberman S, Bryden NA, Polansky MM, and Anderson RA. Effect of chromium and guar on sugar-induced hypertension in rats. *Clin. Neph.* 44:170–177, 1995.

44a. Preuss HG, Talpur N, Fan AY, Bagchi M, and Bagchi D. Treating impaired glucose tolerance and diabetes mellitus via nutrition and dietary supplement. *The Original Internist.* 8:4, 17–23, 42–44, December 2001.

45. Preuss HG, Zein M, Knapka J, and DiPette, D. Effects of heavy sugar eating on three strains of Wistar rats over their lifespan. *J. Am. Coll. Nutr.* 17:36–47, 1998.

46. Reaven GM. Role of insulin resistance in human disease (Banting Lecture 1988). *Diabetes.* 37:1595–1607, 1988.

47. Reiser S, Handler HB, Gardner LB, Hallfrisch JG, Michaelis OE, and Prather ES. Isocaloric exchange of dietary starch and sucrose in humans. II. Effect on fasting blood insulin, glucose, and glucagon and on insulin and glucose response to a sucrose load. *Am. J. Clin. Nutr.* 32:2206–2216, 1979.

48. Ryan M, McInerney D, Owens D, Collins P, Johnson A, and Tomkin GH. Diabetes and the Mediterranean diet: a beneficial effect of oleic acid on insulin sensitivity, adipocyte glucose transport and endothelium-dependent vasoreactivity. *Quart. J. Med.* 93:85–91, 2000.
49. Sartor G, Schersten B, Carlstrom S, Melandor A, Nordan A, and Persson G. Ten-year follow-up of subjects with impaired glucose tolerance: prevention of diabetes by tolbutamide and diet regulation. *Diabetes.* 29:41–49, 1980.
50. Shanmugasundaram ER, Rajeswari G, Baskaran K, Kumar BRR, Shanmugasundaram KR, and Ahmath BK. Use of Gymnema sylvestre leaf extract in the control of blood glucose in insulin-dependent diabetes mellitus. *J. Ethnopharmacol.* 30:281–294, 1990.
51. Stearns DM, Wise JP Sr, Patierno SR, and Wetterhahn KE. Chromium (III) picolinate produces chromosome damage in Chinese hamster ovarian cells. *FASEB J.* 9:1643–1648, 1995.
52. Sun Y, Ramirez J, Woski SA, and Vincent JB. The binding of trivalent chromium to low-molecular-weight chromium-binding substance (LMWCr) and the transfer of chromium from transferrin and chromium picolinate to LMWCr. *J. Biol. Inorg. Chem.* 5:129–136, 2000.
53. The Diabetes Control and Complications Trial Research Group. The effect of intensive treatment of diabetes on the development and progression of long-term complications of insulin dependent diabetes mellitus. *N. Engl. J. Med.* 329:977–986, 1993.
54. Therapeutic Research Faculty. *Natural Medicines. Comprehensive Database.* Stockton CA, 2001.
55. Timar O, Sestier F, and Levy E. Metabolic syndrome X: a review. *Can. J. Cardiol.* 16:779–789, 2000.
56. Tuomilehto J, Lindstrom J, Eriksson JG, Valle TT, Hamalainen H, Ilanne-Parikka P, Keinanen-Kiukaanniemi S, Laakso M, Louheranta A, Rastas M, Salminen V, and Uusitupa M. Finnish Diabetes Prevention Study Group: Prevention of type 2 diabetes mellitus by changes in lifestyle among subjects with impaired glucose tolerance. *N. Engl. J. Med.* 344:1390–1392, 2001.
57. Virtanen SM and Aro A. Dietary factors in the aetiology of diabetes. *Ann. Med.* 26:469–478, 1994.
58. Wheeler ML. Nutrition management and physical activity as treatments for diabetes. *Primary Care; Clinics in Office Practice.* 26:857–868, 1999.
59. Yudkin J. Sucrose, coronary heart disease, diabetes, and obesity. Do hormones provide a link? *Am. Heart J.* 115:493–498, 1988.
60. Zammit VA, Waterman IJ, Topping D, and McKay G. Insulin stimulation of hepatic triacylglycerol secretion and the etiology of insulin resistance. *J. Nutr.* 131:2074–2077, 2001.
61. Zimmet PZ. Primary prevention of diabetes mellitus. *Diabetes Care.* 11:258–262, 1988.

5 Nutritional Management of Metabolic Disorders

Virginia Utermohlen

CONTENTS

0-8493-0945-X/03/$0.00+$1.50

DISEASES OF THE ENDOCRINE CONTROL
OF METABOLISM

INTRODUCTION: THE ROLE OF HORMONES IN REGULATING FUEL METABOLISM, METABOLIC RATE, AND APPETITE

The endocrine control of metabolism is designed to provide both rapid and long term adaptations to food intake and metabolic demand. Hormones can provide this control by altering the flux of fuels through the different metabolic pathways (e.g., the glucocorticoids), or by altering cell sensitivity to other hormones and neurotransmitters (e.g., thyroid hormones). This chapter will focus on glucocorticoids and thyroid hormones. However, these hormones act to regulate fuel flux in concert with epinephrine, glucagon, insulin, and growth hormone. Therefore, we will also give a brief overview of the effects of these hormones.

Brief Overview of the Metabolic Interrelationships among Fuels

The carbon backbones of glucose, fatty acids and amino acids can all potentially be used to form ATP through oxidation to carbon dioxide. These backbones can be stored as glycogen, fats, and proteins, and can be converted to fuel as needed. Among the important roles of hormones are to:

- Ensure the build-up and preservation of stores when fuel input exceeds demand
- Coordinate interconversion of potential fuels should they be needed
- Regulate oxidation rates in response to demand

For example, after a meal, insulin ensures the entry of:

- Glucose into muscle, to be used either as fuel or for storage as glycogen
- Amino acids for the formation of protein, which can serve both functional roles and as a source of carbon backbones to be used as fuel
- Glucose into fat cells, and its metabolism to form triglycerides

When there is a demand for fuel, glycogen is broken down to glucose, triglycerides to free fatty acids and glycerol, and protein can be converted to glucose (gluconeogenesis) as needed. Figure 5.1 shows the interconversion of carbon backbones to provide the fuels. Note that certain conversions can occur only in one direction. Thus gluconeogenesis requires the conversion of oxaloacetate from the citric acid cycle to phosphoenolpyruvate through activation of phosphoenolpyruvate carboxykinase (PEP-CK). The phosphoenolpyruvate can then either be converted to pyruvate for use as fuel, or to glucose, for export to the rest of the body.

When glucose is in short supply, or where there is a need for continuous activity (for example, in the heart or in postural [slow-twitch] muscles), cells preferentially use ketone bodies for most of their oxidative needs, then fatty acids, and only thereafter, glucose. However, the brain is an obligate user of glucose, unless it has become acclimated to use ketone bodies for half of its fuel needs (for example, in starvation). This acclimation is a process that requires several days to a week or two

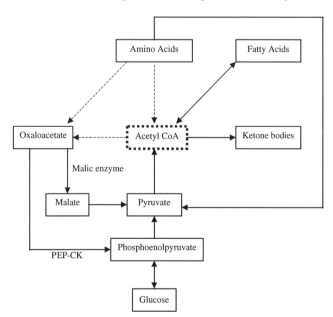

FIGURE 5.1 Interconversion of macronutrients/major tissue components. The dotted arrows indicate that different amino acids participate in different branches of these interconversions, for example, by entering the citric acid cycle at different points: tyrosine enters at fumarate, isoleucine and methionine enter at succinyl-CoA via proprionyl CoA, while valine enters directly at succinyl-CoA; and histidine, proline and arginine enter at α-ketoglutarate via glutamate. PEP-CK: phosphoenolpyruvate carboxykinase.

to become fully operative. Therefore under conditions of early starvation and of metabolic stress, hormonal adaptations are designed to shunt available carbon backbones from amino acids to the formation of glucose to fuel the brain, while peripheral utilization of glucose is diminished, and breakdown of fatty acids and utilization of ketone bodies is increased.

Role of Pituitary, Thyroid, Adrenal, and Pancreatic Hormones in Regulating Fuel Metabolism

Table 5.1 shows the effects of the pituitary, thyroid, adrenal, and pancreatic hormones on fuel metabolism and metabolic rate. As can be seen, two major anabolic hormones — insulin and growth hormone — and a large number of hormones (e.g., glucagon, cortisol, epinephrine) respond to metabolic needs by increasing catabolism, and thereby fuel availability.

These hormones can also be classified according to the timing of their release. Thus, insulin and glucagon are secreted rapidly and in direct response to need. Insulin is secreted when the pancreatic beta cells sense that blood glucose and amino acid levels are rising or are high. The presence of glucose and amino acids in the intestinal lumen causes gastrointestinal inhibitory peptide to be secreted, which in turn induces insulin secretion. Similarly, glucagon is secreted when the alpha cells directly sense that blood glucose and amino acid levels are low. Epinephrine can also be released rapidly, as anyone who has had a sudden scare can appreciate.

By contrast, it takes time for the increase in metabolic rate induced by thyroid hormone to make itself evident. A single dose of thyroid hormone is followed by a "latent period" of 6 to 10 hours, while its effect lasts for several days to a week. Similarly, growth hormone takes several hours before it has maximum effect on plasma free fatty acids. The effects of cortisol are also delayed, but can last for several hours.

Role of Pituitary, Thyroid, and Adrenal Hormones in Regulating Appetite

Regulation of appetite is a complex and incompletely understood process, in which both hormones and blood fuel levels play a role. Abnormally high or low levels of the pituitary, adrenal, and thyroid hormones lead to changes in appetite. Such changes pose significant problems for people with pathological levels of these hormones.

The ventromedial nucleus of the hypothalamus is the area of the brain about which most is known; it serves as a satiety center, which signals to the rest of the brain that food consumption is no longer required to maintain homeostasis. Elevated blood glucose levels activate this nucleus, provided sufficient insulin is available to allow glucose to move into its cells.* By responding to fuel availability as it is "perceived" by peripheral cells, this nucleus can modify appetite in response to blood

---

* Note that unlike other parts of the central nervous system, the ventromedial nucleus of the hypothalamus requires insulin for glucose transport. As a consequence, in conditions of low insulin availability, for example in untreated Type I diabetes mellitus, the patient feels hungry even though blood glucose levels are high.

TABLE 5.1
Effects of Selected Hormones on Fuel Flux

	Epinephrine	Glucagon	Insulin	Cortisol	Growth Hormone	Thyroid Hormones
Metabolic Rate	Marked increase	Increased	[a]	Marked increase	Increased	Marked increase
Lipolysis	Increased	Increased	Decreased	Both increase and decrease[b]; permissive effect on the lipolysis due to epinephrine and growth hormone	Increased	
Lipogenesis	Decreased	Decreased	Increased	Decreased	Decreased	Decreased
Insulin secretion/ insulin resistance	Secretion inhibited through α-adrenergic pathways (major effect); stimulated through β-adrenergic pathways	Secretion stimulated	Secretion inhibited	Increased insulin resistance except in heart muscle	Increased insulin resistance in muscle	Secretion stimulated[d]
Glycogen formation	Decreased	Decreased	Increased	Increased	Increased in liver	Decreased
Gluconeogenesis	Increased	Markedly increased	Decreased	Increased	Increased	Increased[c]
Protein synthesis		Decreased	Increased	Increased in liver, decreased at other sites	Increased	Increased or decreased[c]
Protein catabolism	Increased	Markedly increased	Decreased	Increased peripherally	Decreased	Increased or decreased[c]

[a] While insulin does not directly affect metabolic rate, it can increase the respiratory quotient (RQ), through an increase in fat deposition. Fat deposition is a process that is energetically expensive, with RQ = 5.

[b] Lipolysis is enhanced in the extremities and lipogenesis is increased in the trunk and face.

[c] In children and at low doses, thyroid hormone enhances protein synthesis, in conjunction with growth hormone; in adults, thyroid hormone enhances protein catabolism and gluconeogenesis.

[d] Thyroid hormones increase insulin secretion indirectly, by increasing the metabolic rate. Hyperthyroidism can cause a diabetic state over time, probably due to the demand on insulin secretion secondary to the increased metabolic rate.

glucose and insulin levels, and in response to insulin resistance. The levels of both blood glucose and insulin are, in turn, regulated by the levels of the other hormones. Table 5.1 indicates the effects of these hormones on insulin secretion and resistance, and on glucose availability.

In general, hormones that increase metabolic rate and glucose utilization, or which cause insulin resistance, increase appetite. Thus, excessive secretion of cortisol or thyroid hormones will dramatically increase appetite, while hypothyroidism and hypocorticalism are associated with anorexia.

At the same time, certain neurotransmitters can affect both appetite and the release of hormones. In general, any substance that increases thyroid hormone level or effect will raise appetite and metabolic rate, while any substance that decreases thyroid hormone level and effect will lower appetite and metabolic rate. Thus serotonin decreases appetite and may also decrease hypothalamic release of thyrotropin releasing hormone (TRH).

THE THYROID

Thyroid Hormones

The thyroid secretes two major hormones, thyroxine (T4) and triiodothyronine (T3), as well as a number of minor iodinated compounds which are low in biological activity, including reverse T3 (rT3). Tyrosine is the amino acid precursor of the thyroid hormones.

The Thyroid Gland

The thyroid gland is formed of two lobes, each of which consists of highly vascularized follicles innervated by both the sympathetic and parasympathetic systems. The lumen of each follicle is filled with a colloid material, primarily of thyroglobulin, secreted by the surrounding epithelial cells.

Production of Thyroid Hormone

In order to release thyroid hormone, the epithelial cells take up thyroglobulin by endocytosis and hydrolyze the peptide bonds that bind the hormones to the thyroglobulin, a process that is controlled by thyroid stimulating hormone (TSH). The hormones are then released into the circulation. In the periphery, thyroxine (T4) is deiodinated to form triiodotyrosine (T3), to the extent that 85% of the circulating T3 is derived from T4.

Thyroid hormones are present in the circulation either in the free form or bound to proteins, primarily thyroxine-binding prealbumin. The free forms are responsible for the metabolic activity of the hormones.

Control of Thyroid Hormone Levels

T3 serves as a negative feedback regulator for the release of thyroid hormone from the thyroid. Low levels of T3 lead to the release of thyrotropin releasing hormone

(TRH) from the hypothalamus. TRH, in turn, causes release of thyroid stimulating hormone (TSH) by the anterior pituitary. TSH, in its turn stimulates the thyroid to release T4, T3, and rT3 through hydrolysis of thyroglobulin. High levels of free T3 or T4, or of insulin-like growth factor 1 (IGF-1) cause the hypothalamus to release somatostatin, which inhibits the release of TSH. T4 can also inhibit TSH release through its conversion to T3 in the pituitary.

Stress and morphine suppress, and cold enhances, norepinephrine release in the brain. Norepinephrine, acting through α receptors, causes the hypothalamus to release TRH. Serotonin also appears to down regulate hypothalamic secretion of TRH.

Roles of Thyroid Hormones

Thyroid hormones bind to regulatory genes in cell nuclei; the higher the binding affinity, the greater the biologic activity, with T3 being about 10 times more active than T4.

The primary effect of thyroid hormones in cells is to increase oxygen consumption, that is, the cells' metabolic rate, through a number of different mechanisms, which have not yet been elucidated. The hypermetabolic effects of the thyroid hormones are found in all cells except those of the adult brain, anterior pituitary, gonads, uterus, and reticuloendothelial system. One mode of action is through an increase in the number of Na^+/K^+ ATPase pumps in the cell membrane. These pumps are ubiquitous and are the major users of the cell's energy supplies.

A second mechanism for the increase in metabolism is through an increase in sympathetic nervous system activity, with increases in both catecholamine secretion and in β-adrenergic receptors. This latter increase amplifies the effect of catecholamines on metabolic rate (Silva, 1995).

Fuel is needed to sustain this increase in metabolic rate. Consequently, thyroid hormones increase the intestinal absorption rate of dietary carbohydrates. However, it may be difficult to eat enough to compensate for the elevated metabolic rate and glucose utilization. At the same time, liver glycogen stores are depleted because of impaired synthesis. Liver glycogen depletion enhances glucagon secretion, and consequently gluconeogenesis and protein catabolism (Kabadi, 1992). Fat catabolism is also enhanced, through activation of hormone-sensitive lipase by catecholamines. Thus the body "burns" both dietary and body carbon backbones.

Other Effects of Thyroid Hormones

Other important effects of the thyroid hormones include:

- Enhancement of gene transcription for:
  - Growth hormone, which promotes protein synthesis and positive nitrogen balance; these processes may partially counteract the protein catabolism due to increased gluconeogenesis
  - The calcium pump in the muscle's sarcoplasmic reticulum, enabling more rapid on–off responses to muscle activation and relaxation
  - The myosin heavy chain α, which has greater myosin ATPase activity than myosin heavy chain β

- NADP malate dehydrogenase ("malic enzyme"), which converts malate derived from oxaloacetate to pyruvate; pyruvate is used for a host of reactions (Figure 5.1)
- The insulin responsive muscle/fat glucose transporter gene
- Cell LDL receptors, with resultant drops in plasma LDL cholesterol levels
- The conversion of carotene to vitamin A
- Decrease in the transcription of the genes coding for the TSH subunits; this inhibition provides for feedback inhibition of T3 formation
- Long-term modulation of developmental processes in the fetus, infant, and child

Definitions of Hyper- and Hypothyroidism

Hyperthyroidism can be defined as the condition associated with increased circulating thyroid hormone levels, and hypothyroidism as the condition associated with decreased levels of circulating thyroid hormone levels. Subclinical hyperthyroidism is characterized by serum T3 and T4 levels within normal limits, coupled with decreased TSH levels and subtle hemodynamic changes (Faber et al., 2001), possibly decreased muscle mass (Lovejoy et al. 1997), lipid abnormalities (Tsimihodimos et al., 1999), and increased bone destruction (Krakauer and Kleerekoper, 1992), suggesting that the person is hyperthyroid at the level of tissues.

Hyperthyroidism and Thyrotoxicosis

Causes

Hyperthyroidism is the term used to describe a variety of conditions all characterized by elevated thyroid hormone levels. *Thyrotoxicosis* is the term used for the condition where elevated thyroid hormone levels are accompanied by clinical signs and symptoms.

These conditions can be divided into:

- Immunologically driven:
  - Graves' disease, caused by overproduction of a thyroid stimulating IgG (TSI); Graves' disease is the most common cause of hyperthyroidism in the U.S.
  - Lymphocytic thyroiditis
  - Subacute granulomatous thyroiditis
- Driven by excessive production of TSH or other stimulator:
  - TSH-producing adenoma
  - Pituitary resistance to suppression of TSH production
- Driven by human chorionic gonadotropin (HCG):
  - Trophoblastic tumors
  - Hyperemesis gravidarum

- Driven by autonomous production of thyroid hormone (i.e., not responsive to feedback inhibition):
  - Toxic multinodular goiter
  - Thyroid carcinoma
  - Ovarian teratoma with active thyroid tissue (struma ovarii)
- Driven by hyper-responsiveness to TSH:
  - Hyperfunctioning solitary thyroid adenoma
- Driven by excessive dietary iodine intake:
  - Jodbasedow syndrome

There are also a number of rare syndromes associated with genetic defects at virtually any step along the pathway of thyroid hormone secretion and action, from the hypothalamus to the peripheral cell. The vast majority of these defects cause thyroid function and tissue responsiveness to decrease.

Signs and Symptoms of Hyperthyroidism, with Their Nutritional Consequences

The signs and symptoms of hyperthyroidism can be divided into those associated with excessive hormone levels, and those associated with the conditions that cause hyperthyroidism. For the purposes of this discussion, we will examine only the signs and symptoms of hyperthyroidism itself, which are primarily associated with hyper-metabolism and increased sensitivity to sympathetic nervous system activation. Table 5.2 lists these signs and symptoms with their nutritional consequences.

As noted above, it has become increasingly apparent that subclinical hyperthyroidism, with values for T3 and T4 within the normal range, but with depressed TSH levels, is associated with milder forms of the complications associated with frank hyperthyroidism. The clinician should be alert to the possibility of hyperthyroidism in patients who present with these problems.

Medical Management of Hyperthyroidism

Treatment of thyrotoxicosis depends on the cause of the disease and the location of the abnormality. For example, thyroiditis is treated through immunosuppression, while Graves' disease can be treated by suppressing thyroid hormone secretion, through antithyroid drugs (propylthiouracil and methimazole), radioactive iodine administration, or surgery to decrease the amount of active thyroid tissue. The aspects of thyrotoxicosis that are due to excessive activation of the sympathetic nervous system are amenable to treatment with β-blockers. The ophthalmic complications of Graves' disease require specific treatment.

Nutritional Management of Hyperthyroidism

While hyperthyroidism itself cannot be treated through nutritional means, its complications require nutritional care.

TABLE 5.2
Common Signs and Symptoms of Hyperthyroidism with Their Nutritional Correlates

Sign/Symptom	Nutritional Correlates	Nutritional Management
Weight loss coupled with increased hunger	Metabolic output in excess of intake	Increase caloric intake
Fatigue	Difficulty with the activities of daily living including shopping, cooking, and eating	Increase caloric intake
Nervousness, anxiety, and emotional lability	Difficulty engaging in therapeutic measures, e.g., failure to take medications, failure to understand dietary advice	Reassurance; provision of clear, unambiguous instructions
Tremor	Difficulty handling cooking and eating utensils	Adjust diet choices
Hair loss, nail loss, and moist smooth skin	Together with mental changes, may lead to lack of self-care	
Excessive perspiration	Dehydration	Include fluid intake in diet prescription
Rapid intestinal transit time with increased gastrointestinal hypermotility, hyperdefecation, and diarrhea	Malabsorption	For diarrhea: high fiber diet, inclusion of psyllium gum products in medical regimen; For malabsorption:
Malabsorption		Replenishment with fat-soluble vitamins, calcium, magnesium, and B vitamins
High output heart failure; particularly, cardiac decompensation in persons with cardiovascular disease	Sodium intolerance	Cautious reduction in sodium intake; water intake and loss, and blood pressure should be monitored carefully; treatment of the hyperthyroidism results in improvement
Bone demineralization	Increased release of calcium from bone	Increase dietary calcium and vitamin D

Nutritional Depletion

Hypermetabolism and diarrhea can cause persons with thyrotoxicosis to become severely nutritionally depleted. Animal experiments suggest that this depletion may be more severe the higher the omega-6 polyunsaturated fatty acid content of the diet (Deshpande and Hulbert, 1995), possibly through the effects of these fatty acids in increasing liver malic enzyme activity.

Both hypermetabolism and diarrhea can be rapidly and dramatically reversed by adequate medical management of the hyperthyroidism. Once this reversal has occurred, nutritional depletion can be treated.

Nutritional depletion may also be due to malabsorption consequent to the diarrhea, and to fuel overproduction and tissue loss.

Management of Nutritional Depletion due to Malabsorption

The diarrhea of thyrotoxicosis is accompanied by steatorrhea. Steatorrhea causes malabsorption of:

- Calcium, magnesium, and other divalent cations
- Fat-soluble vitamins, A, D, E, and K

In addition, there may be malabsorption of some water-soluble vitamins, particularly folate and thiamin. Thus, all patients with thyrotoxicosis and diarrhea should receive multivitamin and mineral supplements. No specific amounts can be specified, so a standard preparation that provides daily recommended intakes should be adequate.

Management of Nutritional Depletion due to Fuel Overproduction

Nutritional replenishment can be accomplished orally, unless other medical or surgical problems exist precluding the use of the gastrointestinal tract. Calculation of caloric requirements should be based on the person's normal appropriate weight for height. A normal balanced diet can be used, coupled with supplements to ensure the required caloric intake, provided the degree of malnutrition is not too severe and the introduction of the caloric increase occurs over a week or two.

Refeeding Syndrome

If the degree of malnutrition is moderate to severe (BMI less than 17), the chances that the patient will develop refeeding syndrome increase, so careful attention should be paid to caloric, electrolyte, and water load. Refeeding syndrome usually occurs within the first 5 days of the start of caloric repletion. It consists of glucose intolerance, fluid overload with electrolyte imbalance, diarrhea, congestive heart failure, and cardiac arrhythmias (Klein and Jeejeebhoy, 1997).

With malnutrition and malabsorption, patients are usually markedly potassium, phosphate, and magnesium depleted, although blood concentrations of these cations tend to remain normal, because of the parallel shrinkage in total body fluid content. At the same time, Na^+/K^+ ATPase pumps systems are hypoactive and cell membranes are leaky. The person may be thiamin depleted, and therefore unable to move glycolysis past the formation of pyruvate, as a result of which lactic acid can accumulate. On refeeding, the provision of carbohydrates activates insulin release.

Insulin responses may be hyperactive due to prior suppression that occurred in the presence of hypermetabolism, starvation, and accelerated gluconeogenesis.

Under these circumstances, carbohydrate and fluid loads can lead to

- Hypokalemia, due to rapid movement of potassium into cells
- Phosphate depletion with muscle weakness and nerve function impairment, arrhythmias, cardiopulmonary failure, seizures, coma, and death
- Hyperinsulinemia, with renal sodium and water retention, which contributes to the likelihood of congestive heart failure

To decrease the chances of a patient developing refeeding syndrome, refeeding should begin with a dose of multivitamins and minerals, and proceed to the slow addition of calories and small frequent feedings, with careful attention to fluid, electrolyte, and cation balance.

Long-Term Management

Long-term management of hyperthyroidism may be complicated by the persistence or recurrence of hyperthyroidism, to be managed as described above. More commonly, the patient may develop hypothyroidism, leading to the potential for excessive weight gain that continues over time (Dale et al., 2001). Patients who were obese prior to treatment, who have Graves' disease, who experienced weight loss, and who became hypothyroid are susceptible to this complication.

Resolution of the hyperthyroidism does not resolve the ocular problems of Graves' disease. These consist of proptosis, diplopia, difficulties with upward and eventually lateral gaze, difficulty closing the eyes, scleral and corneal desiccation and abrasions, blurred vision, decreased visual acuity with field defects, and impaired color vision. These have nutritional consequences insofar as the visual disturbances, discomfort, and disfigurement may render shopping, cooking, serving food, and eating difficult. The clinician should make careful and sympathetic inquiry into the person's eating-related activities and feelings about eating to determine problems and to develop possible solutions.

Drug–Nutrient Interactions in Hyperthyroidism

The most common side effects of antithyroid drugs include rash, urticaria, joint pains, fever, and decreased white blood cell counts, none of which have notable nutritional consequences. Rarer side effects without direct nutritional consequences include clotting problems due to thrombocytopenia and hypoprothrombinemia, agranulocytosis and aplastic anemia, and lupus-like syndromes with vasculitis. Note, however, that patients who develop agranulocytosis or aplastic anemia should engage in cooking practices which minimize exposure to food-borne pathogens.

Rarer side effects, which do have nutritional consequences, include decreased sense of taste, diarrhea, hepatitis, cholestatic jaundice, and the formation of anti-insulin antibodies leading to hypoglycemia. Management of each of these complications should be individualized.

Thyroid Storm

Thyroid storm is a life-threatening condition which cannot be identified by laboratory tests, but can be anticipated in persons with severe thyrotoxicosis who decompensate as a result of severe illness, sepsis, injury, or major surgery (Tietgens and Leinung, 1995).

The mechanism for the development of thyroid storm varies from a decrease in available binding globulin to an increase in inflammatory cytokines such as interleukin 1 and tumor necrosis factor alpha.

Characteristic signs and symptoms include extreme elevation of body temperature, extreme agitation and restlessness, with the potential for developing psychosis, and even coma, cardiac decompensation with hypotension, congestive heart failure, and abdominal distension with signs of an acute abdominal crisis. Supportive management should begin immediately upon diagnosis and includes nutritional support with fluid, glucose, electrolytes, vitamins (thiamin), and minerals, accompanied by careful monitoring.

Hypothyroidism

Causes

Primary hypothyroidism can be divided into goitrous and atrophic hypothyroidism. Goitrous hypothyroidism may be due to:

- Dietary deficiency of iodine; note that people with extremely limited diets may develop iodine deficiency in places where dietary deficiency is not endemic (Pacaud et al., 1994)
- Inflammatory and autoimmune processes: Hashimoto's thyroiditis, subacute thyroiditis, silent lymphocytic thyroiditis/postpartum thyroiditis, gluten enteropathy (Sategna-Guidetti et al, 2001), dermatitis herpetiformis (Reunala and Collin, 1997), Down's syndrome (Micklefield and Wenzel, 2000)
- Hereditary defects in the synthesis of thyroid hormones and in the ability to respond to these hormones
- A number of drugs (propylthiouracil, methimazole, iodides, amidarone, tamoxifen, and lithium)
- Excessive consumption of goitrogens (e.g., "cabbage goiters")

Atrophy of the thyroid gland can occur:

- Following treatment for hyperthyroidism
- Following external radiation
- With excessive formation of thyroid-stimulating hormone receptor-binding antibodies
- With the atrophic thyroiditis that follows autoimmune attack
- With age, particularly in women

There are a number of naturally occurring goitrogens in edible plants. These goitrogens are usually not toxic in people who have adequate access to iodine and who eat a varied diet. However, food faddists and vegetarians can develop a form of hypothyroidism called "cabbage goiters." Vegetables of the Brassicaceae family (e.g., turnips, cabbage, and rutabagas) contain a compound, progoitrin, which can be converted to goitrin (L-5-vinyl-2-thiooxazolidine), a heat-labile compound with antithyroid activity. Cooking can destroy preformed goitrin, but heat-stable pro-goitrin is activated in the intestine, probably by intestinal bacteria. As a result, consumption of significant amounts of these vegetables, whether cooked or not, can lead to hypothyroidism (Van Etten and Tookay, 1983). Populations living in iodine-deficient areas who consume cassava (which contains thyrotoxic cyanates) may be at greater risk of developing hypothyroidism than populations living in similar areas who do not consume cassava (Biassoni et al., 1998).

Fetal alcoholism and congenital hypothyroidism share many characteristics, including microcephaly and mental retardation. These characteristics are probably due to a decrease in thyroid hormone receptor levels in the fetal brain (Scott et al., 1998).

Signs and Symptoms of Hypothyroidism, and Their Nutritional Correlates and Management

Signs and Symptoms

The signs and symptoms of hypothyroidism are, for the most part, consequences of the decreased metabolic rate and lack of responsiveness to β-adrenergic stimulation, coupled with unchecked α-adrenergic responsiveness. Thus, weight gain, lethargy and fatigue, and slowing of the thought processes are characteristic findings. Table 5.3 lists the common signs and symptoms of hypothyroidism with their nutritional correlates and suggestions for nutritional management. Silent hypothyroidism may compromise efforts at weight loss and should be considered when people on a weight-loss regimen fail to respond (Simmons et al., 2000). Silent hypothyroidism should also be considered when people present with lipid abnormalities (Tsimihod-imos et al., 1999).

In addition to the problems listed in Table 5.3, severe hypothyroidism may cause reversible panhypopituitarism, hyperprolactinemia and precocious puberty, effusions (abdominal, pleural, or pericardial), arthritis, and carpal tunnel syndrome.

The most severe form of hypothyroidism is myxedema coma, which constitutes a medical emergency (Jordan, 1995). Myxedema coma is characterized by decreased neural responsiveness, hypothermia, hypoventilation and bradycardia, hypoglyce-mia, and hyponatremia, and is most often precipitated by infection. Note that the usual accompaniments of infection (fever, tachycardia, sweating) are absent, so the clinician must have a high index of suspicion.

Specific nutritional management of hypothyroidism is directed at supplying iodine to persons who are iodine deficient, and at reduction of nutritional risk factors for cardiovascular disease. There is evidence that thyroid function may be improved by correcting the patient's hypozincemia (Bucci et al., 1999). Nutritional manage-ment of the other consequences of hypothyroidism is palliative, and medical man-agement of the underlying cause is critical to the patient's well-being.

TABLE 5.3
Common Signs and Symptoms of Hypothyroidism with Their Nutritional Correlates

Sign/Symptom	Nutritional Correlates	Nutritional Management
Weight gain	Intake in excess of metabolic output	Decrease caloric intake
Lethargy and fatigue	Difficulty with the activities of daily living including shopping, cooking, and eating	Help with activities of daily living
Slowing of thinking processes	Difficulty engaging in therapeutic measures, e.g., failure to take medications, failure to understand dietary advice	Help with activities of daily living
Depression	Loss of interest in life, and in interaction with others, failure to care for self and to engage others in caring for self	Adjust diet choices to increase enjoyment; help with activities of daily living
Hair loss and dry skin	Together with mental lethargy and depression, may lead to lack of self-care	Help with activities of daily living
Paresthesias, delayed deep tendon reflex relaxation phase	Loss of sensation in hands and slowing of reflexes may lead to self-injury, for example while cooking	Give advice concerning simple techniques for meal preparation; where feasible, have someone other than the patient prepare meals
Constipation		High fiber diet, inclusion of psyllium gum products in medical regimen. Note that there is some evidence that fiber (bran) may inhibit thyroxine absorption.
Hypercholesterolemia due to decreased hepatic LDL receptors	Increased risk for cardiovascular disease	Consumption of a low fat/high fiber diet. See note about inhibition of thyroxine absorption by fiber.
Carotenemia	Undesirable skin discoloration	Reduction of dietary carotenoid intake.
Metrorrhagia	Iron depletion	Iron supplementation. Note that iron supplements, either on their own or formulated within a vitamin pill, decrease thyroxine absorption.
Decreased stomach acid secretion with malabsorption	Iron deficiency; vitamin B12 deficiency	Supplementation; note caution concerning malabsorption of thyroxine when taken with iron supplements.
Non-pitting edema (myedema), bradycardia, diastolic hypotension, hyponatremia	Myxedema *should not be* confused with pitting edema.	Sodium restriction may be detrimental

Medical Management of Hypothyroidism

Hypothyroidism is best managed by thyroxine replacement. If the patient's pituitary function is intact, treatment is directed at normalizing TSH levels. If the patient has central hypothyroidism, due to pituitary or hypothalamic disease (secondary hypothyroidism), cortisol as well as thyroxine should be administered.

Nutritional Management of Hypothyroidism

Nutritional management of hypothyroidism should be coordinated with medical management. As noted in Table 5.3, nutritional problems associated with hypothyroidism are the consequence of problems relating to weight gain, self-care, mental depression, and depressed gastrointestinal and cardiovascular function.

While precise medical management of hypothyroidism with hormone replacement is conceptually simple, in practice patients and their physicians may experience difficulties in regulating thyroid status. Particularly in those with mild hypothyroidism, determination of the correct dosage of levothyroxine can be difficult. As a consequence, patients under treatment may continue to experience one or more of the problems noted in Table 5.3. Perhaps the most common problems are difficulty in maintaining normal weight and the development of cardiovascular disease.

Dietary gluten elimination may reverse hypothyroidism in patients with celiac disease (gluten enteropathy; Sategna-Guidetti et al., 2001).

Failure to lose weight despite adherence to a proper regimen should alert the clinician to the possibility of hypothyroidism, which should be diagnosed and treated appropriately (Simmons et al., 2000).

The nutritional management of myxedema coma includes correction of hyponatremia (primarily by water restriction, but, when severe, by the cautious administration of saline) and the correction of hypoglycemia with glucose administration.

Drug–Nutrient Interactions in Hypothyroidism

Levothyroxine should be taken on an empty stomach and without other medication, because absorption is better and more predictable in these states (Utermohlen, 1999). Iron preparations and multivitamins containing iron interfere with levothyroxine absorption, as do aluminum hydroxide preparations, and possibly calcium supplements (Utermohlen, 1999). Note that a number of drugs used to treat the complications of hypothyroidism also increase levothyroxine requirements. These include sertraline, estrogens, lovastatin, rifampin, dilantin, carbamazepine, phenobarbital, sucralfate, and cholestyramine.

Glucocorticoid Disease: Hyper- and Hypocorticalism, and Drug-Induced Glucocorticoid Disease

Definition of Glucocorticoid Disease

Glucocorticoid disease is caused by abnormal levels of the glucocorticoids, principally cortisol. These can be the result of excessive or too little endogenous production of cortisol, or can result from the administration of adrenocorticotropic hormone (ACTH) or exogenous drugs with corticosteroid activity.

Control of Cortisol Biosynthesis

Synthesis of cortisol is controlled by the brain, in part as a response to cortisol levels in the bloodstream. The brain signals the hypothalamus to release corticotropin-releasing hormone (CRH), which in turn signals the basophilic cells of the anterior pituitary to release ACTH. ACTH, in turn, signals the adrenal cortex (zona fasciculata) to produce cortisol. Plasma cortisol levels regulate production of CRH and ACTH by feedback inhibition. Low levels of cortisol signal the hypothalamus to produce CRH, while high levels decrease production of CRH rapidly, and the production of ACTH more slowly, in a delayed fashion.

The brain signals that induce cortisol secretion wax and wane in a circadian pattern, as well as in response to physical and emotional stress. The diurnal rhythm is characterized by pulsatile release of ACTH during sleep, which leads to pulsatile secretion of cortisol from the adrenal cortex. The released cortisol is bound to cortisol-binding globulin, also known as transcortin, which prevents cortisol from being degraded in the liver and thus prolongs its half-life. As a result, ACTH secretion leads to a gradual rise in cortisol levels which peak in the morning hours. Levels then decrease during the day, to reach their lowest concentration in early evening. Physical or emotional stress (e.g., depression) can eradicate this normal rhythm, leading to high levels throughout the day as well as at night. In stress, these high levels fail to be effectively suppressed by exogenous corticosteroid, such as dexamethasone, which suggests that in these conditions feedback inhibition of ACTH secretion is limited or abolished.

Hypercorticalism and Drug-Induced Glucocorticoid Disease

Signs and Symptoms of Hypercorticalism

Hypercorticalism, also known as Cushing's syndrome, is characterized by a constellation of signs and symptoms listed in Table 5.4. Cachexia, rather than truncal and nuchal obesity, is characteristic of patients with elevated corticosteroid levels due to malignancy, particularly small cell carcinoma of the lung. However, as with Cushing's syndrome, this catabolic form of hypercorticalism is also accompanied by hypertension, edema, and hypokalemia.

Conditions Associated with Hypercorticalism

Causes of hypercorticalism include:

- Hypersecretion of ACTH:
  - Hypersecretion of CRH
  - Pituitary tumors: these may be numerous and small, or single and large
  - Non-endocrine tumors which secrete substances with CRH or ACTH activity:
    - Small cell carcinoma of the lung
    - Endocrine tumors, including thyroid medullary carcinoma, carcinoid (thymic and bronchial), islet cell tumor
  - Pheochromocytoma
  - Ovarian tumors

TABLE 5.4
Signs and Symptoms of Hypercorticalism and Their Nutritional Correlates and Management

Sign/Symptom	Cause/Nutritional correlate	Nutritional management
Truncal obesity	Peripheral lipolysis with central deposition of fat	Small frequent meals; portion control; encourage exercise
Increased appetite	Excessive caloric intake	Limit caloric intake to maintenance levels
Moon face	Fat deposition; parotid enlargement	
Insulin resistance	Hyperglycemia; may develop into frank diabetes mellitus	Small frequent meals; portion control; limit carbohydrate intake; encourage exercise
Hypertension	Mineralocorticoid activity of high levels of glucocorticoids leads to tubular loss of potassium and retention of sodium	Ensure adequate intake of potassium; sodium restriction as necessary
Hypokalemic alkalosis	Mineralocorticoid activity of high levels of glucocorticoids leads to tubular loss of potassium and retention of sodium	Ensure adequate intake of potassium
Osteoporosis	Increased bone catabolism and decreased anabolism, due to inhibition of osteoblast replication and function; decreased calcium and phosphate absorption in the gastrointestinal tract (antagonism of vitamin D); increased renal calcium and phosphate excretion	Ensure adequate dietary intakes of calcium, phosphate, and vitamin D[a]; note that biphosphonates may be recommended for people receiving exogenous corticosteroids for more than 3 months
Muscle mass depletion	Protein depletion due to increased gluconeogenesis; failure to regenerate muscle mass due to decreased growth hormone secretion	Ensure adequate protein intake; encourage exercise; growth hormone production may be reversed with a hypocaloric diet (Leal-Cerro et al., 1998)
Poor wound healing	Protein depletion due to increased gluconeogenesis	Ensure adequate protein intake; encourage exercise
Acne	Excess androgen secretion; no known nutritional correlate	Reassure patient that cause is not dietary

[a] Intake of at least the RDAs should be attained; appropriate levels of intake have not been established.

- Primary hypersecretion of glucocorticoids:
  - Adrenal tumor
  - Micronodular adrenal disease
- Exogenous glucocorticoid (corticosteroid) or ACTH administration, either iatrogenic or through self-administration of herbal or other preparations that contain corticosteroids
- Food-dependent hypercorticalism

It is important to note here that many so-called "herbal" or "natural" preparations and remedies, which escape regulation by the Food and Drug Administration under the protection of the Dietary Supplement Health and Education Act of 1994, may contain significant amounts of corticosteroids. The corticosteroids are included in these preparations because the hormone can cause a sense of physical well-being, for example by decreasing the inflammation associated with arthritis.

Food-dependent hypercorticalism is a mild condition which has been described in one patient with adrenal hyperplasia (Gerl et al., 2000), but which may in fact be common and unrecognized. In this condition, the release of gastric inhibitory peptide (GIP) with meals induces the secretion of cortisol.

Surgical and Medical Management of Hypercorticalism

Medical and surgical management of hypercorticalism depends on the cause of the condition.

Endogenous hypercorticalism is most frequently due to tumor, either of the pituitary or of another organ. Surgical removal of tumors should be attempted whenever possible. The prognosis depends on the nature and location of the tumors, and the extent to which they have spread. With undetectable or widely spread tumors, ketoconazole, which blocks adrenal steroid synthesis, can be used to control corticosteroid hypersecretion.

Iatrogenic hypercorticalism cannot be avoided if the corticosteroids are required for management of an underlying condition. However, every effort should be made to minimize dosage and provide the drug as a single morning dose, to parallel the timing of normal physiological secretion.

Self-administration of corticosteroids should be actively discouraged, and any person who develops Cushingoid signs and symptoms should be queried concerning the use of corticosteroids or dietary supplements/herbal preparations that might contain corticosteroids.

Nutritional Management of Hypercorticalism

The nutritional problems posed by hypercorticalism and their management are listed in Table 5.4.

An important though often neglected problem for persons with hypercorticalism is the development of dental caries and periodontal disease, the result of demineralization, poor wound healing, and immunosuppression. These conditions may affect nutrient status by making food consumption difficult and painful. Thorough and regular tooth cleaning, and the use of fluoride rinses or prescription toothpastes are critical to preventing these problems.

Drug–Nutrient Interactions in Hypercorticalism

The following drug–nutrient interactions are of note (Utermohlen, 1999):

Ketoconazole, an antifungal drug which acts by inhibiting steroid synthesis, needs an acidic environment for its dissolution in the stomach, and therefore for its absorption. Patients with either endogenous achlorhydria or acquired suppression of stomach acid secretion (through the use of antacids, H2 histamine blockers, or proton pump inhibitors), may have problems obtaining therapeutically effective doses of this drug.

The consumption of black licorice, even in small amounts (a couple of licorice twists a day) enhances the mineralocorticoid effect of endogenous glucocorticoids. Licorice contains glycyrrhizic acid, a compound which inhibits 11-β-hydroxysteroid dehydrogenase activity. Inhibition of this enzyme blocks the conversion of cortisol to cortisone, the pathway for cortisol inactivation. In addition to licorice candy, chewing tobacco is a major source of licorice.

Hypocorticalism and Addison's Disease

Signs and Symptoms of Hypocorticalism

Hypocorticalism occurs in two forms, acute and chronic, although in most patients the manifestations are usually intermediate between these two forms. The chronic syndrome is characterized by vague feelings of malaise, accompanied by anorexia, weight loss, orthostatic hypotension, and, less frequently, vague abdominal pain, or skin hyperpigmentation. The onset of the acute syndrome is rapid and, if untreated, can quickly lead to hypotensive shock, coma, and death. The onset is characterized by arterial hypotension, hypoglycemia, fever, abdominal pain, confusion, and agitation. The shock that develops is not responsive to the usual pressor agents, but is the consequence of cardiac failure in the face of increased circulating blood volume.

Table 5.5 indicates the signs and symptoms of chronic hypocorticalism which require nutritional care.

Causes of Hypocorticalism

Causes of hypocorticalism include:

- ACTH-dependent (secondary hypocorticalism):
  - Suppression of the hypothalamic–pituitary-adrenal axis, coupled with too rapid withdrawal of chronic administration of corticosteroids or ACTH
  - Lesions of the hypothalamic–pituitary axis
  - Isolated ACTH deficiency
  - Head trauma
  - Sarcoidosis
- ACTH-independent:
  - Destruction of the adrenal gland:
    - Autoimmune; may be associated with multiple endocrine abnormalities and gluten intolerance (Valentino, et al., 1999)

- Infectious disease of the adrenal gland, including tuberculosis, fungal infection, HIV infection
- Adrenal hemorrhage
- Metastatic cancer
- Sarcoidosis
- Amyloidosis
- Adrenoleukodystrophy
- Adrenomyeloneuropathy
- Ketoconazole administration
- Congenital adrenal hyperplasia

Medical Management of Hypocorticalism

Medical management of hypocorticalism depends on whether the adrenal gland is destroyed or not. If the adrenal gland is intact, mineralocorticoid secretion may be preserved, while it is not if the adrenal is destroyed.

Medical management of hypocorticalism, either isolated or coupled with mineralocorticoid deficiency, depends on replacement of the missing hormone(s). The amount of corticosteroid needed for replacement is determined by "titration." That is to say, a good replacement dose is achieved when the patient experiences neither hyper- nor hypocorticalism. This is easier on paper than in practice: needs vary considerably from person to person; stress, either physical or mental, increases cortisol requirements by amounts that are difficult to determine. In particular, the degree of mental stress is difficult to determine. By contrast, physical stress can be obvious: fever, nausea and vomiting, trauma, and surgical procedures, even dental extraction. A rule-of-thumb for treating physical stress is to administer twice the normal dose of hormone for the duration of the stress.

TABLE 5.5

Signs and Symptoms of Hypocorticalism and Their Nutritional Correlates and Management

Sign/Symptom	Cause/Nutritional Correlate	Nutritional Management
Anorexia, weight loss, lethargy	Lack of glucocorticoid effects on the brain	Small frequent meals; ensure adequate caloric intake
Relative hypoglycemia	Depressed gluconeogenesis and lipolysis	Acute hypocorticalism: administration of glucose IV
		Chronic hypocorticalism; small frequent meals; adjustment of insulin dosage in insulin-dependent diabetics
Salt craving, hyponatremia, hyperkalemia, hypotension		Acute hypocorticalism: immediate intravenous correction of sodium and fluid status
		Chronic hypocorticalism, correction is accomplished by fludrocortisone

Estimation of the success of replacement therapy depends on the presence or absence and severity of clinical signs and symptoms of either hyper- or hypocorticalism. Weight gain or loss and an increase or decrease in appetite are sensitive measures that should be interpreted in light of the nutritional measures noted below. For example, weight gain due to deposition of fat in the trunk, and in particular in the supraclavicular area, are due to hypercorticalism rather than a reflection of adequate caloric replacement.

If the adrenal gland is destroyed, mineralocorticoid function is also lost, so replacement with fludrocortisone acetate is required. This drug restores appropriate renal retention of sodium and excretion of potassium. Because of its broad therapeutic index, a single daily dose of the drug is sufficient for most patients, although it is prudent to monitor plasma potassium levels on occasion.

Nutritional Management of Hypocorticalism

The nutritional management of chronic hypocorticalism is outlined in Table 5.5. As with other hormone-dependent conditions where weight loss, anorexia, and gastrointestinal disturbances are prominent, correction of hormonal balance, for example by corticosteroid replacement, leads to improvement in both weight and appetite. In acute hypocorticalism, correction of the hypoglycemia is critical.

During the maintenance phase, after hormonal treatments have begun, caloric requirements should be based on the patient's appropriate weight. The gastrointestinal disturbances are not associated with malabsorption, so no specific replacement is required. The normocytic normochromic anemia that accompanies hypocorticalism is due to hormonal deficiency rather than to nutrient deficiency, so in itself it does not require specific nutritional treatment. However, if hormone replacement does not result in recovery from the anemia, other causes for the anemia should be sought. The patient may have had longstanding anorexia, which may have led to specific nutrient deficits, e.g., of iron or folate/vitaminB12.

With hormone replacement, attention should be paid to the patient's overall nutritional status, especially if the patient develops signs and symptoms of hypercorticalism, which can lead, as noted earlier, to a number of nutritional deficits.

Mineralocorticoid deficiency does not require specific nutritional treatment if the patient is receiving fludrocortisone acetate, although an occasional check of potassium status is recommended. If for some reason the person is not taking this drug, the diet should contain enough sodium to compensate for the renal sodium loss, i.e., about 10 mEq/kg/d. Small children, the elderly, food faddists, and patients with cardiovascular disease, renal disease, or other problems related to salt balance can be treated with fludrocortisone acetate, coupled with the dietary treatment appropriate to their condition. Careful monitoring of blood pressure and cardiovascular functioning should accompany this regimen. It is important to take into account causes of excessive sodium loss, such as sweating, which are easily overlooked.

Drug–Nutrient Interactions in Hypocorticalism

The drug–nutrient interactions in hypocorticalism are those noted for hypercorticalism. Fludrocortisone and prednisone or other cortisol replacement should be taken with the morning meal to ensure appropriate timing and degree of absorption.

DISEASES OF CARBOHYDRATE INTOLERANCE

INTRODUCTION

Definition of Diseases of Intolerance

Adverse reactions to food can be classified as food hypersensitivity, pharmacological intolerance of food, metabolic intolerance of food, and food toxicity (Guarderas, 2001).

Food hypersensitivity refers to the condition where the food elicits an immune reaction. The common term for food hypersensitivity is *food allergy*. For example, a person who develops IgE antibodies to, say, peanut proteins, and who develops anaphylactic shock upon coming into contact with these proteins, has peanut protein hypersensitivity.

By contrast, in food (or food component) intolerance, the immune system becomes involved only secondarily, if it is involved at all. In metabolic intolerance, the primary problem lies in the person's inability to properly metabolize the food component. An example is galactosemia. Persons with this condition are galactose intolerant because they cannot metabolize galactose to glucose. In pharmacological intolerance, normal amounts of a foodstuff can have a pharmacological effect in a susceptible individual. An example is licorice intolerance: individual susceptibility to the mineralocorticoid effect of glycyrrhizic acid (a component of licorice) varies from person to person, but all persons will experience adverse effects at high doses. Finally, certain foods can cause the development of toxic reactions in susceptible individuals. Gluten intolerance is generally considered an example of this type of intolerance.

It is worth noting that diseases of intolerance are very commonly misunderstood by the lay public, often because of a failure to differentiate between intolerance and allergy. Many believe that a person can outgrow an intolerance, or that, by increasing exposure, the body "gets used to it." Once a diagnosis is made, the clinician should explore the patient's and the patient's caregivers' perceptions of intolerance and correct any misconceptions. This process may take repeated visits and prolonged effort.

Types of Clinical Problems Associated with Carbohydrate Intolerance

The problems associated with carbohydrate intolerance depend on whether the problem lies in the absorption or in the metabolism of the carbohydrate in question.

Carbohydrate malabsorption leads to passage of unabsorbed carbohydrate into the colon, which can lead to feelings of bloating, pain due to intestinal distention, the production of intestinal gas, and an osmotic diarrhea. A person in good health can tolerate these problems without developing clinical illness. The problems can become more significant if the person is already ill and at risk for the development of dehydration in the presence of osmotic diarrhea.

The severity of these signs and symptoms depend on:

- The digestive capacity for the carbohydrate in question
- The intestine's absorptive capacity

- The quantity of carbohydrate consumed — the quantity necessary to cause symptoms varies depending on digestive and absorptive capacities
- The rate at which the stomach empties — note that stomach emptying is slowed by undigested carbohydrates in the intestinal lumen
- The amount of water and electrolytes that enter the small intestine and the rate at which they enter, in response to the carbohydrate's osmotic effect
- The extent of increased intestinal mobility and decreased transit time, in response to this osmotic load
- The rate of production of gas and short-chain fatty acids (and therefore the rate of removal of the carbohydrate) by the colonic flora, and the intestine's response to these products

The diagnosis of carbohydrate malabsorption will depend on the timing of the signs and symptoms, and their association with the consumption of a possible offending carbohydrate. The diagnosis also depends on the excretion of an acid stool, coupled with a lack of other causes for the signs and symptoms of the disease. The clinician should be alert to hidden sources of carbohydrate (for example, in medications).

By contrast, metabolic intolerance of a carbohydrate leads to signs and symptoms that depend on the type of carbohydrate and the point in the carbohydrate's metabolic pathway that is affected. Infants are usually affected immediately after birth, so management must be lifelong.

The types of hereditary carbohydrate intolerance and their associated defects are listed in Table 5.6.

General Principles of the Nutritional Management of Diseases of Intolerance

Management of diseases of intolerance is predicated on limiting the dietary intake of the substance that is poorly tolerated. If this substance is not necessary for survival, then elimination is compatible with a normal life, and the problem of management lies simply in recognizing the dietary sources of the substance. If, however, the substance or its metabolites are required for life or health, then it is critical to titrate the amount of the substance in the diet to sail between the Scylla and Charybdis of lack and excess. Thus, the management of lactose intolerance is the elimination of lactose from the diet. By contrast, phenylketonuria cannot be managed by the elimination of phenylalanine from the diet; instead, intake of phenylalanine must be carefully calculated to prevent the development of problems of both deficiency and excess.

Whether complete or partial elimination of a dietary substance is the therapeutic goal, it is critical for the patient and his or her caregivers to be aware of the composition of foods. We can consider the composition at two levels. The first is composition of individual ingredients, and the second is the composition of compound foods. For example, a label for a compound food might list an ingredient called "milk solids." The name of this ingredient does not specifically indicate whether or not it contains lactose, so the person who is lactose or galactose intolerant

TABLE 5.6
Types of Carbohydrate Malabsorption and Their Associated Enzyme Deficiencies

Condition	Enzyme Deficiency	Type of Inheritance	Age at Onset	Signs and Symptoms	Treatment
Sucrose-isomaltose intolerance	Sucrase-isomaltase (α-glycosidase)	Autosomal recessive; note that this intolerance is more common than once thought, and is present at high rates in Inuit.	Usually in infancy, after feeding with sucrose or starch dextrins; rarely may begin as late as puberty; symptoms diminish in adulthood	May range from mild to severe; watery osmotic/fermentive diarrhea, with increased excretion of lactic acid, dehydration, malnutrition, failure to thrive.	Avoidance of sucrose
Trehalose intolerance (trehalose is a carbohydrate present in mushrooms and insects)	Trehalase	Autosomal recessive	Occurs only when very large quantities of trehalose are ingested	Diarrhea and flatulence	None required
Hereditary fructose intolerance[a]	Unknown defect in intestinal transport	Variable	In infancy, when fruit or foods with high fructose corn syrup is first given	Diarrhea and flatulence	Avoidance of fructose-containing foods, including fruits, but also foods containing high fructose corn syrup
Primary adult hypolactasia	Lactase; multiple mechanisms as yet to be determined; transcriptional control in the lactase gene and post-transcriptional control of lactase production in the enterocyte may both be involved.	Autosomal recessive	Early adolescence	Diarrhea and flatulence	Avoidance of lactose, depending on the extent of symptoms;
Congenital lactase deficiency	Defect in the lactase gene, as yet undefined.		Infancy	Diarrhea, flatulence and failure to thrive	Avoidance of lactose

[a] Note that the term fructose intolerance is also applied to deficiency of fructose phosphate aldolase, which is associated with vomiting, hypoglycemia, and failure to thrive, and is associated with a large number of different deficiency alleles.

needs to know *a priori* that this type of dairy product does contain lactose. Unlabeled compound foods, or foods which contain ingredients labeled as "spices" or "flavorings," or foods prepared in restaurants, pose even greater problems, as individual manufacturers and food preparers differ in their choice of ingredients. As store-bought foods become more and more complex, the person with intolerance needs to have more and more sophisticated knowledge of food composition and preparation, or needs to avoid store-bought or restaurant food altogether.

Avoidance of prepared foods in the U.S. is difficult because demands of work outside the home have made less time available for food preparation. In response to the demand for information, a number of disease-specific non-profit organizations inform consumers about foodstuffs, primarily through web pages. However, even ingredients with the same name may differ in composition, depending on where they were produced. This is particularly true for sweeteners, which are usually corn-based if manufactured in the U.S., but may be wheat-based or sugar-based when manufactured outside the U.S. People who are gluten-sensitive should pay particular attention to the source of sweetener, because they must avoid all wheat-based products.

Physicians and patients alike should also be aware of the composition of ingredients used in medications, whether prescription or over-the-counter, or "herbal" or "natural." For many, the word "natural" implies harmless, with the result that what may be toxic ingredients are believed to be harmless when included in a remedy. The clinician should explore the possibility of such a belief.

The effects of consuming a poorly tolerated substance may be immediate or long-term. Immediate reactions include nausea, vomiting, or changes in mentation, all of which may be readily associated with ingestion of the offending substance. Long-term consequences, however, are less apparent, and their effects may not be perceptible until significant damage is done. Because the effects are not as dramatic at the time of ingestion, the patient or caregiver may be tempted to "cheat." The clinician should be aware of this possibility and deal with it compassionately by exploring alternative approaches to satisfying the patient's needs. We should note that even the best-intentioned and most knowledgeable patients and caregivers may take a long time to understand what may be drastic lifestyle changes, and to successfully incorporate them into their daily routine.

Clinicians should also be aware that adolescents and adults with chronic disease may rebel against the strictures of treatment, especially when under stress or peer pressure. The support of others with the same condition may be crucial to overcoming these crises.

GALACTOSEMIA

Definition of Galactosemia

Galactosemia is a genetic condition characterized by elevated blood galactose levels and failure to convert galactose to glucose. Galactosemia is most commonly caused by mutations in the gene for galactose 1-phosphate uridyltransferase (GALT), the enzyme responsible for the second step in converting galactose to glucose and uridinediphosphogalactose (UDP-galactose). Lack of this enzyme leads to accumulation of galactose-1-phosphate, the first step in the process of galactose metabolism, which is catalyzed by galactokinase; to the excessive formation of minor metabolites, such as galactitol (a sugar alcohol); to depletion of myo-inositol; and to functional depletion of sugar nucleotides such as UDP-galactose. With respect to the latter, UDP-galactose is required for the galactosylation of a number of important molecules. The excess galactose-1-phosphate inhibits galactosylation. The two other forms of galactosemia are caused by defects in the galactokinase gene, which prevents galactose-1-phosphate formation, or in the uridine diphosphate galactose4-epimerase gene, responsible for the conversion of UDP-galactose to UDP-glucose (Novelli and Reichardt, 2000).

Note that diabetics have defects in galactose metabolism, although they do not have galactosemia (Birlouez-Aragon et al., 1993) and may benefit from galactose avoidance.

Medical Complications of Uncontrolled Galactosemia

Untreated classic galactosemia leads to vomiting and jaundice in newborns within a short time after the first milk feeding. These manifestations are accompanied by changes in liver function, immune function, renal function, brain function, and pituitary function.

Liver damage is progressive and rapidly leads to cirrhosis, with its associated complications, including edema, portal hypertension, ascites, splenomegaly, and bleeding diatheses.

Galactose and its metabolites impair immune function, and in particular the function of the hexose monophosphate shunt that granulocytes use to produce H_2O_2 for killing pathogens. Thus, untreated infants can die of osteomyelitis, meningitis, gangrene, or from overwhelming sepsis, usually caused by *Escherichia coli.*

Classic galactosemia is associated with a renal Fanconi syndrome, characterized by aminoaciduria, glycosuria, proteinuria, and hypochloremic alkalosis, all of tubular origin. Brain function changes include increased intracranial pressure with bulging of fontanelles (pseudotumor cerebri) and mental retardation. Growth failure, particularly in females, is accompanied by hypergonadotropic hypogonadism, with probable *in utero* oocyte dysgenesis. The types of complications patients develop depend on their molecular genotype, on their ability to oxidize galactose-1-phosphate, and on their ability to lower dietary galactose intake (e.g., Guerrero et al., 2000).

Nutritional Management of Galactosemia, and Complications
Associated with Nutritional Management

Nutritional management of classic galactosemia consists of lifelong elimination of galactose from the diet. Milk, dairy products, and prepared foods containing dairy products are major sources of galactose. In addition, some fruits, such as watermelons and tomatoes, are also sources of galactose (Gross and Acosta, 1991; Berry et al., 1993). Table 5.7 indicates the foods to be avoided. Infant formulas made from soybeans or casein hydrolysates (e.g., Nutramigen®) may be used to replace milk-based formulas. The galactose content in these formulas is very low, though not nil.

TABLE 5.7
Foods to be Avoided by Persons with Galactosemia

Dairy products

  Milks: All forms, including nonfat milk, nonfat dry milk, dry milk, buttermilk, nonfat dry milk solids, milk solids

  Milk derivatives: cream, butter, cheese, dried cheese, sour cream, yogurt, ice cream, sherbet.

  Milk-derived food ingredients: lactose, casein, whey and whey solids, dry milk protein, sodium caseinate, calcium caseinate, lactostearin, lactalbumin, buttermilk solids, hydrolyzed protein (e.g., found in canned/preserved meats) *except* hydrolyzed vegetable protein

  Margarine (check label: most margarines contain milk derivatives)

Organ meats

  Liver, heart, kidney, brains, sweetbreads, pancreas

Plant foods:

  Garbanzo beans, tomatoes, watermelons, dates, papayas, bell peppers, persimmons

Fermented soy products (galactose is released with fermentation)

  Fermented soy sauce, miso, tempe

Condiments

  MSG (monosodium glutamate, which may contain lactose extenders)

Dough conditioners containing caseinate

Most fruits and vegetables, except those listed above, have 5 to 10 mg galactose per 100 grams and should be consumed with caution. Seeds and nuts, and gums and fibers (acacia, agar, carageenan, carob, guar gum, locust bean gum, xanthan gum) do not have available galactose.

Note that laypersons may assume that lactate, lactic acid, and lactylate contain lactose, which they do not; these do not have to be avoided.

Parent education, and eventually education of the child as he or she grows up, is critical to dietary adherence. Monitoring of adherence to the diet may be accomplished through measurement of red blood cell galactose-1-phosphate and urinary galactilol. However, levels of these metabolites are abnormally high even when patients are on the best controlled diet, probably due to endogenous production of galactose-1-phosphate.

A diet lacking in milk and milk products has the potential to be inadequate in calcium and riboflavin. Alternative sources for these nutrients should be included in the diet.

Elimination of dietary galactose prevents the acute manifestations of the condition but does not protect against the development of long-term complications. It has become clear that, despite the best possible compliance with treatment, developmental verbal dyspraxia is a prominent problem (Robertson and Singh, 2000) and mental function deteriorates with age (Schweitzer et al., 1993). Cerebral atrophy is present on MRI. Even those children in whom the IQ remains normal show school problems, including learning and behavioral disabilities. Similarly, ovarian failure is not prevented, and overall growth failure in females is a prominent problem (Guerrero et al. 2000).

The reasons for this failure are not clear, but may lie in:

- The inability to eliminate galactose completely from the diet
- The inability to effectively reduce the levels of galactose and its toxic metabolites despite strict dietary management
- The presence of prenatal damage which cannot be compensated for postnatally

Cataract formation is a long-term consequence of galactosemia. In rats, cataract formation has been overcome by the use of antioxidants such as butylated hydroxytoluene (BHT) and lycopene from tomatoes (which are otherwise proscribed, due to their high galactose content; see Table 5.7) (Pollack et al., 1996/7). Similarly, ovarian toxicity may be limited by antioxidant therapy (Liu et al., 2000).

Recently, the existence of an alternative pathway for galactose oxidation has been confirmed (Berry et al., 2001). Whether it will be possible to use this pathway as a therapeutic approach to galactose disposal remains to be seen.

Lactose Intolerance

Definition and Diagnosis of Lactose Intolerance

Lactose intolerance is defined as a condition in which the inability to split lactose into glucose and galactose in the small intestine leads to symptoms associated with the osmotic retention of water in both the small and the large intestine, and with bacterial digestion of the lactose in the large intestine.

The inability to split lactose is caused by lactase deficiency. Lactase is a brush border enzyme, part of the β-glycosidase complex, which also contains glycoceramidase. This complex can split not only lactose, but a number of other β-glycosides, such as cellobiose and glycosyl-β-ceramides. In humans, the β-glycosidase complex appears in fetal life. Mucosal β-glycosidase activity increases rapidly before birth and remains high in neonates, dropping somewhat during the first year of life.

Lactose intolerance, or hypolactasia, can be classified as primary or secondary. Primary adult hypolactasia is the norm among most populations of the world. Adult hypolactasia is characteristic of most mammals at weaning, when milk is no longer consumed, and the presence of lactase is no longer required. An autosomal dominant mutation for lactase persistence, with multiple alleles, has developed in human populations. This mutation provides selective advantages for those groups that have relied on milk as a significant food and source for calories, protein, and calcium.

Congenital lactase deficiency is an extremely rare condition, characterized by severe diarrhea that develops shortly after the newborn's first feedings. A lactose-free diet permits normal growth and development.

An empirical diagnosis of lactose intolerance can be made by eliminating lactose-containing foods from the diet and then reintroducing lactose into the diet. If elimination of dietary lactose removes symptoms and reintroduction results in the return of symptoms, a presumptive diagnosis of lactose intolerance can be made. However, if there is no improvement or only partial improvement in symptoms with lactose withdrawal, lactose intolerance may still be present but may be associated with other conditions, for example, irritable bowel syndrome (Parker et al., 2001). A number of people with subjective lactose intolerance have irritable bowel syndrome (Vesa et al., 1998) and, conversely, some patients with a primary diagnosis of irritable bowel syndrome actually have lactose intolerance as their primary problem (Bohmer and Tuynman, 2001). The diagnosis by exclusion of irritable bowel syndrome is complicated by the observation that some persons with hypolactasia may also be intolerant of other carbohydrates in large amounts (Teuri et al., 1999).

If symptoms return with dietary challenge, possible diagnoses include not only lactose intolerance, but also allergy to cow's milk. As noted below, lactose intolerance may be secondary and accompany other conditions such as malnutrition and monosaccharide malabsorption.

A more definitive diagnosis of lactose intolerance can be made through an oral load test, in which a known amount of lactose is consumed. Either blood glucose levels or, more commonly, breath hydrogen levels are then measured. Bacterial fermentation of lactose yields large amounts of hydrogen which diffuses into the bloodstream and is released by the lungs as breath hydrogen. A failure to observe an appropriate rise in blood glucose or excessive production of breath hydrogen indicates lactose intolerance.

Jejunal biopsies have been used for the diagnosis of lactose intolerance. However, jejunal biopsies may not be revealing because the distribution of lactase along the intestinal tract is not precisely known and appears to increase as one moves along the tract (Newcomer and McGill, 1966). Thus, a given jejunal sample may show lactase deficiency, yet the person may be able to tolerate lactose.

Incidence, Prevalence, and Genetics of Lactose Intolerance

Lactose intolerance may be congenital or acquired and can be caused by decreased synthesis of lactase or by defective transport of the enzyme to the enterocyte brush border. As noted earlier, congenital lactase deficiency is extremely rare, while adult hypolactasia is the norm among most human populations, except those in whom milk-drinking has traditionally continued into adulthood.

Adult Hypolactasia

Adult hypolactasia is due to a developmental switch in the control of lactase production. The exact genetic mechanism underlying the developmental reduction in lactase in the intestinal brush border is not known. The gene for lactase in hypolactasia is normal, and normal lactase mRNA is produced. However, the amount of lactase present at the brush border is low. These findings and others have prompted the suggestion that adult hypolactasia is the result of reduced expression of lactase (Swallow et al., 2001). Although heterozygotes for hypolactasia appear to have approximately half the enzyme activity found in homozygotes, it is sufficient for handling a lactose load.

Between 80 and 100% of Asians, sub-Saharan Africans, and Native Americans are lactose intolerant. Exceptions occur in nomadic pastoralists, e.g., the Masai of Africa, who are dependent on milk as a food source. (For a table of the world distribution of adult hypolactasia, see Flatz, 1995, pp. 4445–4446.) Lactase persistence into adulthood is probably the "abnormal" condition.

Secondary Lactase Deficiency

Of all the disaccharidases, lactase is the most sensitive to intestinal wall damage. Lactase is located at the tips of the microvilli and is one of the last enzymes to appear as the enterocyte moves from its birthplace in the crypts of Lieberkuhn to the tip of the villus. Thus, lactase deficiency can be acquired in any condition in which the intestinal brush border is damaged. Lactase deficiency appears early in the progression of intestinal disease, becomes more profound than other disaccharidase deficiencies, and is the last deficiency to disappear as recovery progresses.

Parasitic Infestation

Lactose intolerance may be aggravated by parasitic infestation, particularly with *Ascaris lumbricoides* and *Giardia lamblia*. *Ascaris lumbricoides*, a roundworm, secretes an enzyme, ascarase, which destroys the intestinal brush border and leads to significant lactase deficiency as well as generalized malabsorption in young children. Because about one fourth of the world's population harbors *Ascaris*, this infestation may be one of the most common causes of lactose intolerance in children.

Infectious Diarrheas

Infectious diarrheas may lead to the loss of intestinal absorptive surface coupled with abnormal enterocyte replication. Together, these defects can lead to lactase deficiency.

Immunological Disease of the Intestine

Intolerance to cow's milk protein can lead to a loss of enterocyte integrity, growth, and development, and can therefore lead to secondary lactose intolerance. Treatment of cow's milk intolerance and lactose intolerance involves the elimination of dairy products from the diet.

Similarly, active celiac disease (gluten enteropathy), through its effect on villous architecture, can also cause secondary lactose intolerance (Murray, 1999). Therefore, patients with active gluten enteropathy may benefit from the complete elimination of lactose as well as sources of gluten from their diets.

Malnutrition

Malnutrition is a common cause of lactase deficiency because it is associated with failure of enterocyte development, loss of villous structure, and a decrease in the number, length, and function of microvilli.

Malnutrition can occur in many contexts. Clearly, situations where food availability is decreased will lead to malnutrition. However, the clinician should consider malnutrition in patients who are underfed and metabolically stressed. Thus, lactose intolerance should be considered in patients with cancer or with severe injury (e.g., burns) who develop watery diarrhea upon refeeding with a milk-based formula or a diet containing lactose.

The clinician should be alert to hidden sources of lactose, which may become significant for patients with severe intestinal damage. These include fillers in medications and prepared foods.

Other Conditions

Lactase deficiency should be considered in patients who develop watery diarrhea on enteral feedings with lactose-containing formulae.

Patients with exocrine pancreatic insufficiency who receive exogenous pancreatic enzymes may develop generalized disaccharidase deficiency due to digestion of the brush border enzymes. The extent to which these deficiencies are clinically relevant will depend both on the degree of deficiency and on the person's diet (Seetharam et al., 1980).

As noted above, some proportion of patients with irritable bowel syndrome may in fact have hypolactasia, so an elimination trial may be useful.

Nutritional Management of Lactose Intolerance

In principle, nutritional management of lactase deficiency is simple: the avoidance of lactose-containing foodstuffs and medications is sufficient. Healthy people with primary adult hypolactasia differ considerably in the amount of lactose they can digest and can usually determine on their own how much of a dairy product they can eat at a sitting. However, it is worth noting that the intestinal flora may become adapted to the amount of lactose a person regularly consumes by increasing the conversion of lactose to absorbable short chain fatty acids. Therefore a sudden drastic decrease in the amount of lactose consumed can cause diarrhea in people who have been tolerant of, and consuming, small amounts of lactose.

Nutritional management of people with secondary hypolactasia is, in principle, the same as for people with primary hypolactasia. The difficulty in management lies in recognition of the condition.

Complications Associated with Nutritional Management

Dairy foods are the primary source of calcium in the Western diet, so avoidance of dairy products may lead to calcium depletion and deficiency (Infante and Tormo, 2000). Lactose intolerance is quite prevalent among people of European descent who have osteoporosis. Calcium depletion in these individuals occurs because their calcium intake is inadequate and not because of malabsorption of calcium.

There are a number of dairy-based products that can provide calcium without causing diarrhea. Of these the most widely available is yogurt. The bacteria in yogurt split and ferment the lactose, so the amount of lactose is low in yogurt and yogurt products. Yogurt products with live cultures appear to be best tolerated (Pelletier et al., 2001). A number of lactose-free milks and milk products have come on the market, and these can also provide adequate amounts of calcium. Furthermore, pills containing β-galactosidase are available, which can be added to milk prior to consumption.

The are a number of lactose-free formulations for patients receiving enteral feedings. The pharmacy should be notified of a patient's lactose intolerance because many medications contain lactose as a filler.

Complications of Lactose Tolerance

Lactose tolerance may be associated with at least two complications, cataracts and certain types of ovarian cancer, because lactose tolerance allows for high levels of milk consumption. It is possible that high levels of lactose consumption, coupled with the presence of a polymorphism of the gene for galactose-1-phosphate uridyl transferase (GALT) or other defect in the metabolism of galactose, may allow for the appearance of these two complications of galactosemia (e.g., Cramer et al., 2000). Whether persons with a family history of galactosemia, cataracts, or ovarian cancer (particularly of endometroid and clear cell types) should avoid lactose has not been determined.

REFERENCES

Adachi JD and Papaioannou A. Corticosteroid-induced osteoporosis: detection and management. *Drug Saf.* 24:697–624, 2001.

Berry GT, Leslie N, Reynolds R, Yager CT, and Segal, S. Evidence for alternate galactose oxidation in a patient with deletion of the galactose-1-phosphate uridyltransferase gene. *Molec. Gen. Metab.* 72:316–321, 2001.

Berry GT, Nissim I, Lin Z, Mazur AT, Gibson JB, and Segal S. Endogenous synthesis of galactose in normal man and patients with hereditary galactosemia. *Lancet.* 346:1073–1074, 1995.

Berry GT, Palmieri MJ, Gross KC, Acosta PB, Henstenberg JA, Mazur A, Reynolds R, and Segal S. The effects of dietary fruits and vegetables in urinary galactolol excretion in galactose-1-phosphate uridyltransferase deficiency. *J. Inherit. Metab. Dis.* 16:91–100, 1993.

Biassoni P, Ravera G. Bertocchi J, Schenone F, and Bourdoux P. Influence of dietary habits on thyroid status of a nomadic people, the Bororo shepherds, roaming a central African region affected by severe iodine deficiency. *Eur. J. Endocrinol.* 138:681–685, 1998.

Birlouez-Aragon I, Ravelontseheno L, Villate-Cathelineau B, Cathelineau G, and Abitbol G. Disturbed galactose metabolism in elderly and diabetic humans is associated with cataract formation. *J. Nutr.* 123:1370–1376, 1993.

Bohmer CJ and Tuynman HA. The effect of a lactose-restricted diet in patients with a positive lactose tolerance test, earlier diagnosed as irritable bowel syndrome: a 5-year follow-up study. *Eur. J. Gastroenterol. Hepatol.* 13:941–944, 2001.

Bucci I, Napolitano G, Giuliani C. Lio S., Minnucci A, DiGiacomo F, Calabrese G. Sabatino G, Palka G, and Monaco F. Zinc sulfate supplementation improves thyroid function in hypozincemic Down children. *Biological Trace Element Research.* 67:257–268, 1999.

Cramer DW, Greenberg ER, Titus-Ernstoff, Liberman RF, Welch WR, Li E, and Ng, WG. A case-control study of galactose consumption and metabolism related to ovarian cancer. *Cancer Epidemiology, Biomarkers and Prevention.* 9:95–101, 2000.

Dale J, Daykin J, Holder R, Sheppard MC, and Franklyn JA. Weight gain following treatment of hyperthyroidism. *Clin. Endocrin.* 55:233–239, 2001.

Deshpande N and Hulbert AJ. Dietary omega-6 fatty acids and the effects of hyperthyroidism in mice. *J. Endocrin.* 144:431–439, 1995.

Faber J, Wiinberg N, Schifter S, and Mehlsen J. Haemodynamic changes following treatment of subclinical and overt hyperthyroidism. *Eur. J. Endocrin.* 145:391–396, 2001.

Flatz, G. The genetic polymorphism of intestinal lactase activity in adult humans, in *The Metabolic and Molecular Bases of Inherited Diseases.* Scriver CR, Beaudet AL, Sly WS, and Valle D, (Eds.) New York: McGraw-Hill, 1995, pp. 4441–4450.

Gerl H, Rohde W, Bierling H, Schulz N, and Lochs H. Food-dependent Cushing's syndrome of long standing with mild clinical features. *Deutsche Med. Wochenschr.* 125:1565–1568, 2000.

Gross KC and Acosta PB. Fruits and vegetables are a source of galactose: implications in planning the diets of patients with galactosemia. *J. Inherit. Metab. Dis.* 14:253–258, 1991.

Guarderas JC. Is it food allergy? Differentiating the causes of adverse reactions to food. *Postgrad. Med.* 109:125–127, 131–134, 2001.

Guerrero NV, Singh RH, Manatunga A, Berry GT, Steiner RD, and Elsas LJ 2nd. Risk factors for premature ovarian failure in females with galactosemia. *J. Pediatr.* 137:833–841, 2000.

Infante D and Tormo, R. Risk of inadequate bone mineralization in diseases involving long-term suppression of dairy products. *J. Pediatr. Gastroenterol. Nutr.* 30:310–313, 2000.

Kabadi, UM. Is hepatic glycogen content a regulator of glucagon secretion? *Metabolism.* 41:113–115, 1992.

Kelepouris N, Harper KD, Gannon F, Kaplan FS, and Haddad JG. Severe osteoporosis in men. *Ann. Intern. Med.* 123:452–460, 1995.

Khaleeli AA, Edwards RH, Gohil K, McPhail G, Rennie MJ, Round J, and Ross EJ. Corticosteroid myopathy: a clinical and pathological study. *Clin. Endocrinol.* 18:155–166, 1983.

Klein S and Jeejeebhoy KN. The malnourished patient: Nutritional assessment and management, in *Gastrointestinal Disease, 6th edition,* Feldman M, Scharschmidt BF, and Sleisinger M. (Eds.) Philadelphia: WB Saunders, 1997, pp. 235–253.

Krakauer JC and Kleerekoper M. Borderline low serum thyrotropin level is correlated with increased fasting urinary hydroxyproline excretion. *Arch. Intern. Med.* 152:360–364, 1992.

Leal-Cerro A, Venegas E, Garcia-Pesquera F, Jimenez LM, Astorga R, Casanueava FF, and Dieguez C. Enhanced growth hormone (GH) responsiveness to GH-releasing hormone after dietary restriction in patients with Cushing's syndrome. *Clin. Endocrinol.* 48:117–121, 1998.

Liu G, Hale GE, and Hughes CL. Galactose metabolism and ovarian toxicity. *Reprod. Toxicol.* 14:377–84, 2000.

Lovejoy JC, Smith SR, Bray GA, Delany JP, Rood JC, Gouvier D, Windhauser M, Ryan DH, Macchiavelli R, and Tulley R. A paradigm of experimentally induced mid hyperthyroidism: effects on nitrogen balance, body composition, and energy expenditure in healthy young men. *J. Clin. Endocrin. Metab.* 82:765–770, 1997.

Micklefield G and Wenzel IU. Acute diarrhea in an adult patient with Down syndrome. *Zeitschrift für Gastroenterologie.* 38:169–172, 2000.

Murray JA. The widening spectrum of celiac disease. *Am. J. Clin. Nutr.* 69:354–358, 1999.

Newcomer AD and McGill DB. Distribution of disaccharidase activity in the small bowel of normal and lactase-deficient subjects. *Gastroenterology.* 51:481–486, 1966.

Novelli G and Reichardt JKV. Molecular basis of disorders of human galactose metabolism: past, present, and future (minireview). *Molec. Gen. Metab.* 71:62–65, 2000.

Pacaud D, Van Vliet G, Delvin E, Dupuis C, Garel L, Chad Z, Delange F, and Deal C. A third world disease in North America: severe diet-induced iodine deficient goitrous hypothyroidism in a 6 yr old atopic boy. *Clin. Invest. Med.* 17(4 Suppl) B32, 1994.

Parker TJ, Woolner JT, Prevost AT, Tuffnell Q, Shorthouse M, and Hunter JO. Irritable bowel syndrome: is the search for lactose intolerance justified? *Eur. J. Gastroenterol. Hepatol.* 13:219–225, 2001.

Pelletier X, Laure-Boussuge S, and Donnazzolo Y. Hydrogen excretion upon ingestion of dairy products in lactose-intolerant male subjects: importance of live flora. *Eur. J. Clin. Nutr.* 55:509–512, 2001.

Pollack A, Madar Z, Eisner Z, Nyska A, and Oren P. (1996/7) Inhibitory effect of lycopene on cataract development in galactosemic rats. *Metab. Pediatr. Syst. Ophthalmol.* 19–20:31–36, 1996/97.

Reunala T and Collin P. Diseases associated with dermatitis herpetiformis. *Br. J. Dermatol.* 136:315–328, 1997.

Robertson A and Singh RH. Outcomes analysis of verbal dyspraxia in classic galactosemia. *Genet. Med.* 2:142–18, 2000.

Sategna-Guidetti C, Volta U, Ciacci C, Usai P, Carlino A, DeFranceschi L, Camera A, Pelli A, and Brossa C. Prevalence of thyroid disorders in untreated adult celiac disease patients and effect of gluten withdrawal: an Italian multicenter study. *Am. J. Gastroenterol.* 96:751–757, 2001.

Schweitzer S, Shin Y, Jacobs C, and Brodehl J. Long-term outcome in 134 patients with galactosemia. *Eur. J. Pediatr.* 152:36, 1993.

Scott HC, Sun GY, and Zoeller RT. Prenatal ethanol exposure selectively reduces the mRNA encoding alpha-1 thyroid hormone receptor in the fetal brain. *Alcoholism Clin. Exp. Res.* 22:2111–2117, 1998.

Seetharam B, Perrillo R, and Alpers DH. Effect of exocrine pancreatic proteases on intestinal lactase activity. *Gastroenterology.* 79:827–30, 1980.

Segal S and Berry GT. Disorders of galactose metabolism, in *The Metabolic and Molecular Bases of Inherited Diseases.* Scriver CR, Beaudet AL, Sly WS, and Valle D, (Eds.) New York: McGraw-Hill, 1995, pp. 967–1000.

Silva JE. Thyroid hormone control of thermogenesis and energy balance. *Thyroid.* 5:481–92, 1995.

Simmons A, Hewlett B, and Cox JSA. Hypo-thyroidism: does it compromise weight loss? *Int. J. Obesity.* 24 (supplement 1) S163, 2000.

Swallow DM, Poulter M, and Hollox EJ. Intolerance to lactose and other dietary sugars. *Drug Metab. Disp.* 29:513–516, 2001.

Teuri U, Vapataalo H, and Korpela R. Fructooligosaccharides and lactulose cause more symptoms in lactose maldigesters and subjects with pseudohypolactasia than in control lactose digesters. *Am. J. Clin. Nutr.* 71:600–602, 1999.

Tietgens ST and Leinung MC. Thyroid storm. *Med. Clin. North Am.* 79:1, 169–184, 1995.

Tsimihodimus V, Bairaktari E, Tzalla C, Miltiadus G, Liberopoulos E, and Elisaf M. The incidence of thyroid function abnormalities in patients attending an outpatient lipid clinic. *Thyroid.* 9:365–368, 1999.

Utermohlen V. Diet, nutrition, and drug interaction, in *Modern Nutrition in Health and Disease, 9th edition.* Shils ME, Olson JA, Shike M, and Ross AC, (Eds.) Baltimore: Williams and Wilkins, 1999, pp. 1619–1641.

Valentino, R, Savastano, S, Tommaselli, AP, Dorato, M, Scarpitta, MT, Gigante, M, Lombardi, G, and Troncone, R. Unusual association of thyroiditis, Addison's disease, ovarian failure and celiac disease in a young woman. *J. Endocrinol. Invest.* 22:390–394, 1999.

Van Etten CH and Tookay HL. Glucosinolates, in *CRC Handbook of Naturally Occurring Food Toxicants.* Reckcigl M Jr. (Ed.) Boca Raton: CRC Press, 1983, pp. 15–30.

Vesa TH, Seppo LM, Marteau PR, Sahi T, and Korpela R. Role of irritable bowel syndrome in subjective lactose intolerance. *Am. J. Clin. Nutr.* 67:710–715, 1998.

Wang ZI, Berry GT, Dreha SF, Zhao H, Segal S, and Zimmerman RA. Proton magnetic resonance spectroscopy of brain metabolites in galactosemia. *Ann. Neurol.* 50:266–269, 2001.

OTHER RECOMMENDED READING

Braverman L and Utiger R. (Eds.) *Werner's and Ingbar's The Thyroid: A Fundamental and Clinical Text. 7th edition.* Philadelphia: Lippincott-Raven, 1996.

Elsas LJ 2nd and Lai K. The molecular biology of galactosemia. *Genet. Med.* 1:40–48, 1998.

Grunfeld JP. Glucocorticoids in blood pressure regulation. *Horm. Res.* 34:111–113, 1990.

Klein I and Ojamaa K. Thyroid hormone and the cardiovascular system. *N. Engl. J. Med.* 344:501–509, 2001.

Patel YT and Minocha A. Lactose intolerance: diagnosis and management. *Compr. Ther.* 26:246–250, 2001.

Segal S and Berry GT. Disorders of galactose metabolism, in *The Metabolic and Molecular Bases of Inherited Diseases.* Scriver CR, Beaudet AL, Sly WS, and Valle D, (Eds.) New York: McGraw-Hill, 1995, pp. 967–1000.

Semenza G and Auricchio S. Small-intestinal disaccharidases, in *Metabolic and Molecular Basis of Inherited Disease. 7th edition.* Scriver RC, Beaudet AL, Sly WS, and Valle D, (Eds.) New York: McGraw-Hill, 1995, pp. 4451–4480.

6 Nutritional Support and Management of Musculoskeletal Diseases

Murray J. Favus, Tammy O. Utset, and Chin Lee

CONTENTS

Musculoskeletal diseases are common chronic disorders that affect children and adults in North America. Diseases such as vitamin D deficiency rickets and osteomalacia arise directly from nutritional deficiencies, while other diseases of bone and cartilage, although not caused by nutritional disturbances, respond to treatments that include nutritional supplementation. This chapter reviews the common metabolic bone and joint disorders with emphasis on the nutritional aspects of their causes and managements.

0-8493-0945-X/03/$0.00+$1.50
© 2003 by CRC Press LLC

OSTEOPOROSIS

POSTMENOPAUSAL OSTEOPOROSIS

Postmenopausal osteoporosis is the most common etiology of this most common metabolic bone disease. The National Osteoporosis Foundation estimates that approximately 10 million Americans have osteoporosis and another 18 million have low bone mass with increased risk for osteoporosis and fracture.[43] Since 1994, the World Health Organization (WHO) definition of osteoporosis has been widely accepted — a systemic disorder of the skeleton in which loss of bone mass and architectural changes increase the fragility and predisposition to fracture.[26] The WHO definition is based on bone density measurements and defines osteoporosis as a T score of –2.5 (2.5 standard deviations below peak adult bone mass), and osteopenia or low bone mass as a T score between –1.0 and –2.49.[2] Using this definition, postmenopausal women account for 75% of all cases. The high rate of osteoporosis in postmenopausal women is due to the large proportion (80%) of women who lose a significant amount of bone during the decade following the onset of menopause. As a result, the average 50-year-old Caucasian woman has a 40% chance of sustaining at least one osteoporotic fracture during her lifetime.[41] The major contributing factors to bone loss and low bone mass in postmenopausal women may be classified as nonreversible (age, sex, race, familial) and reversible (estrogen status, nutritional, exercise, alcohol, cigarette use).

Prevention of bone loss and the treatment of established osteoporosis may be accomplished using one of several FDA-approved agents (Table 6.1). Estrogen replacement is effective in preventing bone loss and reducing fracture risk for as long as it is taken.[8] However, the average length of time a woman uses estrogen is relatively brief (3 to 5 years) and insufficient to preserve bone mass during most of the postmenopausal years. In recent years, a new group of compounds called selective estrogen receptor modulators (SERMS) have become available. These agents

TABLE 6.1
FDA-Approved Agents in the Treatment of Postmenopausal Osteoporosis

Prevention	Treatment
Estrogen	Estrogen
Alendronate	Calcitonin
Raloxifene	Alendronate
Risedronate	Raloxifene
	Risedronate

Note: Prevention refers to agents used to prevent bone loss; Treatment refers to agents used to treat those with osteoporosis. Agents are listed in the order of time when FDA approval was given.

exert estrogen-like actions in some tissues and have anti-estrogen actions in other estrogen target tissues. The first SERM to be approved by the U.S. Food and Drug Administration (FDA) for treatment in established osteoporosis is raloxifene. Raloxifene is estrogen-like in its actions on bone to decrease bone turnover and reduce vertebral fractures.[11] Raloxifene differs from estrogens in that it does not stimulate breast tissue or the uterine endometrium. As a result, some of the more common symptoms that cause women to discontinue estrogens, such as breast tenderness and menstrual flow, do not occur with raloxifene.[11] Raloxifene stimulates small increases in lumbar spine and proximal femur bone mass, with significant reductions in vertebral and some nonvertebral fractures.[11] Reduction in hip fracture rates has not been demonstrated during raloxifene therapy.

The availability of potent antiresorbing agents called bisphosphonates that specifically inhibit osteoclast-mediated bone resorption has established nonhormonal therapy in postmenopausal osteoporosis. Several large randomized clinical trials have demonstrated efficacy of two bisphosphonates (alendronate; risedronate) in preventing bone loss, increasing bone mass at the lumbar spine, proximal femur, and distal radius, and reducing vertebral, hip, and other nonvertebral fractures in early postmenopausal women and older women with established osteoporosis.[7,21,29,38] Bisphosphonates can reduce the appearance of new vertebral and hip fractures by about 50% in women with established osteoporosis whether or not they have already had one or more fractures.[7,21,29,38]

Synthetic human parathyroid hormone 1–34 (rhPTH 1–34) in small daily doses is the first agent developed to treat osteoporosis that primarily increases bone mass through stimulation of bone formation. Randomized clinical trials conducted in postmenopausal women with osteoporosis with or without prevalent fracture have demonstrated that rhPTH-134 increases spinal and proximal femoral bone mass and decreases vertebral and nonvertebral fracture rates after 18 months. At the time of this writing, FDA approval for use of rhPTH 1–34 in osteoporosis was pending.

Alterations in optimal nutrition may contribute to postmenopausal bone loss and fractures. Low body weight is a risk factor for the development of osteoporosis.[5] In young women, weight loss from extreme caloric restriction, excessive exercise, and conditions of anorexia and bulimia are associated with loss of menses, low bone mass, and low peak bone mass. As peak bone mass is an important predictor of bone mass at the onset of menopause and thereafter, the interruption of skeletal development and mineralization in the early adult years has serious implications later in life. While the relation between weight loss and bone loss has been recognized clinically for many years, recent evidence implicates the adipocyte peptide hormone leptin as a regulator of bone cell function. Current studies in mice show that leptin acts on bone cells indirectly through actions in the central nervous system. If leptin is shown to be an important regulator of bone mass in older women, then additional mechanisms for improving bone mass and reducing fracture risk may emerge.

As 99% of body calcium is located in the skeleton, calcium deficiency would be expected to affect bone mass. There is abundant evidence that dietary calcium restriction can limit bone development in children and cause bone loss in adult men and women. The absorption of dietary calcium is inefficient, with only 20 to 35% of that ingested absorbed. In children and young adults, low intake of calcium prompts an adaptation of the intestine to increase the efficiency of calcium absorption and thereby meet skeletal mineral requirements. This adaptation requires an increase in the renal proximal tubule synthesis of 1,25-dihydroxyvitamin D (1,25-OH-D or calcitriol), the hormonal form of vitamin D that stimulates intestinal calcium absorption. By the sixth decade, men and women lose the ability to increase calcitriol production and the intestinal calcium transport adaptive response during low-calcium diet. As a result, low calcium intake predisposes older adults to negative calcium balance.[24] Because a large portion of the adult population avoids dairy products because they follow a low-fat diet, or avoid high calorie foods, or have lactose intolerance, the average man and women in the U.S. has a calcium intake that is well below the recommended daily allowance (RDA). The self-imposed calcium restriction in the early adult years may not alter bone mass as long as the adaptation mechanism is intact. However, with the loss of the intestinal adaptation process, the long-term effects of habitual low calcium intake may contribute to low bone mineral density and hip fracture. The intestine can be stimulated by calcitriol administration, but a resistance, in terms of intestinal calcium absorption, to calcitriol in older adults may also be present.

The predisposition to osteoporosis by virtue of low calcium intake is aggravated by the increase in urinary calcium losses in postmenopausal women. The mechanism of the increased urinary calcium losses is not known, but a small increase in filtered load of calcium and decreased tubular calcium reabsorption likely play a role. The higher serum parathyroid hormone levels found in older adults, by increasing plasma calcium and therefore the amount of calcium filtered, may contribute to the increase in urinary calcium excretion.

Recommendations for calcium intake for all ages are based on the RDA, but recent studies suggest that optimal calcium intakes may be higher than the current RDA recommendations (Table 6.2). For premenopausal women, calcium intake should be increased during pregnancy to balance the substantial calcium shift from the mother to the growing fetus during pregnancy and for the mother's increased calcium requirements during breast-feeding. The increased dietary calcium recommendations for postmenopausal women are based upon calcium balance studies conducted over a range of calcium intakes performed in premenopausal women and postmenopausal women with or without estrogen. Premenopausal women and postmenopausal women taking estrogen reached neutral calcium balance at an average of 1200 mg calcium intake per day, whereas postmenopausal women not taking estrogen required 1500 mg daily intake to reach neutral calcium balance.

Sodium excretion is a major determinant of renal tubular calcium handling, and high salt intake and urinary excretion are associated with high urine calcium levels. Unlike calcium, intestinal sodium absorption is virtually complete. Therefore, under conditions of stable sodium intake, urine sodium excretion is a good measure of dietary sodium intake. The typical North American diet is excessive in salt, and therefore, the average adult has an excessive sodium intake and urinary excretion. Minimal sodium requirements are about 2 to 3 mEq per day (about 50 to 70 mg); however, in the U.S. population, the excretion of sodium is usually greater than 150 mg and often more than 200 mg per 24 hours. Reduction in salt intake reduces urinary sodium and calcium excretion and thereby improves calcium balance.

TABLE 6.2
RDA and Optimal Daily Intake of Calcium

Age	RDA	Age	Optimal
Birth-6 months	210	Birth–6 months	400
6 months – 1 year	270	6 months – 1 year	600
1– 3 years	500	1 –10 years	800–1200
4– 8	800		
9– 13	1300	11 – 24	1200–1500
14– 18	1300		
19– 30	1000	25–50	1000
31– 50	1000	(women and men)	
51– 70	1200	51–64	1000
		(women on ERT and men)	
		51 and over	1500
		(women not on ERT	
		65 or older	1500
Pregnant or lactating		Pregnant or lactating	1200–1500
14-18	1300		
19–50	1000		

Note: Values are in mg per day. RDA is Recommended Daily Allowance as set by the National Academy of Sciences. Optimal Calcium Intake is from the NIH Consensus Panel on Optimal Calcium Intake, 1994. ERT is estrogen replacement therapy.

Diuretic agents may also influence sodium and calcium excretion. Loop diuretics such as furosemide increase sodium and calcium excretion and may contribute to negative calcium balance. Thiazide diuretics and the related chlorthalidone decrease urine calcium excretion while enhancing sodium excretion and can improve calcium balance.[10] In addition, there is evidence that long-term use of a thiazide agent is associated with lower rate of hip fracture. Therefore, measurements of urine sodium and calcium in a 24-hour urine collected while the patient is following his or her usual diet may permit adjustments in salt intake and choice of diuretic if one is indicated for other medical conditions. For example, if a diuretic is required, then hydrochlorothiazide 50 mg daily, the long-acting chlorthalidone 25 mg per day, or long-acting indapamide 2.5 mg daily may conserve calcium while stimulating sodium and water diuresis.

Dietary protein may also influence urine calcium excretion, and high animal protein intake from meat, fish, and poultry contributes to increased urine calcium excretion and decreased calcium balance. Animal proteins are rich in the sulfur-containing amino acid methionine. Renal handling of sulfur results in increased proton excretion and a lowering of urine pH. The lower methionine content of vegetable proteins is reflected in the urine pH of vegetarians, which is near 7, compared to meat-eaters, who typically have a lower urine pH, in the 5 to 6 range. Because acid loads are buffered in large part by the bicarbonate content in bone, and the process of bone buffering is associated with bone resorption, high intake of animal proteins may contribute to bone loss. In contrast, very low protein intake in the elderly is associated with muscle weakness, falling, and increased risk for hip fracture. Additional studies are required to define optimal animal protein intake and determine whether high protein intake of animal origin contributes to the appearance and severity of osteoporosis. A significant role of chronic acid loading on bone mass was demonstrated by improved calcium balance in postmenopausal women taking potassium bicarbonate.[60] More long-term studies are needed to substantiate this short-term beneficial effect of alkali loading.

The role of vitamin K in bone health remains controversial. Vitamin K is required for post-translational glutamate carboxylation of osteocalcin and other selected bone matrix proteins. Low levels of adequately carboxylated osteocalcin in serum have been associated with higher rates of hip fracture in older adults. Further, low levels of vitamin K are common in the general population due to a limited intake of dark green leafy vegetables and legumes. Vitamin K supplementation in postmenopausal women may reduce bone resorption, decrease urine calcium losses, and improve bone density. The use of vitamin K supplementation in osteoporotic women awaits reliable techniques to measure vitamin K body stores and well-designed randomized controlled trials showing efficacy of vitamin K in reducing fracture risk.

GLUCOCORTICOID-INDUCED OSTEOPOROSIS

Low bone mass, osteoporosis, and fractures are common complications of chronic glucocorticoid therapy.[31] The adverse effects of glucocorticoids on bone are dose related. Short-term glucocorticoid therapy of 3 months or less is not associated with permanent bone loss. However, over 50% of patients treated with 5 mg prednisone daily or greater or inhaled steroids at doses of 1000 micrograms per day or greater for longer than 3 months will have bone mineral density (BMD) of the lumbar spine or proximal femur measurements in the osteopenic (T score of −1.0 to −2.49) or osteoporotic (T score −2.50 and below) range.[1] Chronic glucocorticoid therapy reduces trabecular bone to a greater extent than cortical bone due to a suppression of osteoblastic bone-forming cells and a stimulation of osteoclastic bone resorption. Bone resorption is stimulated through modest increases in circulating parathyroid hormone levels. Glucocorticoid-induced suppression of intestinal calcium absorption and enhancement of urinary calcium excretion contribute to the negative calcium balance and progressive bone loss. As a result of these early effects on calcium metabolism especially in trabecular bone, vertebral bone loss and vertebral compression fractures are common initial presentations. With chronic use, reductions in proximal femur bone mass become evident and fractures of the long bones begin to occur.

Prevention of fractures in high-risk patients is described in an updated version of the American College of Rheumatology Recommendations for the prevention and treatment of glucocorticoid-induced osteoporosis.[2] Patients receiving prednisone 5 mg/d or greater or equivalent medicine for 3 months or longer should undergo measurement of bone mineral density (BMD). A baseline measurement of BMD is indicated when initiating long-term therapy. If baseline measurements are low, then BMD should be repeated at 6-month intervals. Annual measurements are sufficient for those whose baseline BMD is normal and in whom treatment to prevent bone loss has been started.

Prevention of bone loss addresses the multiple actions of glucocorticoids. Calcium supplementation sufficient to increase calcium intake to about 1500 mg/d may overcome suppressed intestinal calcium absorption. Serum vitamin D levels (25-OH-vitamin D; 1,25-OH-D) are normal in patients taking chronic glucocorticoid therapy. Nevertheless, adequate vitamin D intake is also recommended at 400 to 800 IU per day to maintain vitamin D body stores. Large doses of vitamin D (50,000 IU daily) are no longer used because they lead to stimulation of bone resorption. If calcium absorption remains low, calcitriol at 0.5 to 1.0 micrograms per day, can overcome the glucocorticoid-induced inhibition and stimulate calcium absorption. The indications for calcitriol and the effects of treatment can be determined by sequential measurements of 24-hour urine calcium excretions while the patients are eating their usual diets and taking calcium supplements. Calcitriol may worsen hypercalciuria in some patients, and therefore its use may be limited.[2] Although calcium and vitamin D supplementation is advised, there is no evidence that calcium and vitamin D alone can prevent bone loss and fractures. Persistent hypercalciuria in the absence of calcitriol therapy may be controlled by thiazide diuretic therapy.

Bisphosphonates, which have been used extensively in postmenopausal osteoporosis, also prevent bone loss induced by glucocorticoids. Bisphosphonates decrease bone turnover and increase bone mass in the presence of chronic steroid therapy.[58] Of the bisphosphonates available, clinical trials with alendronate 10 mg/d[58] and risedronate 5 mg/d have demonstrated efficacy in men and women with a variety of underlying disorders that require glucocorticoid therapy. In these studies, bisphosphonates significantly increased BMD of the lumbar spine and proximal femur and reduced vertebral fracture rates. For premenopausal women, bisphosphonates should be used with caution, and women should receive counseling on effective contraception.

Calcitonin, whether administered by subcutaneous injection or nasal inhalation, has not been shown to increase bone density or decrease fracture rates.[2] Therefore, calcitonin can at best be considered a second-line treatment.

Glucocorticoids may suppress pituitary gonadotropin secretion and cause hypogonadism in men and women. Hormone replacement therapy should be considered for all postmenopausal women unless there is a contraindication.[2] Estrogen replacement therapy should also be considered in premenopausal women who develop oligo- or amenorrhea during glucocorticoid therapy. Testosterone replacement may be considered in men who become hypogonadal with low serum testosterone.

RICKETS AND OSTEOMALACIA

Defective mineralization of bone matrix is manifest clinically (as the rickets syndrome in children and osteomalacia in adults) as proximal muscle weakness, bone tenderness, difficulty in walking, hypocalcemia, change in posture, and fractures.[9,49] Bone mass measurements in osteomalacia may show low bone mass with T scores of −2.5 and below. However, the histology of osteomalacia differs substantially from that of osteoporosis,[46] and it is critical not to confuse rickets or osteomalacia with osteoporosis. In rickets and osteomalacia, the causes are reversible by treatment with vitamin D; treatment with antiresorptive agents used for osteoporosis may interfere with normal mineralization and worsen the osteomalacia. Vitamin D deficiency and phosphate depletion may each cause the mineralization defect of rickets and osteomalacia.[46] Vitamin D deficiency is diagnosed by low serum 25-OH-vitamin D levels,[9,49] which are usually below 10 ng/mL (mean normal of 25 ng/mL in summer and 15 ng/mL in winter). Hypocalcemia and secondary hyperparathyroidism develop as more unmineralized bone matrix becomes exposed on bone surfaces.[46] Elevated total or bone specific alkaline phosphatase is the hallmark of active rickets or osteomalacia. Serum osteocalcin, a bone marker of osteoblastic activity, can also serve as an indicator of bone cell involvement in osteomalacia. Serum alkaline phosphatase and osteocalcin levels are also useful to follow response to therapy and the healing of the osteomalacia.

VITAMIN D DEFICIENCY

Vitamin D deficiency may develop during the course of a variety of disorders (Table 6.3). Serum 25-hydroxyvitamin D (25-OH-vitamin D) levels, which are good estimates of vitamin D body stores, are reduced in all conditions.[9,49] It is estimated that 7 to 10 years are required to deplete vitamin D body stores. Therefore, in the absence of adequate intake or absorption, vitamin D deficiency develops over months and years. With time, depletion of vitamin D stores decreases 25-OH-vitamin D and lowers 1,25-dihydroxyvitamin D [$1,25(OH)_2D$] production and serum levels. As a result, intestinal calcium absorption declines, and bone mineralization is delayed, with accumulation of unmineralized osteoid. Extensive osteoid coverage of the bone surfaces prevents bone resorption, reduces bone turnover, and permits serum calcium levels to fall, notwithstanding an increase in parathyroid hormone secretion. Ultimately, the clinical syndrome of rickets (children) or osteomalacia (adults) appears.[9,49]

Vitamin D deficiency is common in children and can be reversed by the oral administration of vitamin D (1000 IU per day) until symptoms clear.[49] More rapid healing can be accomplished using higher doses for a shorter period of time. Declining skin vitamin D_3 synthesis with age results in low vitamin D body stores in the elderly. The decline may be large enough to cause reduction in bone density. Therefore, exogenous sources of vitamin D must be used, either in the form of multivitamin tablets (that contain 100 to 200 IU per tablet), food products such as milk (about 100 IU vitamin D per 8 ounces), or cereals. Surveys of elderly populations in the U.S. and many European countries have revealed widespread vitamin D deficiency.[30] In the U.S., 30 to 40% of older adults have serum 25-OH-vitamin D levels of 10 ng/mL and below.[30] Elevated levels of serum alkaline phosphatase (total or bone specific) or of osteocalcin are important diagnostic markers of active osteomalacia.[9,30] Repletion of vitamin D stores can be accomplished by administration of 50,000 IU of vitamin D in capsule form once a week until the serum 25-OH-vitamin D levels have returned to the normal range.[9,30] Sustained use of this dose, however, may cause vitamin D intoxication with hypercalcemia. In most cases of vitamin D intoxication, serum 25-OH-vitamin D levels are 200 ng/mL and above.[30]

TABLE 6.3
Causes of Vitamin D Deficiency

Malabsorption syndromes	Inadequate sunlight exposure
Celiac sprue	Unsupplemented breast-fed infants
Pancreatic insufficiency	Total parenteral nutrition
IBD with ileal resection	Chronic anticonvulsant therapy
Small bowel bypass for obesity	Aging and insufficient skin synthesis of vitamin D
Blind loop syndrome	Primary biliary cirrhosis
Post-gastrectomy	

Note: All conditions are associated with low serum 25-OH-vitamin D levels indicative of vitamin D deficiency. IBD is inflammatory bowel disease.

MALABSORPTION SYNDROMES

Vitamin D deficiency and osteomalacia may complicate the course of intestinal and liver disorders that interfere with the absorption of vitamin D (Table 6.3). When these diseases occur in adults, the clinical presentation and laboratory tests are the same as in nutritional vitamin D deficiency described above.[33] Early evaluation of these conditions should include analysis of serum calcium, phosphate, magnesium, and 25-OH-vitamin D. Serum alkaline phosphatase or osteocalcin and parathyroid hormone levels should also be measured if serum calcium or vitamin D levels are low. Repletion with oral vitamin D should be initiated by administration of 50,000 IU of vitamin D as described above.[33] Serum 25-OH-vitamin D measurements should be repeated at intervals of 4 to 6 weeks to assess adequacy of repletion. Some patients will not absorb enough oral vitamin D and will require parenteral vitamin D (50,000 IU) administered intramuscularly weekly until repletion has been achieved.[33] Thereafter, once-monthly maintenance dosing may be sufficient.

PHOSPHATE DEPLETION SYNDROMES

The deposition of calcium phosphate, ultimately in the form of apatite, on bone protein constitutes skeletal mineralization. The driving force for mineralization is provided by the supersaturated concentration product of calcium and phosphate in the plasma. The blood level of phosphate is largely a function of phosphate intake and absorption, and of renal phosphate reabsorption. Thus, phosphate intake is a key determinant of the mineralization process.[46] Phosphate is abundant in the diet, occurring in most food groups, i.e., meats, dairy products, and breads and grains. Phosphate is well absorbed, much of it absorbed by passive diffusion. When phosphate intake is low, renal $1,25(OH)_2D$ production increases, with elevated serum $1,25(OH)_2D$ levels stimulating active phosphate transport in the jejunum. Therefore, in otherwise healthy subjects with normal intestinal absorptive function, hypophosphatemia is the result of diminished renal phosphate conservation. When sodium-dependent phosphate (Na/P) reabsorption in the renal proximal tubule is defective, urine phosphate excretion increases and serum phosphate concentrations are lowered.[18,23] As a result, the calcium × phosphate product is undersaturated, and bone mineralization is slowed or halted. Reduced Na/P transporter function may be the result of a mutation of the transporter gene or of the PHEX gene, which may regulate NA/P function. This occurs in familial X-linked hypophosphatemia (XLH).[18,23] A similar defect also occurs as a random, nonfamilial genetic mutation. A rare cause of acquired osteomalacia is a small mesodermal-derived tumor that produces an inhibitor of the Na/P transporter. This condition, referred to as oncogenic osteomalacia, causes progressive hypophosphatemia and hypocalcemia, as well as osteomalacia. Removal of the tumor is followed by an intense remineralization of bone and resolution of the osteomalacia.

Treatment of Hypophosphatemic Osteomalacias

Large doses of vitamin D, effective in the treatment of vitamin D deficient osteomalacia, do not heal hypophosphatemic osteomalacia because serum 25-OH-vitamin D levels are normal in this condition. XLH is associated with low serum calcitriol; therefore, calcitriol (0.5 to 1.0 micrograms per day) in divided doses is current therapy.[18] Sustained increases in serum phosphate have been associated with improvement in mineralization and skeletal growth. Because the renal phosphate transport defect cannot as yet be corrected, serum phosphate must be increased. Multiple daily oral doses of sodium phosphate tend to restore serum phosphate levels to normal. Initial dosage is usually sodium phosphate (250 mg in tablet form administered 4 to 5 times per day and at bedtime). The maximal dose that can be tolerated is limited by the appearance of loose stools or diarrhea, which is a common complication of oral phosphate therapy. Oral phosphate therapy is monitored by measuring fasting serum phosphate and bone mineralization markers (serum alkaline phosphatase, osteocalcin). Serum and urine calcium levels are normal in XLH, but both may increase in response to calcitriol. Therefore, fasting serum calcium and 24-hour urine calcium excretion should be measured periodically. Short stature and deformities of the lower extremities are common in untreated XLH. Early effective treatment may improve growth and avoid curvature of the lower extremities.[18] For those children who fail to respond to calcitriol and phosphate alone, growth hormone therapy has improved linear growth.

PRIMARY HYPERPARATHYROIDISM

The hypercalcemia and hypercalciuria of primary hyperparathyroidism are due to increased bone resorption and intestinal calcium hyperabsorption.[28,45] As bone resorption is the dominant process, restricting dietary calcium may not significantly lower urine calcium excretion,[28] nor will calcium restriction normalize serum calcium. In fact, restriction of dietary calcium may promote bone loss, because renal tubular reabsorption of calcium may already be at a maximum, reached by only modest elevations of serum calcium.[27,28] In patients with high serum $1,25(OH)_2D$ levels and increased intestinal calcium absorption, high intakes of calcium will only further increase serum calcium and thus worsen the hypercalciuria.[27] Therefore, calcium intake should be adequate and neither restricted nor increased. Optimal calcium intake is in the 800 to 900 mg/d range, but may have to be adjusted depending on the level of urinary calcium excretion. Limiting sodium intake is one way of lowering urine calcium excretion without undue restriction of calcium intake. In some studies,[12,22,27] patients who form kidney stones were found to have levels of calcium intake, intestinal calcium absorption, and serum $1,25(OH)_2D$ that were higher than normal, but this was not borne out by other studies.[12] Thiazide diuretic agents, which are effective in reducing urine calcium excretion in normocalcemic hypercalciuric states, raise serum calcium levels in primary hyperparathyroidism and therefore are contraindicated. Thus, the management of patients with primary hyperparathyroidism with nephrolithiasis must be surgical removal of the enlarged glands.

Serum phosphate is low in about one third of patients due to parathyroid hormone-stimulated inhibition of tubular phosphate reabsorption. However, the phosphate levels rarely decline to below 2.3 mg/dL, and therefore, there is no indication for phosphate supplementation. Phosphate therapy to reduce kidney stone formation has not been subjected to randomized, prospective trial, but worsening of renal function may be a concern.

Serum 25-OH-vitamin D levels are lower than normal in hyperparathyroid patients. This may be due to the increased rate of renal conversion of 25-OH-vitamin D to calcitriol. There is no evidence that vitamin D supplementation is beneficial for the control of the hypercalcemia or hypercalciuria in hyperparathyroidism. High vitamin D intake may aggravate the hypercalcemia.

OSTEOARTHRITIS

Osteoarthritis (OA) is the most prevalent form of arthritis. Radiological evidence of osteoarthritis is present by age 55 in the majority of individuals. Before menopause, the rates of OA are similar in both sexes; after menopause, the rate of OA markedly increases in women. Gender, genetics, occupation, obesity, hypermobility, trauma, and congenital variation in joint anatomy are all known contributors to OA risk.[17]

Age is the greatest predisposing factor in osteoarthritis. The water content and the elasticity of articular cartilage deteriorate with age. Such changes result in increased mechanical force on articular cartilage and subchondral bone. While roughening, or fibrillation, of the surface of articular cartilage is a common result of joint use in adults, this traumatic change may trigger an excess of catabolic forces in cartilage. If destructive processes exceed reparative processes, a net loss of articular cartilage will occur over time.

The first event in the pathogenesis of OA is thought to be microtrauma to articular cartilage, which is aggravated by age-related changes in the cartilage. This microtrauma results in the release of inflammatory cytokines such as interleukin-1β and tumor necrosis factor-α by the synovial membrane and chondrocytes. These cytokines activate chondrocytes to produce proteolytic enzymes that degrade the articular cartilage. The major enzymes mediating cartilage destruction in osteoarthritis are the matrix metalloproteinases, zinc-dependent enzymes that are inhibited by tetracycline. Concurrent with the production of matrix metalloproteinases, tissue inhibitors of matrix metalloproteinases are also produced. Chondroprotective cytokines, such as transforming growth factor-β and insulin-like growth factor-1, can offset many effects of pro-inflammatory cytokines in younger individuals, but older individuals become less responsive to growth factors. This poor response to growth factors may be an important reason why a shift in the balance of chondrocyte activity toward cartilage degradation occurs in many older adults. As cartilage is lost, traumatized subchondral bone is stimulated to attempt repair. This repair response is manifested by subchondral sclerosis and the growth of marginal osteophytes. Thus OA is mechanically driven and biochemically complex.[48]

In view of the role of mechanical stressors in the pathogenesis of OA, it is not surprising that epidemiological studies have shown obesity to be a major risk factor for OA of the knee and hands.[17] The increased loading forces that are experienced in the knees of an obese individual constitute a reasonable explanation for the propensity for knee OA in such individuals. Similarly, OA of the hands occurs more frequently in individuals whose jobs require repetitive gripping. However, many women who have not been subjected to industrial trauma also develop OA of the hands. In these individuals the etiology is less clear, but the association with obesity suggests that some metabolic alteration caused by obesity may affect the balance between cartilage damage and repair in joints exposed to normal, everyday stresses.

A significant protective effect of weight loss on incident knee OA was demonstrated in the Framingham Knee Osteoarthritis Study. A weight loss of approximately 12 pounds was associated with a greater than 50% decreased risk of incident knee OA in women at 10-year follow-up.[16] Thus, obese patients should be educated in the importance of weight loss in preventing or minimizing OA, and a weight loss program should be instituted.

The role of estrogens in the development and progression of OA is controversial. The dramatic increase in OA in postmenopausal women argues for the importance of estrogen in maintaining healthy articular cartilage. While some epidemiological studies have found estrogen replacement to have a protective role in the risk of developing OA, others have not.[17] The reason for the discordance in these studies is not known. It seems reasonable, therefore, to state that estrogen replacement in postmenopausal women may decrease the risk of OA, but that the magnitude of any such benefit is not known and further research is needed.

Epidemiological studies suggest that several vitamins may impact the natural history of OA in humans. The Framingham Knee Osteoarthritis Study found no relationship between the incidence of radiologically defined knee OA and micronutrient intake. However, the frequency of progression in cases of established OA was increased threefold among individuals in the lowest tertile of vitamin C intake, compared to individuals in the middle and highest tertiles of vitamin C intake.[34] A threefold increased risk of OA progression was also found in individuals in the middle and lowest tertile of vitamin D intake relative to those in the highest tertile of vitamin D intake.[35] Lesser effects were suggested with beta-carotene. Vitamin E appeared protective against OA progression in men, but not in women.[34] Two animal studies using surgically induced OA have demonstrated a protective effect of vitamin C given prior to and following trauma.[40,59] Several small, short-term studies have found modest symptomatic improvement in OA with vitamin E supplementation; this may be due more to the anti-inflammatory effects of vitamin E than to its disease-modifying potential.[36] Based on these studies, it is reasonable to recommend a daily multivitamin containing 400 to 800 IU of vitamin D and moderate supplemental vitamin C to decrease the risk of progression in patients with established OA.

Food supplements have become an extremely popular lay remedy among individuals with osteoarthritis. The most widely studied of these supplements are glucosamine and chondroitin sulfate, which were popularized in several self-help books. Glucosamine is a hexosamine sugar used by chondrocytes in the synthesis of constituents of cartilage, including glycosaminoglycans and proteoglycans. Chondroitin is a glycosaminoglycan naturally present in articular cartilage. Both compounds are absorbed by the intestine, and glucosamine is absorbed to a greater extent. Glucosamine is commercially available in various chemical forms, but it is the sulfates of glucosamine and chondroitin that have been used in most clinical studies. Typical doses for glucosamine sulfate are 1500 mg/d, and 800 to 1200mg/d for chondroitin sulfate. N-acetyl glucosamine, available commercially as a nutritional supplement, is not absorbed and is unlikely to be of therapeutic benefit. One case of an allergic reaction to glucosamine has been reported, but no other toxic effects have been reported.

A large number of clinical trials have examined the efficacy of glucosamine and chondroitin sulfate for the symptomatic treatment of OA. In a meta-analysis of these trials, McAlindon[37] detected many methodological flaws and publication bias. Nevertheless, the preponderance of data on the combination of glucosamine and chondroitin sulfate indicates a beneficial effect in decreasing OA pain. However, the magnitude of this benefit cannot be estimated due to publication bias. A potential disease-modifying effect has also been suggested. A 3-year, double-blinded, placebo-controlled trial of the effects of glucosamine sulfate on the rate of progressive joint space loss in knee OA demonstrated decreases in the rate of joint space loss in patients on glucosamine sulfate relative to those on placebo.[53] However, this study has been criticized because pain relief in the glucosamine group may have affected positioning on radiographs, leading to the apparent improvement in the radiologically determined joint space. Thus, uncontested disease-modifying effects of glucosamine sulfate have not yet been established. Further studies, using more rigorous radiological methods, need to be done.

Because these products are not regulated as drugs in the U.S. and have been designated as dietary supplements, they do not have to be tested for efficacy, manufacturing purity, or product content. It is not surprising, therefore, that they have been found to be of widely divergent quality by an independent consumer group, ConsumerLab.com (http://www.consumerlab.com). The National Institutes of Health are currently sponsoring a large, multicenter trial of various combinations of glucosamine, chondroitin, and placebo. If the results of the trial support the use of these supplements for either symptomatic or disease-modifying effects, the Federal Drug Administration may be able to regulate them as prescription drugs and impose quality control standards. Until that time, physicians need to make patients aware of ongoing quality problems in these unregulated products.

A variety of other unregulated dietary supplements have gained popularity in the self-treatment of osteoarthritis. Methylsulfonylmethane (MSM), also known as dimethyl sulfone (DMSO2), is purported to be helpful in a variety of arthritic conditions. MSM is a naturally occurring substance in both plants and animals, but its biological role is not known. Review of published literature on MSM indicates veterinary use only. No trials of MSM in human arthritis have been published in peer-reviewed medical journals. Unlike its closely related cousin, dimethyl sulfoxide (DMSO), no toxic effects have been reported in humans.

A popular home remedy for arthritis, S-adenosylmethionine (SAMe), is a by-product of protein degradation and may be used as a natural substrate in the synthesis of neurotransmitters and other compounds. Two small clinical trials from the 1970s and 1980s purported to find a beneficial effect of SAMe in osteoarthritis. Montrone et al.[42] conducted a small, double-blinded, controlled trial of oral SAMe 400 mg three times daily vs. placebo. Both treatment and placebo groups improved significantly by study end, although the improvement was statistically greater in the SAMe group in only four of their seven outcome parameters. Three of 32 patients receiving SAMe reported gastrointestinal upset as an adverse event and required withdrawal from the trial. Because of the lack of adequate studies of the effectiveness or toxicity of this compound, it cannot be recommended to patients. As with glucosamine and chondroitin sulfate, the actual chemical content of nutritional supplements claiming to contain either MSM or SAMe may vary and patients should be warned of the lack of quality control in these products.

In summary, patients with osteoarthritis need to be counseled about the importance of weight loss if obesity is present. They should also ensure adequate dietary intake of calcium and vitamins C and D. Supplementation with glucosamine and chondroitin sulfate may be considered to ameliorate symptoms arising from osteoarthritis, with the understanding that the degree of benefit is not clear, but that toxicities appear low. Supplementation with other nutritional products is not supported by available research.

RHEUMATOID CACHEXIA

Cachexia is a feature of many rheumatic diseases including the vasculitides, systemic lupus erythematosus, and rheumatoid arthritis. Of these, weight loss in patients with rheumatoid arthritis (RA), termed "rheumatoid cachexia," has been the most carefully studied.

Most rheumatoid arthritis patients may not appear grossly malnourished, but anthromorphometric studies show cachexia present in 67% of RA patients. Thirty percent of RA patients are below the 5th percentile in muscle mass. Metabolic studies of RA patients reveal two simultaneous processes: an elevated metabolic rate compared to healthy control patients, and a relative anorexia, that is, their appetite is not adequate to keep pace with their metabolic rate. As a result, a reduction in body cell mass (BCM) occurs.[56]

Loss of BCM is a powerful predictor of death in chronic medical diseases in general. A 40% loss of BCM predicts death. Well-controlled RA patients have an average 13% loss in BCM, or about one third of the dispensable body cell mass. RA patients also have significantly reduced life expectancy, with death occurring from a wide range of ailments.[20] This loss of BCM may be an indicator of frailty among RA patients, leading to premature demise when other medical disease is superimposed on RA. The degree of BCM loss correlates with the severity of RA.[56]

Loss of BCM in RA is multifactorial, including contributions from chronic inactivity and medications. However, the degree of BCM loss correlates significantly with the production of tumor necrosis factor-α (TNF-α) and to a lesser degree with interleukin-1β (IL-1β) production.[56] These pro-inflammatory cytokines are major mediators of joint inflammation in RA, but have also been found to play causal roles in the cachexia that occurs in sepsis. TNF-α appears to play a role in the modulation of insulin-sensitivity, whereas IL-1β mediates central anorexia.[3] A standard animal model of rheumatoid arthritis, adjuvant-induced arthritis in mice, reproduces the anorexia and hypermetabolism of RA patients. The animal model also exhibits strong correlations of weight loss with levels of TNF-α and IL-1β under conditions of controlled feeding.[57]

Nutritional supplementation has not been shown to be an effective intervention in RA patients with cachexia, even if anorexia is present. However, in patients with clearly inadequate intake, it seems sensible to provide nutritional supplements to minimize the severity of weight loss. Progressive resistance exercise programs can help to reverse the protein catabolism seen in RA patients.[51] Because cytokines clearly are playing a large role in the weight loss in RA patients, the most important intervention in rheumatoid cachexia is controlling the inflammatory disease. Fortunately, many of the new therapeutic agents for RA inhibit TNF-α and IL-1. These agents are likely to be powerful tools that not only treat RA, but also prevent the loss of body mass, which contributes to frailty in RA patients.

SCLERODERMA AND NUTRITION

Systemic sclerosis, or scleroderma, is a generalized disorder of connective tissue characterized by induration and thickening of the skin, Raynaud's phenomenon and other vascular abnormalities, musculoskeletal manifestations, and visceral involvement, including the gastrointestinal tract, lungs, heart, and kidneys. The underlying pathogenesis of scleroderma has not been fully elucidated, but the prevailing hypothesis is that activation of the immune system and the endothelial cells, possibly by an environmental factor, results in the release of a variety of cytokines from T-cells, macrophages, and platelets, which triggers extracellular matrix production by fibroblasts.[61]

The gastrointestinal tract is the third most common organ system involved in scleroderma.[61] Esophageal dysfunction occurs in 80% of patients with scleroderma, and half are symptomatic.[50] In contrast, involvement of the lower intestinal tract is less common and typically asymptomatic.[63] However, 10 to 25% of individuals with intestinal involvement experience problems related to malabsorption,[50] and secondary malnutrition can pose a major problem in patients with advanced systemic sclerosis.

Upper gastrointestinal involvement, particularly disordered peristalsis of the lower two thirds of the esophagus, can present as dysphagia and odynophagia. Some individuals with esophageal involvement compensate by swallowing smaller quantities of food, rather than improving mastication of solid foods. This subtly reduces their overall dietary intake.[61] Such chronic attempts to minimize symptoms can inadvertently contribute to the development of malnutrition.

Systemic sclerosis involving the lower gastrointestinal tract can result in malabsorption from bacterial overgrowth of the small intestine, abnormalities of the intestinal absorptive surface, and pancreatobiliary insufficiency.[62]

Bacterial overgrowth is the most important cause of malabsorption in systemic sclerosis and often a reflection of underlying intestinal dysmotility.[62] The presence of an abnormal bacterial flora is believed to result from stasis within the dilated, atonic loops of small bowel. It is postulated that bacterial flora causes deconjugation of bile salts, which inhibits proper digestion and absorption of dietary fats and nutrients, resulting in steatorrhea and malabsorption. Micronutrient deficiency, such as vitamin B12 deficiency, can also occur because of competition with enteric bacteria for vitamins in the atonic gut. The most consistent laboratory abnormality in scleroderma bowel disease is an elevated 72-hour stool fat content. Many of these patients will not describe clinical symptoms of steatorrhea despite the presence of an abnormally high fat content in their stool. There are well-documented cases of marked improvement in the clinical status of the patients as well as a correction of the abnormal laboratory studies following treatment with broad spectrum antibiotics, such as ciprofloxacin, amoxicillin, tetracycline, oral vancomycin, or metranidazole.[50]

Submucosal collagen deposition and intestinal fibrosis can also lead to reduced permeability of the intestinal wall and further contribute to malabsorption independent of bacterial overgrowth.[32] This is demonstrated by the D-xylose absorption and glucose tolerance tests for passive mucosal uptake that can be abnormal in those with systemic sclerosis even in the absence of bacterial overgrowth. Additional evidence of intestinal mucosal dysfunction and malabsorption comes from low levels of folic acid among some individuals with systemic sclerosis.[62]

Abnormal pancreatobiliary secretion has been described in individuals with systemic sclerosis. Reduced pancreatic secretion of digestive enzymes, such as lipase and trypsin, has been shown to occur in almost a third of unselected individuals with systemic sclerosis. Such pancreatobiliary dysfunction is believed to be mild in most scleroderma patients, but 10 to 15% of patients with pancreatobiliary dysfunction may experience clinically significant symptoms of malabsorption.[62]

Diffuse fibrosis of the colon and hypomotility of the gastrointestinal tract can lead to complaints of constipation or infrequent bowel movements in those with scleroderma. Increased consumption of dietary fiber, traditionally advocated for improvement of constipation in normal individuals, may not be an ideal remedy for those with scleroderma. Individuals with scleroderma who increased their dietary fiber intake have developed acute abdominal pain and distention requiring hospitalization. One explanation is that supplemental fiber aggravates an underlying pseudo-obstruction that has resulted from the colonic atony of scleroderma bowel disease. If dietary fiber intake is increased, it should be introduced and increased very gradually. Alternative methods, including osmotic agents such as lactulose, should be considered, but no known controlled studies exist to demonstrate the safety and benefit of these interventions in patients with scleroderma.[19]

A deficiency of antioxidants has been suggested as a predisposing factor in the endothelial injury occurring in systemic sclerosis, but this has not been examined in clinical studies. Early studies demonstrated low circulating levels of selenium and ascorbic acid in individuals with scleroderma. However, more recent studies have failed to show a significantly lower level of these micronutrients in patients with systemic sclerosis compared to controls. The benefits of adding vitamins A, D, E, and other factors to the diet have not been extensively evaluated. Available studies generally are based on small, uncontrolled human trials. Therefore, no specific recommendations for vitamin supplementation can be made in the management of scleroderma patients.[32,67]

Dietary contribution of essential fatty acids has been evaluated to assess beneficial effects in the management of scleroderma. One rationale for the use of fatty acids is that derivatives of these fatty acids, such as vasoactive prostaglandins, might ameliorate the chronic ischemia–reperfusion lesions characteristic of scleroderma. However, the results of studies using fatty acid supplementation have been mixed. In an open trial of four women who had had systemic sclerosis for 13 to 15 years, supplementation of omega-6 fatty acids in the form of evening primrose oil for 1 year was followed by clinical improvements. These included relief of pain in the extremities, improvement of telangectasia and skin texture, and healing of ulcers.[65] However, a double-blind study of combined supplementation with the omega-3 and omega-6 fatty acids showed no objective evidence of benefit with regard to vascular symptoms after 6 months.[64]

Antibiotics to control bacterial overgrowth and dietary supplements may bring about improvement of malabsorption and malnutrition; however, severe refractory cases of systemic sclerosis usually require long-term parenteral nutritional support. The complication and mortality rates for chronic parenteral support have been as high as 40%.[62] One study with more encouraging results of parenteral nutritional support included 15 patients with severe scleroderma bowel disease and a history of intestinal pseudo-obstruction, malabsorption, and malnutrition. These patients had good weight gain. There was an overall improvement in the hematocrit and serum albumin levels and a reduction in the symptoms of vomiting while on home central venous hyperalimentation (HCVH). The authors concluded that HCVH appears to be a useful adjunct in the management of patients with severe debilitating scleroderma bowel disease, despite its high cost.[44]

Gastrointestinal involvement by scleroderma poses major challenges to the clinician. The use of proton-pump inhibitors to control esophageal disease and cyclical use of antibiotics for bacterial overgrowth represent the best opportunities to intervene in scleroderma patients with weight loss. Weight loss in the scleroderma patient should alarm the clinician to take immediate steps to prevent further weight loss before severe inanition develops.

GOUT

Gout is a metabolic disease with a constellation of clinical features that most often occur in the presence of chronic or intermittent hyperuricemia. The accumulation and deposition of monosodium urate monohydrate crystals can present as recurrent inflammatory arthritis, asymptomatic collections of urate in the form of tophi, interstitial deposition of urate crystals in the renal parenchyma (urate nephropathy), or nephrolithiasis.[6]

Gout is the most common inflammatory arthritis in men. Women, however, are also affected, with the vast majority of cases arising after menopause. The incidence and prevalence of gout can range from 0.60 to 70 cases per 1000 and from 0.9 to 4.0 per 1000, respectively, depending on the population and the degree of hyperuricemia.[55]

The first episode of an acute gouty attack usually occurs between the 4th and 5th decades in men and in the 6th and 7th decades in women.[55] Acute inflammatory monoarticular arthritis is the initial presentation in a majority of patients, and the classic involvement of the first metatarsophalangeal joint, termed *podagra*, ultimately affects 75 to 90% of individuals.[52] Other common sites in the order of frequency include the instep of the foot, ankles, heels, wrists, fingers, and elbows, including the olecranon bursae. Less commonly, gout can initially present as a polyarticular arthritis. The onset of an acute gouty attack typically evolves over 8 to 12 hours, resulting in an intensely exuberant inflammation of one or more joints, and may be accompanied by fevers and chills. Without medical intervention, an acute episode can persist for several weeks before spontaneously remitting and entering a quiescent phase known as the intercritical period. The average time before the occurrence of a second gouty attack is 1 year. In the absence of treatment, the duration of the intercritical period may shorten and progress to more frequent episodes of longer duration. Recurrent flares in the face of long-standing hyperuricemia can lead to unrelenting symptomatic disease and insidious tophus formation.[6]

The identification of negatively birefringent needle-shaped monosodium urate crystals by arthrocentesis from an involved site or aspiration of suspected tophi establishes the diagnosis of gout. Even during the intercritical period, aspiration of the knee or the first metatarsophalangeal joint will usually reveal uric acid crystals and confirm the diagnosis of gout.[47]

Uric acid is the final breakdown product of purine metabolism. It is derived from the metabolism of either endogenous or exogenous substrates. The two pathways of purine metabolism (Figure 6.1) include the *de novo* and the *salvage* pathways. In the *de novo* pathway, nonpurine precursors are incorporated onto a ribose-5-phosphate backbone that then forms 5-phosphoribosyl-1-pyrophosphate (PRPP) and, in a series of steps, results in the formation of nucleotides. The nucleotides are used to synthesize necessary cellular components, such as DNA and RNA, or to catabolize the free bases, guanine and hypoxanthine. These bases may be converted to xanthine and then to uric acid in a reaction catalyzed by the enzyme xanthine oxidase. In contrast, the *salvage* pathway involves the recycling of preformed purine bases resulting from the degradation of endogenous or exogenous cellular nucleic acids. Salvaged purine bases are preferentially re-utilized for nucleotide formation, but some are oxidized to hypoxanthine and guanine, and then converted to uric acid. Uric acid is the end product of purine metabolism in humans, and it is eliminated via renal excretion or intestinal uricolysis. The kidney excretes approximately two thirds of the uric acid, while the remainder is degraded to ammonia and carbon dioxide.[6]

Humans tend to develop hyperuricemia because they lack uricase, the enzyme responsible for degradation of uric acid into a more soluble metabolite. Hyperuricemia is defined as a serum uric acid level greater than 6.8 mg/dL, because this

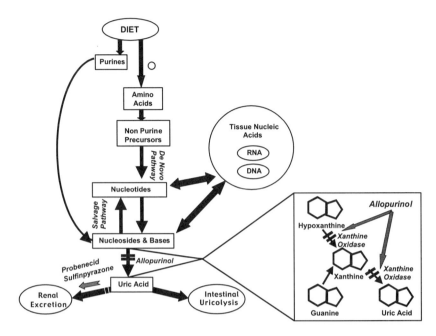

FIGURE 6.1 The *de novo* and salvage pathways of purine metabolism. The sites of action of hypouricemic drugs are shown. (Adapted from *Clinical Medicine*, JA Spitell Jr., Ed., Vol. 9, Chapter 10, Harper and Row Publishers, 1986. With permission.)

concentration is the solubility threshold at 37°C. The laboratory definition of a "normal" uric acid level is based on a population average rather than on a serum saturation value. As a result, many individuals considered to be within the "normal range" of serum urate are actually supersaturated with uric acid and are at increased risk of developing gout. Uric acid overproduction or underexcretion may result in hyperuricemia. Nearly 90% of individuals with gout underexcrete urate, while the remaining 10% of patients overproduce uric acid.[6]

The body produces much of its purine load, but purines are also consumed in the diet. Foods rich in nucleic acid such as organ meats, including liver and kidney, along with "sweetbreads," such as thymus and pancreas, are examples of foods that can contribute to hyperuricemia.[66] About half of RNA and a quarter of DNA purines from the diet are excreted in the urine in the form of uric acid. However, the amount of purine consumed in a typical diet may not by itself explain the degree of hyperuricemia seen in gout, inasmuch as diet accounts for only 1.0 mg/dL of the uric acid in the serum of patients with normal renal function. Therefore, changes in diet are likely to have only a modest effect on serum urate. In patients with impaired renal function, however, a high dietary intake of purines is of greater significance and contributes significantly to the development and severity of gouty arthropathy.[52]

Not all purine-rich foods contribute equally to the serum urate pool because the level of uricogenic bases, adenine and guanine, may vary, making some vegetables and grains more uricogenic than others (Table 6.4). Additionally, foods that contain appreciable amounts of hypoxanthine may contribute more to the serum urate pool

TABLE 6.4
Purine Content of Various Foods

Foods Highest in Purines (150 – 825 mg/100 g)

Anchovies (363 mg/100 g)	Liver (calf/beef – 233 mg/100 g)
Brains	Mackerel
Kidney (beef – 200 mg/100 g)	Meat extracts (160–400 mg/100g)
Game meats	Sardines (295 mg/100 g)
Gravies	Scallops
Herring	Sweetbreads (825 mg/100 g)

Foods High in Purines (50 –150 mg/100 g)

Asparagus	Mushrooms
Breads & cereals, whole grain	Oatmeal
Cauliflower	Peas, green
Eel	Poultry-chicken/duck/turkey
Fish, fresh & saltwater	Shellfish – crab/lobster/oysters
Legumes, beans/lentils/peas	Spinach
Meat – beef/lamb/pork/veal	Wheat germ and bran
Meat soups and broths	

Foods Lowest in Purines (0 – 50 mg/100 g)

Beverages – coffee/tea/sodas	Gelatin
Breads and cereals except whole grain	Milk
Cheese	Nuts
Eggs	Sugars, syrups, sweets
Fats	Vegetables (except those listed above)
Fish roe	Vegetable and cream soups
Fruits and fruit juices	

From Pennington, J.A.T. Bowes & Church's *Food Values of Portions Commonly Used* . 17th Edition. Lippincott; 1998, pp. 391. With permission.

than foods with similar amounts of adenine, since adenine can be degraded to end products other than uric acid.[52]

Individuals who drink excessive amounts of alcohol have long been observed to have higher frequencies of hyperuricemia and gout. Initial studies have demonstrated that alcohol intake increases serum uric acid levels by reducing excretion of urate in gout patients. Also, alcohol was thought to contribute to hyperuricemia by providing readily absorbable purines. More recent studies have shown that ethanol intake contributes to urate synthesis by increasing the turnover of adenine nucleotides.[15] Excessive alcohol intake also leads to dehydration and transient increases in serum uric acid, which may then provoke gouty diathesis. Gout and lead nephropathy may often occur when illegally distilled alcohol, or "moonshine," is consumed, because this liquor is often contaminated with lead. The major reason for the occurrence of "saturnine gout," the severe gout seen in lead poisoning, is decreased uric acid clearance,[4] but obesity, ethanol consumption itself, hypertension, and heredity may be important factors that contribute to the development of gout among those who drink moonshine.[54]

Medications can also predispose to hyperuricemia and gout. For instance, thiazide diuretics provoke hyperuricemia by enhancing tubular reabsorption of sodium and urate by the kidneys. Furosemide causes hyperuricemia by suppressing tubular urate excretion. Low doses of salicylates, pyrazinamide, ethambutol, and nicotinic acid all promote hyperuricemia. Cyclosporin and tacrolimus dramatically escalate the severity of hyperuricemia and gout in transplant patients.[6]

A clear correlation exists between obesity and hyperuricemia. In one study, only 3.4% of individuals with a relative weight below the 20th percentile were hyperuricemic, as compared with 11.4% of those above the 80th percentile.[39] Weight loss also appears to correlate with a reduction in serum uric acid levels and in the frequency of gouty attacks.[13] Insulin resistance also seems to correlate with the level of hyperuricemia, yet no direct relationship between diabetes and gout or hyperuricemia is known to exist.[39]

Several nutritional recommendations can be made to patients with gout. Complete elimination of purine-rich foods from the diet is unrealistic and impractical. A more reasonable approach is to educate patients with gout on specific types of purine-rich foods that should be reduced (Table 6.4). Although no definitive recommendations exist for the dietary management of gout, dietary modification should be attempted because of the relationship of hyperuricemia with diet, obesity, alcohol intake, and hydration status. These suggestions include: 1) limiting foods containing 150 mg or more purines per 100-g serving; 2) moderating the daily intake of protein with most coming from low-fat dairy products; 3) reducing fats to approximately 30% of daily caloric intake; 4) avoiding large, heavy meals, with a goal of gradual weight loss to maintain ideal body weight; 5) eliminating or restricting alcohol consumption to less than 100 g/d; and 6) consuming fluids liberally — at least 2 to 3 quarts/d.[66]

The appropriate management of gout and underlying hyperuricemia becomes more important when one considers associated morbid conditions. "Syndrome X" is the name given to describe the common co-occurrence of insulin resistance, hyperlipidemia, hypertension, atherosclerosis, and coronary heart disease; it is a condition epidemiologically linked to gout and hyperuricemia. Although no direct causal effect has been established, the presence of gout or hyperuricemia may indicate the existence of these serious conditions.[39]

Hyperlipidemia, typically hypertriglyceridemia, is not uncommon in those with hyperuricemia or gout. Between 25 and 60% of individuals with gout are reported to have hyperlipidemia,[39] and hyperuricemia is found in up to 82% of those with hypertriglyceridemia.[6] Although hyperuricemia alone does not appear to be an independent risk factor for coronary heart disease, it is predictive for the development of hypertension and coronary artery disease, as well as for cardiovascular-associated morbidity and mortality.[25] Hypertension is present in up to 44% of patients with gout. In one study of 4575 subjects, gout developed in 3.1% of hypertensive vs. 0.9% of normotensive subjects who had been followed for 8 years. Clinicians should therefore consider evaluating individuals with hyperuricemia or gout for underlying hypertension, hyperlipidemia, and cardiovascular disease.[39]

In view of the strong epidemiological links between gout, hyperuricemia, and Syndrome X, dietary modification should go beyond the goals of lowering serum urate concentration and address other co-morbidities.[52] A diet containing moderate levels of purines, low in saturated fats, high in grains, fresh fruits and vegetables, with lean meats, fish, and poultry as sources of protein, may help to optimize the management of gout and associated conditions.

Because dietary purine limitation and lifestyle changes alone may not achieve normouricemia in most patients, pharmacological management is necessary to prevent recurrences of gout.[39] This can be achieved with chronic anti-inflammatory agents, such as small doses of either colchicine or NSAIDs, or with the utilization of medications to lower serum urate levels. Antihyperuricemic drugs such as allopurinol, a xanthine oxidase inhibitor which decreases the rate of uric synthesis, or probenecid and sulfinpyrazone, which are uricosuric agents (Figure 6.1), are indicated for persons with frequent bouts of acute gouty attacks (> 2 per year), tophaceous gout, nephrolithiasis, or if erosive bony disease is present. Uricosuric agents are contraindicated in individuals with nephrolithiasis, as the increased concentration of urate in the excreted urine may provoke further stone formation.[14] Many patients who discontinue long-term antihyperuricemic agents will experience recurrence of tophi or acute gout attacks; they therefore need lifelong therapy.[6]

With the exception of the occasional patient who has an unusual diet that is very high in purines, the two lifestyle changes likely to have the greatest impact on serum urate levels and the severity of gouty disease in patients with intact renal function are reductions in body weight/adiposity and in the amount of alcohol consumed. Whenever possible, treating physicians should also discontinue drugs that cause or exacerbate hyperuricemia. Many individuals afflicted with gout need therapy with antihyperuricemic agents; however, awareness and appropriate changes in aggravating medications, in body weight, diet, and alcohol habits can optimize the management of gout and its associated co-morbid conditions.

REFERENCES

1. Adinoff, A.D. and Hollister, J.R. Steroid-induced fractures and bone loss in patients with asthma. *N. Engl. J. Med.* 309: 265–268, 1983.
2. American College of Rheumatology Ad Hoc Committee on glucocorticoid-induced osteoporosis. Recommendations for the prevention and treatment of glucocorticoid-induced osteoporosis: 2001 update. (2001) *Arth. Rheum.* 44: 1496.
3. Argiles, J.M. and Lopez-Soriano, F.J. Catabolic proinflammatory cytokines. *Metab. Care.* 1: 245–251, 1998.
4. Ball, G.V. and Sorensen, L.B. Pathogenesis of hyperuricemia in saturnine gout. *N. Engl. J. Med.* 280: 1199–1202, 1969.
5. Bauer, D.C., Browner, W.S., Cauley, J.A., Orwoll, E.S., Scott, J.C., Black, D.M., Tao, J.L., and Cummings, S.R. Factors associated with appendicular bone mass in older women. *Ann. Int. Med.* 118:657–665, 1993.
6. Becker, M.A. *Arthritis and Allied Conditions.* Lippincott Williams and Wilkins. Philadelphia, 2001, Chapter 114.

7. Black, D.M., Cummings, S.R., Karpf, D.B., Cauley, J.A., Thompson, D.E., Nevitt, M.C., Bauer, D.C., Genant, H.K., Haskell, W.L., Marcus, R., Ott, S.M., Torner, J.C., Quandt, S.A., Reiss, T.F., and Ensrud, K.E. Randomised trial of effect of alendronate on risk of fractures in women with existing vertebral fractures. *Lancet.* 348:1535–1541, 1996.

8. Cauley, J.A., Seeley, D.G., Ensrud, K., Ettinger, B., Black, D., and Cummings, S.R. Estrogen replacement therapy and fractures in older women. *Ann. Int. Med.* 122: 9–16, 1995.

9. Chapuy, M.-C. and Meunier, P.J. (1997) Vitamin D insufficiency in adults and the elderly, in *Vitamin D.* Feldman, D., Glorieux, F.H., and Pike, J.W. (Eds.) Academic Press, San Diego, 1997, Chapter. 43.

10. Coe, F.L., Parks, J.H., Bushinsky, D.A., Langman, C.B., and Favus, M.J. Chlorthalidone promotes mineral retention in patients with idiopathic hypercalciuria. *Kidney Int.* 33:1140–1146, 1988.

11. Delmas, P.D., Bjarnason, N.H., Mitlak, B.H., Ravoux, A.C., Shah, A.S., Huster, W.J., Draper, M., and Christiansen, C. Effects of raloxifene on bone mineral density, serum cholesterol concentrations, and uterine endometrium in postmenopausal women. *N. Engl. J. Med.* 337:1641–1647, 1997.

12. Dent, C.E. Calcium intake in patients with primary hyperparathyroidism. *Lancet* 2:330, 1966.

13. Dessein, P.H., Shipton, E.A., Stanwix, A.E., Joffe, B.I., and Ramokgadi, J. Beneficial effects of weight loss associated with moderate calorie/carbohydrate restriction, and increased proportional intake of protein and unsaturated fat on serum urate and lipoprotein levels in gout: a pilot study. *Ann. Rheum. Dis.* 59: 539–543, 2000.

14. Emmerson, B.T. The management of gout. *N. Engl. J. Med.* 334: 445–551, 1996.

15. Faller, J. and Fox, I.H. Ethanol-induced hyperuricemia: evidence for increased urate production by activation of adenine nucleotide turnover. *N. Engl. J. Med.* 307: 1598–1602, 1982.

16. Felson, D.T., Zhang, Y., Anthony, J.M., Naimark, A., and Anderson, J.J. Weight loss reduces the risk for symptomatic knee osteoarthritis in women. *Ann. Int. Med.* 116: 535–539, 1992.

17. Felson, D.T., Lawrence, R.C., Dieppe, P.A., Hirsch, R., Helmick, C.J., Jordan, J.M., Kington, R.S., Lane N.E., Nevitt, M.C., Zhang, Y., Sowers, M., McAlindon, T., Spector, T.D., and Poole, A.R. Osteoarthritis: new insights. Part 1: the disease and its risk factors. *Ann. Int. Med.* 133:635–646, 2000.

18. Glorieux, F.H. Hypophosphatemic vitamin D-resistant rickets, in *Primer on the Metabolic Bone Diseases and Disorders of Mineral Metabolism.* Favus, M.J. (Ed.) Lippincott Williams & Wilkins. Philadelphia, 1996, Chapter 64.

19. Gough, A., Sheeran, T., Bacon, P., and Emery, P. Dietary advice in systemic sclerosis: the dangers of a high fibre diet. *Ann. Rheum. Dis.* 57: 641–642, 1998.

20. Guedes, C., Dumont-Fischer, D., Leichter-Nakache, S., and Boissier, M.C. Mortality in rheumatoid arthritis. *Rev. Rheum. Engl. Ed.* 66: 492–498, 1999.

21. Harris, S.T., Watts, N.B., Genant, H.K., McKeever, C.D., Hangartner, T., Keller, M., Chesnut, C.H. 3rd, Brown, J., Eriksen, E.F., Hoseyni, M.S., Axelrod, D.W., and Miller, P.D. Effects of risedronate treatment on vertebral and nonvertebral fractures in women with postmenopausal osteoporosis. A randomized controlled trial. *JAMA.* 282: 1344–1352, 1999.

22. Hodgkinson, A. Biochemical aspects of primary hyperparathyroidism: an analysis of 50 cases. *Clin. Sci.* 25:231, 1963.

23. Holm, I.A., Nelson, A.E., Robinson, B.G., Mason, R.S., Marsh, D.J., Cowell, C.T., and Carpenter, T.O. Mutational analysis and genotype-phenotype correlation of the PHEX gene in X-linked hypophosphatemic rickets. *J. Clin. Endocrinol. Metab.* 86:3889–3899, 2001.

24. Ireland, P. and Fordtran, J.S. Effect of dietary calcium and age on jejunal calcium absorption in humans studied by intestinal perfusion. *J. Clin. Invest.* 52:2672–2681, 1973.

25. Johnson, R.J., Kivlighn, S.D., Kim, Y.G., Suga, S., and Fogo, A.B. Reappraisal of the pathogenesis and Consequences of hyperuricemia in hypertension, cardiovascular disease and renal disease. *Am. J. Kid. Dis.* 33: 225–234, 1999.

26. Kanis, J.A., Melton, L.J. 3rd, Christiansen, C., Johnston, C.C., and Khaltaev, N. The diagnosis of osteoporosis, *J. Bone Miner. Res.* 8:1137–1141, 1994.

27. Kaplan, R.A., Haussler, M.R., Deftos, L.J., Bone, H., and Pak, C.Y. The role of 1,25-dihydroxyvitamin D in the mediation of intestinal hyperabsorption of calcium in primary hyperparathyroidism and absorptive hypercalciuria. *J. Clin. Invest.* 59:756–760, 1977.

28. Klugman, V.A., Favus, M.J., and Pak, C.Y.C. Nephrolithiasis in primary hyperparathyroidism, in *The Parathyroids: Basic and Clinical Concepts.* Bilezikian, J.P., Marcus, R., and Levine, M.A. (Eds.) Academic Press, San Diego, 2001, Chapter 27.

29. Liberman, U.A., Weiss, S.R., Broll, J., Minne, H.W., Quan, H., Bell, N.H., Rodriguez-Portales, J., Downs, R.W. Jr., Dequeker, J., and Favus, M. Effect of oral alendronate on bone mineral density and the incidence of fractures in postmenopausal osteoporosis. The alendronate phase III osteoporosis treatment study group. *N. Engl. J. Med.* 333: 14337–1443, 1995.

30. Lips, P. Vitamin D deficiency and secondary hyperparathyroidism in the elderly: consequences for bone loss and fractures and therapeutic implications. *Endocrinol. Rev.* 22:477–501, 2001.

31. Lukert, B.P. Glucocorticoid-induced osteoporosis, in *Primer on the Metabolic Bone Diseases and Disorders of Mineral Metabolism.* Favus, M.J., (Ed.) Lippincott Williams & Wilkins, Philadelphia, 1996, Chapter 55.

32. Lundberg, A.C., Akesson, A., and Akesson, B. Dietary intake and nutritional status in patients with systemic sclerosis. *Ann. Rheum. Dis.* 51:1143–1148, 1992.

33. Mawer, E.B. and Davies, M. Bone disorders associated with gastrointestinal and hepatobiliary disease, in *Vitamin D.* Feldman, D., Glorieux, F.H., and Pike, J.W. (Eds.) Academic Press, San Diego, 1997, Chapter 51.

34. McAlindon, T.E., Jacques, P., Zhang, Y., Hannan, M.T., Aliabadi, P., Weissman, B., Rush, D., Levy, D., and Felson, D.T. Do antioxidant micronutrients protect against the development and progression of knee osteoarthritis? *Arth. Rheum.* 39: 648–656, 1996.

35. McAlindon, T.E., Felson, D.T., Zhang, Y., Hannan, M.T., Aliabadi, P., Weissman, B., Rush, D., Wilson, P.W.F., and Jacques, P. Relation of dietary intake and serum levels of vitamin D to progression of osteoarthritis of the knee among participants in the Framingham Study. *Ann. Int. Med.* 125: 353–356, 1996.

36. McAlindon, T. and Felson, D.T. Nutrition: risk factors for osteoarthritis. *Ann. Rheum. Dis.* 56: 397–400, 1997.

37. McAlindon, T.E., LaValley, M.P., Gulin, J.P., and Felson, D.T. Glucosamine and chondroitin for treatment of osteoarthritis. *JAMA.* 283: 1469–1475, 2000.

38. McClung, M.R., Geusens, P., Miller, P.D., Zippel, H., Bensen, W.G., Roux, C., Adami, S., Fogelman, I., Diamond, T., Eastell, R., Meunier, P.J., and Reginster, J.Y. (Hip Intervention Program Study Group.) Effect of risedronate on the risk of hip fracture in elderly women. *N. Engl. J. Med.* 344:333–340, 2001.

39. McGill, N.W. Gout and other crystal-associated arthropathies. *Bailliere's Best Prac. Res.* 14: 445–460, 2000.

40. Meacock, S.C.R., Bodmer, J.L., and Billingham, M.E.J. Experimental OA in guinea pigs. *J. Exp. Path.* 71:279, 1990.

41. Melton, L.J. III, Chrischilles, E.A., Cooper, C., Lane, A.W., and Riggs, B.L. How many women have osteoporosis? *J. Bone Miner. Res.* 7:1005–1010, 1992.

42. Montrone, F., Fumagalli, M., Sarzi Puttini, P., Boccassini, L., Santandrea, S., Volpato, R., Locati, M., and Caruso, I. Letter: double-blinded study of s-adenosyl-methionine versus placebo in hip and knee arthrosis. *Clin. Rheum.* 4: 484–485, 1985.

43. National Osteoporosis Foundation. 1996 and 2015: osteoporosis prevalence figures. State-by-state report. Washington, D.C., 1997.

44. Ng, S.C., Clements, P.J., Berquist, W.E., Furst, D.E., and Paulus, H.E. Home central venous hyperalimentation in fifteen patients with severe scleroderma bowel disease. *Arth. Rheum.* 32: 212–216, 1989.

45. Pak, C.Y.C., Oata, M., Lawrence, E.C., and Snyder, W. The hypercalciurias: causes, parathyroid functions, and diagnostic criteria. *J. Clin. Invest.* 54: 387–400, 1974.

46. Parfitt, A.M. (1997) Vitamin D and the pathogenesis of rickets and osteomalacia, in *Vitamin D.* Feldman, D., Glorieux, F.H., and Pike, J.W. (Eds.) Academic Press, San Diego, Chapter 41.

47. Pascual, E., Batlle-Gualda, E., Martinez, A., Rosas, J., and Vela, P. Synovial fluid analysis for diagnosis of intercritical gout. Ann. Int. Med. 131: 756–759, 1999.

48. Pelletier, J.P., Martel-Pelletier, J., and Abramson, S.B. Osteoarthritis, an inflammatory disease: potential implication selection of new therapeutic targets. *Arth. Rheum.* 44: 1237–1247, 2001.

49. Pettifor, J.M. and Daniels, E.D. Vitamin D deficiency and nutritional rickets in children, in *Vitamin D.* Feldman, D., Glorieux, F.H., and Pike, J.W. (Eds.) Academic Press, San Diego, 1997, Chapter 42.

50. Poirier, T.J. and Rankin, G.B. Gastrointestinal manifestations of progressive systemic scleroderma based on a review of 364 cases. *Am. J. Gastroenterol.* 58: 30–44, 1972.

51. Rall, L.C., Rosen, C.J., Dolnikowski, G., Hartman, W.J., Lundgren, N., Abad, L.W., Dinarello, C.A., and Roubenoff, R. Protein metabolism in rheumatoid arthritis and aging. Effects of muscle strength training and tumor necrosis factor. *Arth. Rheum.* 39: 1115–1124, 1996.

52. Rall, C.L. and Roubenoff, R. *Encyclopedia of Human Nutrition.* Associated Press. San Francisco, 978–982, 1998.

53. Reginster, J.Y., Deroisey, R., Rovati, L.C., Lee, R.L., Lejeune, E., Bruyere, O., Giacovelli, G., Henrotin, Y., Dacre, J.E., and Gossett, C. Long-term effects of glucosamine sulphate on osteoarthritis progression: a randomized, placebo-controlled clinical trial. *Lancet.* 357: 251–256, 2001.

54. Reynolds, P.P., Knapp, M.J., Baraf, H.S., and Holmes, E.W. Moonshine and lead. *Arth. Rheum.* 26: 1057–1064, 1983.

55. Roubenoff, R. Gout and hyperuricemia. *Rheum. Dis. Clinics N. Am.* 16:539–550, 1990.

56. Roubenoff, R., Roubenoff, R.A., Cannon, J.G., Kehayias, J.J., Zhuang, H., Dawson-Hughes, B., Dinarello, J.G., and Rosenberg, I.H. Rheumatoid cachexia: cytokine-driven hypermetabolism accompanying reduced body cell mass in chronic inflammation. *J. Clin. Invest.* 93: 2379–2386, 1994.

57. Roubenoff, R., Freeman, L.M., Smith, D.E., Abad, L.W., Dinarello, C.A., and Kehayias, J.J. Adjuvant arthritis as a model of inflammatory cachexia. *Arth. Rheum.* 40: 534–539, 1997.

58. Saag, K.G., Emkey, R., Schnitzer, T.J., Brown, J.P., Hawkins, F., Goemaere, S., Thamsborg, G., Liberman, U.A., Delmas, P.D., Malice, M.P., Czachur, M., and Daifotis, A.G. Alendronate for the prevention and treatment of glucocorticoid-induced osteoporosis. *N. Engl. J. Med.* 339:292–299, 1998.

59. Schwartz, E.R., Leveille, C., and Oh, W.H. Experimentally induced osteoarthritis in guinea pigs: effect of surgical procedure and dietary intake of vitamin C. *Lab. Anim. Sci.* 31: 683–687, 1981.

60. Sebastian, A.S.T., Harris, S.T., Ottaway, J.H., Todd, K.M., and Morris, R.C. Jr. Improved mineral balance and skeletal metabolism in postmenopausal women treated with potassium bicarbonate. *N. Engl. J. Med.* 330:1776–1781, 1994.

61. Seibold, J.R. *Kelley's Textbook of Rheumatology.* W.B. Saunders Co. St. Louis, 2001, Chapter 83.

62. Sjogren, R.W. Review: gastrointestinal motility disorders in scleroderma. *Arth. Rheum.* 37:1265–1279, 1994.

63. Stafford-Brady, F.J., Kahn, H.J., Ross, T.M., and Russell, M.L. Advanced scleroderma bowel: complications and management. *J. Rheum.* 15: 869–874, 1988.

64. Stainforth, J.M., Alison, M.L., and Goodfield, M.J.D. Clinical aspects of the use of gamma linolenic acid in systemic sclerosis. *Acta Derm. Venereol.* 76: 144–146, 1996.

65. Strong, A.M., Campbell, A., and Thomson, J. The effect of oral linoleic acid and gamma-linolenic acid. *Br. J. Clin. Pract.* 39:444–445, 1985.

66. Touger-Decker, R. *Krauses's Food, Nutrition and Diet Therapy: Nutritional Care in Rheumatic Diseases.* W.B. Saunders Co. Philadelphia, 1996, Chapter 40.

67. Werbach, M.R. *Textbook of Nutritional Medicine.* Third Line Press, Inc. Tarzana, CA, 1999, 400–402

7 Nutritional Support and Management of Renal Disorders

T. Alp Ikizler

CONTENTS

INTRODUCTION

The kidney has several important functions including excretion of metabolic waste products, synthesis of hormones, degradation of peptides and low-molecular-weight proteins, and regulation of the composition of body fluids. In the course of kidney failure, many metabolic functions fail or are diminished in effectiveness. This in turn affects body homeostasis unfavorably. The requirements and utilization of different nutrients also change significantly in the course of kidney failure. These changes ultimately cause renal failure patients to be at higher risk for protein-calorie malnutrition, a condition that is an important predictor of poor outcome. Understanding the nutritional principles and methods of improving nutritional status for the management of these patients is, therefore, of paramount importance.

Chronic kidney disease usually worsens over time. Accordingly, the risk for adverse outcomes and complications, such as protein-calorie malnutrition, also increases. It is essential, therefore, to understand the different stages of kidney failure. A classification system for kidney disease based on an extensive literature review has been recently published.[34] However, to provide a better understanding of this chapter, these stages and their acronyms are briefly defined as follows. Chronic renal failure (CRF): progressive loss of renal function, applied to patients not yet on dialysis and classified as mild, moderate, and severe; end-stage renal disease (ESRD): patients with minimal or no renal function, applied to patients just prior to initiation of dialysis (glomerular filtration rate <15 ml/min) and to patients on renal replacement therapy (treated ESRD, either by hemodialysis or peritoneal dialysis).

NUTRIENT METABOLISM IN RENAL FAILURE

PROTEIN METABOLISM

In general, the minimal daily protein requirement is that protein intake which maintains a neutral nitrogen balance and prevents malnutrition; this has been estimated as approximately 0.6 g/kg in healthy individuals, with a safe level of protein intake equivalent to the minimal requirement plus 2 standard deviations, or approximately 0.75 g/kg/d. Dietary protein intake above this level results in nitrogenous waste products via increased amino acid oxidation. When the glomerular filtration rate (GFR) is decreased, these products accumulate, leading to renal failure. Chronic renal failure patients respond by decreasing their protein intake. Indeed, anorexia is a hallmark of advanced renal failure. Several studies have indicated that CRF patients spontaneously restrict their dietary protein intake (DPI), with levels less than 0.6 g/kg/d when glomerular filtration rate is less than 10 ml/min.[17]

Decreased dietary nutrient intake may also be related to factors other than the accumulation of toxins. Patients with renal failure secondary to diabetes mellitus are likely to be more prone to decreased dietary nutrient intake because of dietary restrictions, gastrointestinal symptoms such as gastroparesis, nausea and vomiting, bacterial overgrowth in the gut, and pancreatic insufficiency. Depression, which is commonly seen in CRF patients, is also associated with anorexia. In addition, CRF patients are usually prescribed a large number of medications, particularly sedatives, phosphate binders, and iron supplements, which are also associated with gastrointestinal complications. Finally, the socioeconomic status of the patients, their lack of mobility, as well as their age, are other predisposing factors for decreased dietary protein intake.

While there is clear-cut indication that DPI is substantially decreased in patients with advanced renal failure, there is a compensatory adjustment in protein turnover in these patients. Studies in patients with severe CRF suggest that their whole-body protein catabolism is decreased, thereby improving the net nitrogen balance to some extent.[10]

Amino Acid Metabolism

Chronic renal failure patients have well-defined abnormalities in their plasma and to a lesser extent in their muscle amino acid profiles. Commonly, essential amino acid (EAA) concentrations are low and nonessential amino acid (NEAA) concentrations are high. There are multiple factors associated with this abnormal profile. The progressive loss of renal tissue where several amino acids are metabolized is an important factor that alters the plasma and muscle amino acid concentrations. Specifically, glycine and phenylalanine concentrations are elevated, and serine, tyrosine, and histidine concentrations are decreased. Plasma and muscle concentrations of valine, leucine, and isoleucine (so-called branched-chain amino acids) are reduced in chronic dialysis patients, with valine displaying the greatest reduction. In contrast, plasma citrulline, cystine, aspartate, methionine, and both 1- and 3-methylhistidine levels are increased. While decreased dietary protein intake is a possible factor in an abnormal essential amino acid profile, certain abnormalities occur even in the presence of adequate dietary nutrient intake. This indicates that the uremic milieu has an additional effect on amino acid profiles. Indeed, it has been suggested that metabolic acidosis, commonly seen in uremic patients, plays an important role in the increased oxidation of branched-chain amino acids.[33] Further, dialytic losses of amino acids are another factor contributing to abnormal amino acid profile in dialysis patients.[16] There are inevitable losses of amino acids during both hemodialysis (HD) and peritoneal dialysis (PD), ranging from 5 to 8 grams of amino acids per hemodialysis session and 5 to 12 g/d of amino acids during peritoneal dialysis.

Carbohydrate Metabolism

While several aspects of carbohydrate metabolism are impaired in renal failure, the most prominent abnormality is insulin resistance leading to glucose intolerance. A post-receptor defect in insulin responsiveness of tissues is the most likely cause of insulin resistance and associated glucose intolerance in uremia.[8] Hyperparathyroidism, usually seen in chronic renal failure, is thought to be, at least partly, responsible for the decreased tissue responsiveness to insulin, caused by inhibition of insulin secretion by pancreatic β-cells. However, it is not clear to what extent this insulin resistance affects protein metabolism in CRF. Following initiation of dialysis, there is marked improvement in glucose tolerance in renal failure patients.

LIPID METABOLISM

Dyslipidemia is quite common in renal failure patients and is often multifactorial. Abnormalities in lipid profiles can be detected in patients once renal function begins to deteriorate. This suggests that uremia is associated with lipid disorders. The presence of nephrotic syndrome or other co-morbidities, such as diabetes mellitus and liver disease, as well as the use of medications that alter lipid metabolism, all contribute to the dyslipidemia seen in renal failure.

Nephrotic syndrome, a common renal disorder, results from the urinary loss of albumin and is characterized by hypoalbuminemia, hyperlipidemia, and edema formation. The abnormalities in the lipid profiles of these patients necessitate special attention.[24] Specifically, there are high serum levels of total cholesterol and triglycerides, mostly of the low-density lipoprotein (LDL), very low-density lipoprotein (VLDL), and intermediate-density protein (IDL) fractions. There is also a decrease in high-density lipoprotein (HDL). These abnormalities are similar to what is observed in atherosclerosis. These metabolic abnormalities may, in turn, worsen the progression of renal disease. While decreasing urinary albumin losses is the mainstay of therapy for these patients, dietary fat restriction and pharmacological therapy in cases not responding to diet may be warranted.

In hemodialysis patients, the most common abnormalities are elevated serum triglycerides and VLDLs, and decreased low- and high-density lipoproteins. The increase in triglycerides is thought to be related to an increase in apoCIII, an inhibitor of lipoprotein lipase. In addition, there is a defect in lipolysis (postprandial lipoprotein metabolism), resulting in the accumulation of chylomicron remnants. A substantial number of chronic hemodialysis patients also have elevated lipoprotein (a) (Lp(a)) levels.

Patients on PD exhibit higher concentrations of serum cholesterol, triglyceride, LDL cholesterol, and apoB, even though the mechanisms that alter the lipid metabolism are similar to what occurs in CHD patients. This is thought to be related to increased protein losses through the peritoneum and glucose load from the dialysate. They also exhibit higher concentrations of Lp(a). A study also showed that PD patients have less pronounced abnormalities in cholesterol transport. Whether these differences in dyslipidemia are clinically significant remains to be clarified.

DYSLIPIDEMIA AND CARDIOVASCULAR RISK IN DIALYSIS PATIENTS

Cardiovascular death is the leading cause of mortality in chronic dialysis patients.[42] Hypercholesterolemia and other certain abnormalities in the lipid profile have been associated with increased risk of atherosclerosis and cardiovascular events in the general population. However, whether this relationship applies to chronic dialysis patients is not clear. Indeed, on the basis of large, cross-sectional studies, it seems that low rather than high cholesterol concentrations constitute an increased mortality risk for chronic dialysis patients.[31] On the other hand, a large multicenter study has shown that, in a cohort of diabetic patients on CHD, those who died from a cardiovascular event had a higher median cholesterol, higher LDL cholesterol, higher LDL/HDL ratio, and higher apoB concentrations at the time of initiation of dialysis than patients who survived.[41]

This can be explained with the help of recent data that indicate that atherosclerosis develops early, at the stage of mild to moderate chronic renal failure. In addition, widespread prevalence of protein-calorie malnutrition in chronic dialysis patients also complicates the use of serum cholesterol concentrations as a risk factor for atherosclerosis. Nevertheless, it is generally accepted that chronic dialysis patients with known risk factors for atherosclerosis and cardiovascular events should be treated with the appropriate regimen, including lipid-lowering agents when indicated. If ESRD patients have a cholesterol concentration higher than 240 mg/dl, an LDL concentration higher than 130 mg/dl, and if other risk factors are present, then treatment is needed. Whether this approach will influence the overall outcome in these patients remains to be answered with further studies.

METABOLIC AND HORMONAL DERANGEMENTS IN RENAL FAILURE PATIENTS

Multiple metabolic and hormonal abnormalities, related to the loss of renal tissue, as well as renal function, become apparent in CRF patients. Metabolic acidosis, which commonly accompanies progressive renal failure, also promotes malnutrition by virtue of an increase in protein catabolism. Detailed *in vitro* and animal studies have shown that muscle proteolysis is stimulated during metabolic acidosis by an ATP-dependent pathway involving ubiquitin and proteasomes.[33] Correction of metabolic acidosis has been shown to actually improve muscle protein turnover in a small number of CHD and PD patients. Two recent studies in CHD and PD patients also showed improvements in anthropometric measurements and body weight.[11,12] However, other cross-sectional and prospective studies did not demonstrate differences in nutritional parameters, particularly in serum albumin. Therefore, correction of metabolic acidosis may be nutritionally beneficial in chronic dialysis patients, and large-scale studies are still needed.

Several hormonal derangements, including insulin resistance, increased glucagon levels, and secondary hyperparathyroidism, have also been implicated as factors in the development of malnutrition in chronic renal failure. The increase in parathyroid hormone levels may be responsible for promoting protein metabolism in uremia by enhancing amino acid release from muscle tissue. Finally, uremic patients are characterized by low thyroxine and triiodothyronine concentrations. These changes resemble those seen in prolonged malnutrition in other patient populations. It has, therefore, been suggested that the thyroid hormone profile of malnutrition and possibly of renal failure is a maladaptive response to decreased energy intake, representing an effort to preserve overall energy balance.

More recently, it has been suggested that abnormalities in growth hormone and the IGF-1 axis constitute an important factor in the development of malnutrition in uremic patients.[18] Growth hormone is the major promoter of growth in children and exerts several anabolic actions in adults, such as enhanced protein synthesis, increased fat mobilization, and increased gluconeogenesis. IGF-1 is the major mediator of these actions. Plasma concentrations of growth hormone actually increase during the progression of renal failure, probably as a result of its reduced renal clearance. Recent evidence suggests, however, that uremia as such is associated with the development of resistance to growth hormone action at cellular levels. In animal experiments, uremia is characterized by reduced hepatic growth hormone receptor mRNA, as well as reduced hepatic IGF-1 mRNA expression.[4] This blunted response would be expected to attenuate the anabolic actions of these hormones. Interestingly, these abnormalities can also be observed with decreased food intake, as well as in experimental metabolic acidosis. Current evidence suggests an interesting, as yet ill-defined interrelationship between these hormonal, metabolic, and nutritional factors, all involved in the evolution of malnutrition in CRF patients.

MINERAL, VITAMIN, AND TRACE ELEMENT REQUIREMENTS

Sodium and potassium intakes should be restricted in renal failure patients (Table 7.1). For patients with advanced renal failure or patients on chronic dialysis, a daily intake of less than 2 g is recommended for both sodium and potassium. A well-known complication of advanced renal failure is the development of renal osteodystrophy, brought on by a number of factors. As renal function falls to levels less than 20%, hypocalcemia develops due to hyperphosphatemia, and there is decreased renal synthesis of $1,25(OH)_2$vitamin D (calcitriol) and worsening hyperparathyroidism with resistance to peripheral actions of parathyroid hormone (PTH).[39] The resulting conditions, which may range from osteitis fibrosa and osteomalacia to mixed and adynamic bone lesions, are important and produce long-term complications that also affect renal failure patients while on dialysis.

In early renal failure, phosphorus control can be achieved by moderate dietary phosphorus restriction. This usually increases calcitriol to near normal levels. Calcitriol also enhances the absorption of calcium from the gut to correct hypocalcemia. Once GFR is less than 20 to 30 ml/min, phosphorus restriction is not enough to stimulate calcitriol production, and gastrointestinal phosphorus binding agents are required to minimize phosphate absorption. Use of aluminum-containing binders should be avoided as much as possible because, in the long term, aluminum accumulation can predispose the dialysis patients to aluminum-related osteomalacia. Calcium acetate is the most commonly used phosphate binder. It is most effective when given with meals. Since there is a patient-to-patient as well as within-patient variability from meal to meal, the dose frequency and timing should be adjusted for each individual meal. For renal failure patients who have low calcium concentrations and/or high PTH levels (> 300 pg/ml), calcitriol administration should be considered. This may alleviate the symptoms and development of renal osteodystrophy. However, the patients must be monitored closely for hypercalcemia and hyperphosphatemia. Products of plasma calcium and phosphorus above 55 should be avoided, since it may lead to soft tissue and vascular calcifications. Newer phosphate binders (calcium-free cross-linked polyallylamine) and vitamin D_3 analogs do not seem to cause hypercalcemia and may be used to treat hyperphosphatemia and vitamin D_3 deficiency.

TABLE 7.1
Recommended Intakes of Protein, Energy, and Minerals in Renal Failure

	Protein	Energy	Phosphorus	Sodium
Chronic Renal Failure				
Mild to moderate (GFR > 30 ml/min)	No restriction	No restriction	600–800 mg/d	< 2 g/d[a]
Advanced (GFR < 30 ml/min)	0.6–0.75 g/k/d[b]	35 kcal/kg/d[c]	600–800 mg/d[d]	< 2 g/d
Nephrotic Syndrome[e]	0.7–1.0 g/kg/d	35 kcal/kg/d[c]	600–800 mg/d[d]	< 2 g/
End-Stage Renal Disease				
Hemodialysis	> 1.2 g/kg/d	35 kcal/kg/d[c]	600–800 mg/d[d]	< 2 g/d
Peritoneal Dialysis	> 1.3 g/kg/d	35 kcal/kg/d[c]	600–800 mg/d[d]	< 2 g/d
Acute Renal Failure				
Predialysis	1.0–1.2 g/kg/d	35 kcal/kg/d	600–800 mg/d[f]	< 2 g/d
Dialysis	1.2–1.4 g/kg/d	35 kcal/kg/d	600–800 mg/d[f]	< 2 g/d

[a] If hypertensive

[b] With close supervision and frequent dietary counseling;

[c] 30 kcal/kg/d for individuals 60 years and older

[d] Along with phosphate binders

[e] Low cholesterol and high polyunsaturated fat diet

[f] If phosphorus > 5.5 mg/dl

From Ikizler, TA, Nutrition and renal failure, in *Primer on Kidney Diseases, 3rd ed* . Greenberg A. (Ed.) National Kidney Foundation, New York, 2001, 420–425. With permission.

The status of many vitamins is altered in chronic dialysis patients with plasma levels both decreased and increased.[5] Vitamin A concentrations are usually elevated and even small intakes lead to excessive accumulation. There have been several reports of vitamin A toxicity in chronic dialysis patients, who therefore should not receive vitamin A supplementation. The vitamin E level in chronic dialysis patients is not well defined and has been reported as increased, decreased, or unchanged. Therefore, vitamin E supplementation may not be needed, but no adverse effects of vitamin E supplementation have been reported. Some short-term studies have indicated improved lipid peroxidation. Vitamin K supplementation is usually not recommended in chronic dialysis patients unless they are at high risk for developing deficiency, as in the case of prolonged hospitalization with poor dietary intake.

The serum concentrations of the water-soluble vitamins are reported to be low in chronic dialysis patients, mainly because intakes are reduced and their clearances are increased by diffusion during hemodialysis. The use of daily multivitamin prescriptions that are specifically designed for renal failure patients usually alleviates these low concentrations. Nevertheless, it is important to recognize that chronic dialysis patients have higher requirements for vitamin B6, folic acid, and ascorbic acid. Plasma levels should be determined in patients at risk, e.g., those requiring prolonged hospitalization. High-flux and high-efficiency dialyzers may also reduce plasma levels of water-soluble vitamins.

The plasma concentrations of most trace elements depend largely on the degree of renal failure.[43] Although there is a list of trace elements whose body fluid concentrations are altered in chronic dialysis patients, only a few of these compounds are thought to be important in this patient population. Serum aluminum concentrations are important in as much as high aluminum levels are associated with dialysis dementia and aluminum-related bone disease. The first reports on aluminum intoxication described patients who were dialyzed with untreated water in regions where the soil is rich in minerals and/or in industrial areas where rigorous environmental precautions are not observed. The dialysis water is treated appropriately in most developed countries. The risk of aluminum intoxication, however, continues in many developing countries. Another source of aluminum is the use of phosphate binders that contain aluminum hydroxide. In chronic dialysis patients with poor control of phosphate intake, prolonged use may be a risk for aluminum intoxication; therefore, these patients' aluminum concentrations should be monitored carefully and frequently, and maintained below 50 μg/L. Finally, concurrent use of aluminum-containing phosphate binders and citrate-containing preparations is contraindicated, because citrate increases aluminum absorption and predisposes the patient to acute aluminum intoxication.

Selenium deficiency has been associated with cardiovascular disease through increased peroxidative damage to the cells. Decreased concentrations of selenium have also been observed in chronic dialysis patients, probably secondary to inadequate dietary intake. However, whether selenium supplementation to correct concentrations would be beneficial is not well defined. Similarly, low concentrations of zinc have been reported in chronic dialysis patients. Zinc deficiency is associated with impotence and anorexia. However, the beneficial effects of supplemental zinc therapy have not been confirmed in dialysis patients.

INDICES OF NUTRITIONAL STATUS IN RENAL FAILURE PATIENTS

Although practical methods to assess nutritional status are imperative, the appropriate interpretation of nutritional markers in renal failure patients remains a challenge. Several markers utilized for nutritional purposes are influenced by many non-nutritional factors. In CRF patients, relatively simple biochemical measures reflecting the visceral protein stores, such as serum albumin, creatinine, and BUN, as well as more complex and not commonly used parameters, such as transferrin, prealbumin, and insulin-like growth factor 1, have been proposed as nutritional markers (Table 7.2). Serum albumin is the most extensively examined nutritional index in almost all patient populations, probably due to its easy availability and strong association with outcome. Serum albumin concentration may, however, also be affected by problems other than malnutrition. Specifically, serum albumin is a negative acute-phase reactant; its serum concentration decreases sharply in response to inflammation and therefore may not necessarily reflect changes in nutritional status in acutely or chronically ill patients.[25] Serum albumin concentration in CRF patients may also be affected by other factors, such as proteinuria, extravascular fluid volume and other illnesses, i.e., liver disease.

TABLE 7.2
Nutritional Indices in Renal Failure Patients with Advantages and Disadvantages

Biochemical Parameters

Serum albumin (< 4.0 g/dl)

Easy to measure	Negative acute phase reactant
Good predictor of outcome	Long (20 days) half-life

Serum transferrin (< 200 mg/dl)

Readily available	Dependent on iron stores
Early response	Negative acute phase reactant

Serum prealbumin (< 30 mg/dl)

Good predictor of outcome	Falsely elevated in renal failure
Good and early response to nutritional support	Negative acute phase reactant

Serum IGF-1 (< 200 ng/dl)

Good association with other markers	Not readily available for clinical use
Short half-life	Not validated in large-scale studies

Body Composition Techniques

Anthropometric measures

Useful if followed longitudinally in the same patient	Crude marker and large variations
	Operator dependent

Bioelectrical impedance analysis

Easy to measure	Affected by fluid status
Good predictor of outcome	Not clinically validated in large studies

DEXA

Good association with other methods	Affected by fluid status
	Expensive and not readily available
	Operator dependent

Dietary Assessment

Protein catabolic rate (< 1.0 g/kg/d)

	Related to short-term dietary intake
	No well-established association with other nutritional markers

Subjective Global Assessment

Includes objective data (disease state, weight changes)	Heavy reliance on clinical judgment
Easy applicability	Inability to tailor a specific nutritional intervention.

From Ikizler TA and Himmelfarb J. Nutritional complications in chronic hemo- and peritoneal dialysis patients, in *Complications of Dialysis* , Lameire N and Mehta RL (Eds.) Marcel Dekker, New York, 2000, 405–426. With permission.

Anthropometric studies can be used for body composition analysis in CRF patients. More reliable and accurate methods of body composition analysis, such as prompt neutron activation analysis, which measures total body nitrogen content, and dual-energy x-ray absorptiometry (DEXA), require expensive equipment only available in specialized centers. Subjective Global Assessment is a recently proposed method to evaluate the nutritional status of chronic renal failure patients. Its advantage is that it combines objective data (disease state, weight changes), several manifestations of poor nutritional status, and the clinical judgment of the involved physician. The limitations are heavy reliance on the clinical judgment and inability to tailor a specific nutritional intervention. Its use as a standard nutritional tool in renal failure is yet to be determined.

Estimation of dietary protein intake can also be used as a marker of overall nutritional status in the CRF patient. Although dietary recall is a direct and simple measure of dietary protein intake, it does not accurately estimate actual intake. Therefore, other means of measuring dietary protein intake, such as 24-h urine urea nitrogen excretion in CRF patients or protein catabolic rate calculations in dialysis patients, have been suggested as useful methods to estimate protein intake.[32]

$$\text{Dietary Protein Intake}/6.25 = \text{Urea Nitrogen Appearance} + 0.031 \text{ g N/kg body weight} \qquad (7.1)$$

However, it should be noted that these indirect estimations of dietary protein intake are valid only in stable patients where nitrogen balance is assumed to be neutral, and may easily overestimate the actual intake in catabolic patients where endogenous protein breakdown can lead to a high urea nitrogen appearance.

EXTENT OF MALNUTRITION IN RENAL FAILURE PATIENTS

Virtually every study that has evaluated the nutritional status of CRF patients has reported some degree of malnutrition in this population, estimated to range from approximately 20 to 60%, depending on the nutritional parameter utilized. In chronic renal failure patients not yet on maintenance dialysis, mild to severe malnutrition by subjective global assessment is reported in 44% of patients. Using the same method, the prevalence of moderate to severe malnutrition is reported to be 30% in chronic hemodialysis patients and 40% in peritoneal dialysis patients.

ASSOCIATION OF NUTRITION AND OUTCOME IN RENAL FAILURE

A number of studies have documented the increased mortality and morbidity in renal failure patients who suffer from malnutrition. In a comprehensive study of prevalent chronic dialysis patients, low serum albumin was identified as the most powerful indicator of mortality.[31] Even serum albumin concentrations of 3.5 to 4.0 g/dl (considered a normal value by most laboratories) resulted in an increased relative risk of death as compared to 4.0 g/dl or higher. In addition, decreases in serum creatinine (an indicator of muscle mass) and percent ideal body weight were also associated with increased risk of death in this patient population. Similar observations can be made for incident dialysis patients. Specifically, low serum creatinine and albumin concentrations at the time of initiation of maintenance dialysis are associated with increased risk of mortality and morbidity over the subsequent years on hemodialysis.

FACTORS AFFECTING THE NUTRITIONAL STATUS OF RENAL FAILURE PATIENTS

Considering the magnitude of the problem, it is likely that multiple factors play important roles in the evolution of malnutrition in renal failure patients. Many of these factors act simultaneously in the progression from suboptimal nutrition to apparent malnutrition. These factors are listed in Table 7.3. In the previous sections of this chapter, we have discussed the importance of decreased dietary nutrient intake, amino acid and protein metabolism abnormalities, and metabolic and hormonal derangements and their potential contribution to development of PCM in renal failure patients. There are several other dialysis-related factors that further exacerbate PCM in these patients.

DOSE OF DIALYSIS

One of the most important factors that affect the nutritional status of dialysis patients is the extent of dialysis that is delivered to CHD and PD patients.[20] The relationship between dialysis dose and dietary protein intake was first reported following the completion of the National Cooperative Dialysis Study.[36] In this randomized study of different dialysis doses, it was noted that patients with the shortest dialysis time, hence the lowest dialysis dose, had the lowest dietary protein intake. Subsequent studies showed a significant linear relationship between dose of dialysis and protein catabolic rate (a marker for dietary protein intake). Hakim et al. reported a prospective 4-year study where the dose of dialysis was increased intentionally in 130 CHD patients.[13] They observed statistically significant differences in serum albumin, serum transferrin, and PCR measurements between patients at the lowest vs. the highest quartile of dialysis dose.

Similar conclusions with regard to the dose of dialysis are reported in PD patients in several studies. Most recently, the results of a large, multicenter study suggested a positive relationship between adequacy of dialysis and nutritional status in PD patients.[7] It was reported that decreasing serum albumin concentrations and wors-

---

TABLE 7.3
Factors Associated with Decreased Nutritional Status of Chronic Renal Failure Patients

Decreased dietary protein and calorie intake (anorexia, frequent hospitalizations, multiple medications)

Co-morbidities (diabetes mellitus, gastrointestinal diseases, ongoing inflammatory response)

Increased resting energy expenditure

Inadequate dialysis dose

Bioincompatible hemodialysis membranes

Losses of nutrients (amino acids and/or proteins) during dialysis

Hormonal and metabolic derangements (metabolic acidosis, hyperparathyroidism, insulin and growth hormone resistance)

ening nutrition, according to subjective global assessment, were predictive of worsening mortality and increasing hospitalizations, and the estimates of adequacy of dialysis and nutritional status are positively correlated in these patients.

All available evidence in chronic renal failure patients confirms the close association between dialysis dose and nutrition. It is important to note, however, that the specific level of optimal dose of dialysis, after which no further improvement in nutritional status is observed, has not been established. Several prospective studies are under way to evaluate this question.

DIALYSIS BIOINCOMPATIBILITY

Another well-defined cause of inappropriate protein catabolism in dialysis patients is the adverse consequence (or enhanced inflammatory response) resulting from contact between blood and foreign material during hemodialysis, i.e., the effects of bioincompatibility. It is now well established that the type of dialysis membrane used affects the protein metabolism in CHD patients. Specifically, bioincompatible membranes that vigorously activate the complement system also induce net protein catabolism, as compared to dialysis membranes that do not activate this inflammatory response. In a prospective randomized study of 159 new hemodialysis patients, randomized to either a low-flux biocompatible membrane or a low-flux bioincompatible membrane, Parker et al. measured the effects of biocompatibility on several nutritional parameters, including estimated dry weight, serum albumin, and IGF-1 over 18 months.[35] They reported that the biocompatible group had a mean increase in their dry weight by 4.36 ± 8.57 kg at the end of the study, whereas no change in average weight was observed in the bioincompatible group. In addition, the biocompatible group had an earlier (6 months vs. 12 months) and more marked increase in serum albumin concentrations compared to the bioincompatible group, as well as consistently higher IGF-1 values.

The mechanism by which the biocompatibility and activation of the complement pathway enhances protein catabolism is not clear. Production of cytokines, such as interleukin-1 and tumor necrosis factor-alpha (TNF-α), may induce muscle protein degradation and excess amino acid release. Complement activation has been shown to result in increased transcription of TNF-α, and in a study by Canivet and co-workers, increased serum TNF-α concentrations were reported in CHD patients dialyzed with a complement activating membrane.[2]

ROLE OF INFLAMMATION

Recently, markers of chronic inflammation have also been associated with adverse clinical outcome in chronic renal failure patients. One of these markers, C-Reactive Protein (CRP), is a significant predictor of mortality as well as morbidity in both chronic hemodialysis and peritoneal dialysis patients. Inflammation, more correctly termed systemic inflammatory response syndrome, is a complex combination of physiological, immunological, and metabolic effects that occur in response to a variety of stimulators arising from tissue injury or a disease process. Metabolic effects of inflammation include fever, anorexia, immune system modulation, and complement activation. Markers of inflammation, as with markers of PCM, are highly prevalent in CRF patients and patients on chronic dialysis.[37,40] Because of the high prevalence of these two entities in the same patient populations, the relationship between PCM and inflammation is actively being pursued. A recent study by Stevinkel et al. demonstrates this relationship in CRF patients not yet on maintenance dialysis.[40] Of 109 patients with advanced chronic renal failure, 44% had moderate to severe protein-calorie malnutrition by subjective global assessment, and 32% had signs of inflammation by elevated CRP concentrations.[40] Most important, 53% of the patients with malnutrition had signs of inflammation, and 72% of the patients with inflammation had signs of malnutrition. These observations suggest that the two conditions may have a common etiology and that there may exist a cause-and-effect relationship such that chronic inflammation can cause PCM in CRF and ESRD patients.

Etiologies of Inflammation-Related Malnutrition

When one considers the metabolic effects of chronic inflammation, the nutritional consequences are evident. Anorexia or suppression of nutrient intake is a well-established metabolic effect of inflammation. Animal studies have also shown an increase in skeletal muscle protein breakdown with TNF (with or without IL-1) administration.[30] In surgical patients, sepsis is associated with an increase in whole-body catabolism.[1] Thus, the combined presence of decreased nutrient intake at the time of increased protein breakdown worsens the overall nitrogen balance, predisposing the patient to accelerated protein-calorie malnutrition. Another well-known effect of inflammation is the activation of the complement system. An important corollary to this observation is that activation of the complement caused by the use of bioincompatible hemodialysis membranes may add to the burden of disease in these patients. Chronic inflammation can also cause cytokine-mediated hypermetabolism. Indeed, increased resting energy expenditure (REE) is observed in most of the chronic inflammatory states, such as congestive heart failure, rheumatoid arthritis, and most forms of cancer. While the exact mechanism of the increased REE in these patients is not well delineated, a common observation is the close association with the increased concentrations of pro-inflammatory cytokines and increased REE. Recent studies by Ikizler et al. have shown that end-stage renal disease patients on chronic dialysis also have increased REE.[21]

Several other indirect effects of chronic inflammation can also predispose renal failure patients to PCM. Chronic inflammation induces a decrease in voluntary activity and the disease initiating the inflammation may require bed rest. It is well known that a prolonged decrease in muscular activity is associated with muscle weakness, muscular atrophy, and negative nitrogen balance, all leading to loss of lean body mass. Finally, there are certain hormonal derangements observed during chronic inflammation. These include disruption of the growth hormone and IGF-1 axis, leading to decreased anabolism and increased leptin concentrations, which may induce anorexia.

STRATEGIES FOR TREATMENT OF MALNUTRITION IN RENAL FAILURE PATIENTS

A list of general measures to prevent or treat malnutrition in different stages of ESRD is presented in Table 7.4. Figure 7.1 shows a proposed diagram of a common treatment protocol for approach and treatment of malnourished chronic renal failure patients.

An important issue that needs to be considered is that the basic nutritional requirements of renal failure patients are altered compared with healthy individuals (Table 7.2). While the "minimal" daily protein requirement may be as low as 0.75 g/kg/d in healthy individuals, CRF patients have been shown to require a protein intake of 1.4 g/kg/d to maintain a positive or neutral nitrogen balance during non-dialysis days, and even this intake may not be adequate for dialysis days. In general, a minimum protein intake of 1.2 g/kg/d is considered safe for both chronic hemo- and peritoneal dialysis patients.[28] Energy requirements are also increased in renal failure patients due to increased REE. In hemodialysis patients, the REE is raised by 5 to 10%. For chronic dialysis patients, a minimum of 30 to 35 kcal/kg/d of energy intake is usually suggested. These recommendations apply to stable patients and may have to be increased at the time of concurrent illnesses, especially during hospitalizations.

PROTEIN RESTRICTION IN CHRONIC RENAL FAILURE PATIENTS

Dietary protein restriction has been recommended as a therapeutic approach for delaying the progression of chronic renal failure. Based on the information in previous sections of this chapter, it is important to reassess the applicability of this

---

TABLE 7.4
Interventions to Prevent or Treat Malnutrition In Renal Failure

Chronic Renal Failure Patients (Moderate to Severe)
   Close supervision and nutritional counseling (especially for patients on protein restricted diets)
   Initiation of dialysis or renal transplant in advanced chronic renal failure patients with apparent protein-calorie malnutrition despite vigorous attempts

Chronic Dialysis Patients
   Continuous dietary counseling
   Appropriate amount of dietary protein and calorie intake
   Optimal dose of dialysis
   Use of biocompatible hemodialysis membranes
   Nutritional support in chronic dialysis patients who are unable to meet their dietary needs
      Oral supplements
      Tube feeds (if medically appropriate)
      Intradialytic parenteral nutritional supplements for hemodialysis patients
      Amino acid dialysate for peritoneal dialysis patients

Growth Factors (Experimental)
   Recombinant human growth hormone
   Recombinant Human Insulin-like Growth Factor-1

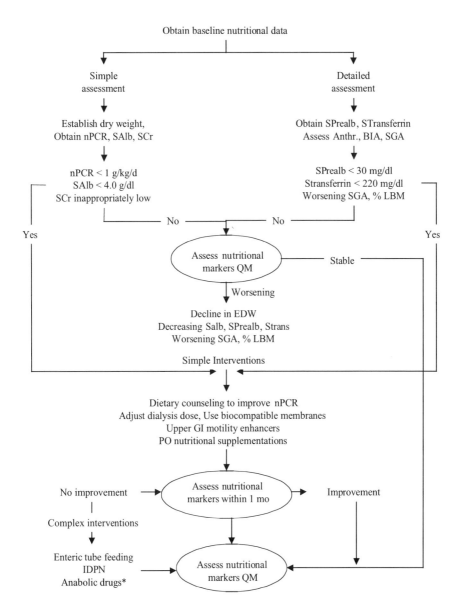

FIGURE 7.1 Proposed algorithm for assessment of nutritional status and treatment of malnutrition in renal failure patients. nPCR: Normalized protein catabolic rate; SAlb: serum albumin; SPrealb: serum prealbumin; SCr: serum creatinine; Anthr: anthropometric measurements; BIA: bioelectrical impedance analysis; SGA: subjective global assessment; LBM: lean body mass; QM: every month; EDW: estimated dry weight; GI: gastrointestinal; IDPN: intradialytic parenteral nutrition. *Mostly experimental. (From Ikizler, T.A. and Hakim, R.M., Renal failure and parenteral nutrition, in *Clinical Nutrition: Parenteral Nutrition*, 3rd ed., Rombeau, J.L. and Rolandelli, R.H. (Eds.) W.B. Saunders Company, Philadelphia, 2001, 366–391. With permission.)

approach. Interestingly, the results of several recent studies on this subject have also been conflicting. The results of the largest randomized clinical trial, The Modification of Diet in Renal Disease (MDRD), did not demonstrate a benefit of dietary protein restriction on progression of renal disease.[26] On the other hand, three recent meta-analyses indicate that such diets may be beneficial in slowing the progression, albeit to a small extent. A comprehensive meta-analysis suggested that dietary protein restriction reduced the rate of decline in GFR only by 0.53 ml/min per year.[23] If such diets are to be used, it is important to make sure patients do not develop malnutrition. Based on the MDRD study, protein-restricted diets with or without essential amino acid or their keto-analog supplementation will maintain nutritional status, provided the patients are carefully followed. However, close supervision, the involvement of dedicated dietitians, and heavy emphasis on appropriate calorie intake, as in the MDRD study, is not usually possible in the majority of patients with chronic renal failure. Such dietary interventions should, therefore, be attempted only in highly motivated and closely supervised patients.

CONVENTIONAL NUTRITIONAL THERAPY

Considering the catabolic nature of chronic dialysis, it is clear that chronic dialysis patients should be continually encouraged to maintain an adequate protein and calorie intake. Most of these patients tend to continue their predialysis diets while on chronic renal replacement therapy. It is, therefore, especially important to ensure that the protein and calorie intake of these patients meet the increased requirements after initiation of dialysis. Frequent comprehensive dietary counseling by an experienced dietitian is important, as is detection of early signs of malnutrition. Similar efforts should be spent not only in outpatient settings, but also when the patients are hospitalized, inasmuch as these patients have even lower protein and calorie intakes.

ENTERAL AND INTRADIALYTIC PARENTERAL NUTRITION

Dietary counseling to improve nutritional status may be unsuccessful in optimizing the dietary intake in certain subgroups of malnourished dialysis patients. For these patients, other forms of supplementation, such as enteral (including oral protein, amino acid tablets, and energy supplementation, nasogastric tubes, percutaneous endoscopic gastroscopy or jejunostomy tubes) and intradialytic parenteral nutrition (IDPN) need to be considered. Several recent reports have recommended use of IDPN in malnourished chronic dialysis patients, if enteral nutritional supplementation has not been effective. The early studies by Heidland and Kult, as well as several subsequent studies, reported positive effects of intradialytic infusions of nutrients on several nutritional parameters.[3,15] In contrast, other studies were not able to show any benefit of IDPN.[19] All of these studies had drawbacks in design and the choice of patient populations. Therefore, no definitive recommendations could be offered. Of the few studies that suggested a positive impact on parenteral nutrition, Cano et al., in a randomized, controlled study, reported improvements in multiple nutritional parameters with IDPN in a group of 26 malnourished CHD patients.[3] In a retrospective analysis of more than 1500 CHD patients treated with IDPN, Chertow et al. have reported a decreased risk of death with the use of IDPN, particularly in patients with serum albumin concentrations below 3.5 g/dl and serum creatinine concentrations below 8 mg/dl, who showed substantial improvements in these parameters following use of IDPN.[6] This treatment mode is, therefore, probably most useful in patients with moderate to severe malnutrition.

Studies using amino acid dialysate (AAD) as a nutritional intervention in malnourished PD patients have also provided conflicting results. In studies which suggested benefit from AAD, serum transferrin and total protein concentrations increased and plasma amino acid profiles tended toward normal with 1 or 2 exchanges of AAD.[22] On the other hand, an increase in BUN concentration associated with exacerbation of uremic symptoms, as well as metabolic acidosis, remains a complication of AAD. In a more recent study, Jones et al. reported significant improvements in serum albumin and prealbumin concentrations in malnourished CAPD patients, particularly in those who had serum albumin concentrations in the lowest tertile.[22] These results are consistent with reports that suggest these interventions are most useful in CHD patients with severe malnutrition.

Overall, the available evidence suggests that IDPN and AAD may be useful in the treatment of malnourished chronic dialysis patients and offers an alternative method of nutritional intervention in dialysis patients in whom oral or enteral intake cannot be maintained. However, most studies evaluating IDPN are retrospective, uncontrolled, and short term. Furthermore, there are no data to show that aggressive nutritional supplementation through the gastrointestinal tract is inferior to parenteral supplementation in dialysis patients. Until a controlled study comparing various forms of nutritional supplementation in similar patient groups is completed, one should be cautious in choosing very costly nutritional interventions.

EXPERIMENTAL NUTRITIONAL THERAPIES

The availability of recombinant forms of certain anabolic agents has made it possible to utilize them at pharmacological doses to promote net anabolism in multiple patient populations. Several preliminary studies have suggested that rhGH administration, especially when given along with nutritional supplementation, results in significant improvements in nitrogen balance, serum albumin, serum transferrin, and IGF-1 concentrations.[38] Recombinant human IGF-1 has also been proposed as an anabolic agent, but the side effect profile of this agent, at least as observed in CRF patients, may impede its widespread use at this time.[14] Interestingly, the combined use of rhGH and rhIGF-1 in healthy subjects has yielded the most efficient anabolic action, with the least side effects.[29] Whether long-term use of these agents in malnourished CHD and PD patients would result in improved nutritional parameters and better outcome is not yet known.

NUTRITION IN ACUTE RENAL FAILURE

The nutritional hallmark of acute renal failure is excessive catabolism. Multiple studies have shown that the protein catabolic rate in ARF patients requiring dialytic support is very high and can be massive at times.[9] Several factors have been suggested as potential causes. Sepsis and trauma can initiate a sequence of catabolic events through specific cytokines, including interleukins and tumor necrosis factor. This, in turn, induces increased whole-body protein breakdown. In addition, ARF patients may also suffer from diminished utilization and incorporation of available nutrients.

Nutritional indices are important for identifying the malnourished ARF patient and designing appropriate nutritional support and assessing the response. In ARF, the major process contributing to poor nutritional status is the metabolic response to ongoing morbidity or catabolism, whereas in other states malnutrition is largely a response to chronic starvation. The nutritional markers that correlate best with efficacy of nutritional therapy and patient outcome may differ considerably in these two disease states, and have not been well delineated in the ARF patient population. Urea nitrogen, serum albumin, prealbumin, and transferrin are all influenced by many factors. Similarly, anthropometry or other traditional measures of body composition find limited application in ARF patients because these patients may experience major shifts in body water.

The actual requirements for protein and energy supplementation in ARF patients are not well defined (Table 7.1). In clinical practice, the patients' actual nutritional needs are frequently not determined. Measurement of urea nitrogen appearance rate, which reflects protein catabolism, may be cumbersome in clinical settings, as is the measurement of energy expenditure. The fluid distribution and the fat free mass may be considerably altered in ARF patients, especially during the oligo-anuric phase. In the presence of diminished utilization and clearance due to diminished renal function, excessive protein supplementation will result in increased accumulation of end-products of protein and amino acid metabolism, i.e., higher BUN concentrations. Large quantities of nutrients to meet the patients' needs require higher levels of fluid infusion and may bring about fluid overload. Aggressive nutrition may also cause hyperglycemia, hyperlipidemia, hyper- or hyponatremia, and abnormalities in amino acid profiles in ARF patients.[27] Although most of these abnormalities can be managed by complementary dialytic support, the initiation, intensity, as well as the dose of dialysis treatment are controversial. In some cases, even dialysis cannot overcome the development of these abnormalities in patients treated with parenteral nutrition.

ACKNOWLEDGMENTS

This work is supported in part by NIH Grant #RO1 45604, RO1 DK 54413, FDA Grant # 000943.

GLOSSARY

CRF   Chronic renal failure
ESRD   End-stage renal disease
GFR   Glomerular filtration rate
DPI   Dietary protein intake
CHD   Chronic hemodialysis

PD   Peritoneal dialysis
PCM   Protein calorie malnutrition
MDRD   Modification of diet in renal disease
TNF   Tumor necrosis factor
IL-1   Interleukin 1
CRP   C-reactive protein
REE   Resting energy expenditure
IGF-1   Insulin-like growth factor 1
IDPN   Intradialytic parenteral nutrition
AAD   Amino acid dialysate
ARF   Acute renal failure
BUN   Blood urea nitrogen
LDL   Low-density lipoprotein
VLDL   Very low-density lipoprotein
HDL   High-density lipoprotein

REFERENCES

1. Bistrian, B.R., Role of the systemic inflammatory response syndrome in the development of protein-calorie malnutrition in ESRD, *Am. J. Kidney Dis.* 32, S113–7, 1998.
2. Canivet, E., Lavaud, S., Wong, T., Guenounou, M., Willemin, J. C., Potron, G., and Chanard, J., Cuprophane but not synthetic membrane induces in serum tumor necrosis factor-alpha levels during hemodialysis, *Am. J. Kidney Dis.* 23, 41–46, 1994.
3. Cano, N., Labastie-Coeyrehourq, J., Lacombe, P., Stroumza, P., Costanzo-Dufetel, J. D., Durbec, J.-P., Coudray-Lucas, C., and Cynober, L., Perdialytic parenteral nutrition with lipids and amino acids in malnourished hemodialysis patients, *Am. J. Clin. Nutr.* 52, 726–730, 1990.
4. Chan, W., Valerie, K.C., and Chan, J.C.M., Expression of insulin-like growth factor-1 in uremic rats: Growth hormone resistance and nutritional intake, *Kidney Int.* 43, 790–795, 1993.
5. Chazot, C. and Kopple, J.D., Vitamin metabolism and requirements in renal disease and renal failure, in *Nutritional Management of Renal Disease*, Kopple, J.D. and Massry, S.G. (Eds.) Williams & Wilkins, Baltimore, 1997, 415–478.
6. Chertow, G.M., Ling, J., Lew, N.L., Lazarus, J.M., and Lowrie, E.G., The association of intradialytic parenteral nutrition with survival in hemodialysis patients, *Am. J. Kidney Dis.* 24, 912–920, 1994.
7. Churchill, D.N., Adequacy of peritoneal dialysis: how much dialysis do we need?, *Kidney Int.* 48, S2-S6, 1997.
8. DeFronzo, R.A., Alvestrand, A., Smith, D., Hendler, R., Hendler, E., and Wahren, J., Insulin resistance in uremia, *J. Clin. Invest.* 67, 563–568, 1981.

9. Druml, W., Nutritional support in acute renal failure, in *Nutrition and the Kidney*, 2nd ed., W.E. Mitch and S. Klahr (Eds.) Little Brown and Company, Boston, 1993, 314–345.

10. Goodship, T.H.J., Mitch, W.E., Hoerr, R.A., Wagner, D.A., Steinman, T.I., and Young, V.R., Adaptation to low-protein diets in renal failure: Leucine turnover and nitrogen balance, *J. Am. Soc. Nephrol.* 1, 66–75, 1990.

11. Graham, K.A., Reaich, D., Channon, S.M., Downie, S., Gilmour, E., Passlick-Deetjen, J., and Goodship, T. H., Correction of acidosis in CAPD decreases whole body protein degradation, *Kidney Int.* 49, 1396–1400, 1996.

12. Graham, K.A., Reaich, D., Channon, S.M., Downie, S., and Goodship, T.H.J., Correction of acidosis in hemodialysis decreases whole body protein degradation, *J. Am. Soc. Nephrol.* 8, 632–637, 1997.

13. Hakim, R.M., Breyer, J., Ismail, N., and Schulman, G., Effects of dose of dialysis on morbidity and mortality, *Am. J. Kidney Dis.* 23, 661–669, 1994.

14. Hammerman, M.R., Insulin-like growth factor I treatment for end-stage renal disease at the end of the millennium, *Curr. Opin. Nephrol. Hypertens.* 9 (1), 1–3, 2000.

15. Heidland, A. and Kult, J., Long-term effects of essential amino acids supplementation in patients on regular dialysis treatment, *Clin. Nephrol.* 3, 234–239, 1975.

16. Ikizler, T.A., Flakoll, P.J., Parker, R.A., and Hakim, R.M., Amino acid and albumin losses during hemodialysis, *Kidney Int.* 46, 830–837, 1994.

17. Ikizler, T.A., Greene, J., Wingard, R.L., Parker, R.A., and Hakim, R.M., Spontaneous dietary protein intake during progression of chronic renal failure, *J. Am. Soc. Nephrol.* 6, 1386–1391, 1995.

18. Ikizler, T.A. and Hakim, R.M., Nutrition in end-stage renal disease, *Kidney Int.* 50, 343–357, 1996.

19. Ikizler, T.A. and Hakim, R.M., Renal failure and parenteral nutrition, in *Clinical Nutrition: Parenteral Nutrition*, 3rd ed., Rombeau, J.L. and Rolandelli, R.H. (Eds.) W.B. Saunders Company, Philadelphia, 2001, 366–391.

20. Ikizler, T.A. and Schulman, G., Adequacy of dialysis, *Kidney Int. Suppl.* 62, S96–100, 1997.

21. Ikizler, T.A., Wingard, R.L., Sun, M., Harvell, J., Parker, R.A., and Hakim, R.M., Increased energy expenditure in hemodialysis patients, *J. Am. Soc. Nephrol.* 7, 2646–2653, 1996.

22. Jones, M., Hagen, T., Boyle, C.A., Vonesh, E., Hamburger, R., Charytan, C., Sandroni, S., Bernard, D., Piraino, B., Schreiber, M., Gehr, T., Fein, P., Friedlander, M., Burkart, J., Ross, D., Zimmerman, S., Swartz, R., Knight, T., Kraus, A., Jr., McDonald, L., Hartnett, M., Weaver, M., Martis, L., and Moran, J., Treatment of malnutrition with 1.1% amino acid peritoneal dialysis solution: results of a multicenter outpatient study, *Am. J. Kidney Dis.* 32, 761–769, 1998.

23. Kasiske, B.L., Lakatua, J.D., Ma, J.Z., and Louis, T.A., A meta-analysis of the effects of dietary protein restriction on the rate of decline in renal function, *Am. J. Kidney Dis.* 31, 954–961, 1998.

24. Kaysen, G.A., The nephrotic syndrome: nutritional consequences and dietary management, in *Nutrition and the Kidney*, 3rd ed., W.E. Mitch and S. Klahr (Eds.) Lippincott–Raven, Philadelphia, 1998, 201–212.

25. Kaysen, G.A., Stevenson, F.T., and Depner, T. A., Determinants of albumin concentration in hemodialysis patients, *Am. J. Kidney Dis.* 29, 658–668, 1997.

26. Klahr, S., Levey, A. S., Beck, G. J., Caggiula, A. W., Hunsicker, L., Kusek, J. W., Striker, G., et al., The effects of dietary protein restriction and blood-pressure control on the progression of chronic renal disease, *N. Engl. J. Med.* 330, 877–884, 1994.

27. Kopple, J.D., The nutrition management of the patient with acute renal failure, *J. Parenteral Enteral Nutr.* 20, 3–12, 1996.

28. Kopple, J.D., National kidney foundation K/DOQI clinical practice guidelines for nutrition in chronic renal failure, *Am. J. Kidney Dis.* 37 (1 Suppl 2), S66–70., 2001.

29. Kupfer, S.R., Underwood, L.E., Baxter, R.C., and Clemmons, D.R., Enhancement of the anabolic effects of growth hormone and insulin-like growth factor I by use of both agents simultaneously, *J. Clin. Invest.* 91, 391–396, 1993.

30. Ling, P.R., Schwartz, J.H., and Bistrian, B.R., Mechanisms of host wasting induced by administration of cytokines in rats, *Am. J. Physiol.* 272, E333–9, 1997.

31. Lowrie, E.G. and Lew, N.L., Death risk in hemodialysis patients: the predictive value of commonly measured variables and an evaluation of death rate differences between facilities, *Am. J. Kidney Dis.* 15, 458–482, 1990.

32. Maroni, B., Steinman, T.I., and Mitch, N.E., A method for estimating nitrogen intake of patients with chronic renal failure, *Kidney Int.* 27, 58–61, 1985.

33. Mitch, W.E. and Goldberg, A.L., Mechanism of muscle wasting: the role of ubiquitin-proteasome pathway, *N. Engl. J. Med.* 335, 1897–1905, 1997.

34. National Kidney Foundation, K/DOQI, Clinical Practice Guidelines for Chronic Kidney Disease: Evaluation, Classification, Stratification, *Am. J. Kidney Dis.* 39, Suppl. 1, S1–S2, 2002.

35. Parker III, T.F., Wingard, R.L., Husni, L., Ikizler, T.A., Parker, R.A., and Hakim, R.M., Effect of the membrane biocompatibility on nutritional parameters in chronic hemodialysis patients, *Kidney Int.* 49, 551–556, 1996.

36. Parker, III,T.F., Laird, N.M., and Lowrie, E.G., Comparison of the study groups in the National Cooperative Dialysis Study and a description of morbidity, mortality, and patient withdrawal, *Kidney Int.* 23 (Suppl. 13), S42-S49, 1983.

37. Qureshi, A.R.V., Alvestrand, A., Danielsson, A., Divino-Filho, J.C., Gutierrez, A., Lindholm, B., and Bergstrom, J., Factors predicting malnutrition in hemodialysis patients: a cross-sectional study, *Kidney Int.* 53, 773–782, 1998.

38. Schulman, G., Wingard, R.L., Hutchinson, R.L., Lawrence, P., and Hakim, R.M., The effects of recombinant human growth hormone and intradialytic parenteral nutrition in malnourished hemodialysis patients, *Am. J. Kidney Dis.* 21 (5), 527–534, 1993.

39. Slatopolsky, E., Brown, A., and Dusso, A., Pathogenesis of secondary hyperparathyroidism, *Kidney Int. Suppl.* 73, S14–9, 1999.

40. Stenvinkel, P., Heimburger, O., Paultre, F., Diczfalusy, U., Wang, T., Berglund, L., and Jogestrand, T., Strong association between malnutrition, inflammation, and atherosclerosis in chronic renal failure, *Kidney Int.* 55, 1899–1911, 1999.

41. Tschope, W., Koch, M., Thomas, B., and Ritz, E., Serum lipids predict cardiac death in diabetic patients on maintenance hemodialysis. Results of a prospective study. The German Study Group Diabetes and Uremia, *Nephron.* 64, 354–358, 1993.

42. U.S. Renal Data System, Excerpts from the USRDS 2001 Annual Data Report, *Am. J. Kidney Dis.* 38 (Suppl 3), S1- S248, 2001.

43. Vanholder, R., Cornelis, R., Dhondt, A., and Ringoir, S., Trace element metabolism in renal disease and renal failure, in *Nutritional Management of Renal Disease*, Kopple, J.D. and Massry, S.G. (Eds.) Williams & Wilkins, Baltimore, 1997, 395–414.

8 Nutrition and Vision

John R. Trevithick and Kenneth P. Mitton

CONTENTS

INTRODUCTION

Most of us take vision for granted. If we encounter difficulty seeing, we react, looking around the head of the person sitting in front of us at the theatre, or straining to see the white line on the highway as we drive in fog or snowstorm. Such experiences can remind us just how important our vision is.

As we age, chronic, progressive, degenerative diseases act to impair our vision. Degenerative changes to the eye lens result in cataract, affecting nearly half of those aged 75 years or more, and macular degeneration, which affects about one fifth of people over 75. The major cause of blindness worldwide, cataract, is treated surgically in developed countries (with a success rate of greater than 99%) by implantation of a plastic intraocular lens (IOL). Implicated as a possible contributing cause of cataract, age-related macular degeneration is not treated surgically. While some surgeons (Jorge Vasco-Posada in Medellin, Colombia) have reported excellent results in individual cases, Russian studies using similar techniques have not shown efficacy by statistical analysis.

Recent experimental treatment of neovascularization by photodynamic therapy, although promising, requires follow-up treatments at 6-month intervals. Dry eye, a condition which also increases with age, can result in corneal damage.

This chapter will focus on nutritional strategies that medical personnel can employ to complement more conventional pharmacological strategies. It will also attempt to point out some pros and cons of herbal remedies, which may affect the eye and vision.

PREVENTION STRATEGIES.....OR ACTIVE TREATMENT TO REDUCE RISK OF EYE DISEASE?

The eye lens contains a person's entire life history, since its cells are laid down approximately one cell layer per day, like tree rings. In one notable example, a cataract in the lower quadrant, interior of the left lens of a radiation technician was related to an exposure to radium needles carried in his shirt pocket 20 years previously.[1] In this case, the damage occurred when cells were being laid down. After 20 years, the damaged cells were located in the interior of the lens at the area of the cataractous opacity. This suggests that the earlier radiation damage increased the risk of death in these cells, resulting in the observed opacity.

Because chronic diseases are slow to develop, preventive strategies to reduce their risk should begin well before the first appearance of signs or symptoms. For instance, the risk reduction observed for cataract when a vitamin C supplement is used achieved significance only after 10 years of supplement use.[2] In our study, risk reduction associated with vitamins E and C was observed in people who had used supplements of vitamin E or vitamin C for a minimum of 5 years.[3,4] Because cortical cataract formation appears to involve protein leakage and cell death, the possibility of reversal of an opacity by any active treatment is remote.

Another concern is whether active treatment is feasible. As pointed out above, although photodynamic therapy offers hopeful results in a small percentage of macular degeneration cases, surgery for this condition is not feasible in developed countries.[5–7]

For this reason, nutritional strategies, initiated before or early in middle age, appear to offer the best opportunity to prevent or at least reduce the risk of degenerative eye diseases, such as cataract or macular degeneration.

NUTRITIONAL RISK REDUCTION STRATEGIES FOR VARIOUS OCULAR DISEASES

CATARACT

Although one study (the Italian American Study[8]) showed no effect of supplemental vitamin E in cataract risk reduction, meta-analysis including this and six other studies of antioxidant levels in populations showed approximately 54% risk reduction for people with elevated vitamin E levels.[9]

Vitamin C, a water-soluble antioxidant, has also been shown to lead to a significant reduction of cataract risk, although (as pointed out above) between 5 and 10 years may be required for the effect to become statistically significant. Leske's studies indicate that multivitamins and consumption of several helpings of fruits and vegetables per day also reduce cataract risk.[10,11] These results suggest that subclinical deficiencies of both vitamins and antioxidants may increase cataract risk, and they call attention to the importance of consuming diets with enough fruits and vegetables. Antioxidants occur in many vegetables and fruit sources as flavonoids, polyphenols, and proanthocyanidins, which serve as monomeric or oligomeric precursors of the polymeric lignin, which forms the plastic portion of plant cell walls.

Recent randomized prospective studies have been split on the benefit of antioxidant supplementation. Over a period of 2 years, the REACT trial followed the effect of supplementation in North America and Europe with a cocktail of vitamin E, C, and selenium on cataract progression in patients having an early stage of cataract. The REACT trial showed approximately 55% risk reduction of cataract progression in participants with elevated vitamin E plasma concentrations over a 2-year period in patients having an early cataract.[12] By contrast, in the VECAT study conducted in Melbourne, Australia, vitamin E supplementation of people 55 years of age and older for a period of up to 5 years did not show any significant reduction in risk of either cataract or macular degeneration.[13] A similar lack of effect was observed in the Age Related Eye Diseases Study in patients followed for an average of 6.3 years.[14,15] Given the long time required for cataracts to develop and the several case-control studies that indicate that no significant effects occur until after 5 years of treatment, it is possible that following the participants of the VECAT study for a longer period would have revealed a significant effect. Alternatively, the ultraviolet exposure of Australian participants in the VECAT study may have been much larger than that encountered at the more northern latitude of the REACT study, resulting in less oxidative stress for the latter group. Further studies of the effects of vitamins E and C are indicated to resolve this apparent conflict.

In this connection, the several studies of alcohol consumption and cataract risk reduction deserve comment. Ruth Clayton's group found that people in Glasgow, Scotland, consuming one drink per day had approximately 50% of the cataract risk of non-drinkers.[16] The curve of risk vs. alcoholic beverage consumption was J shaped, however, with cataract risk reduction at two drinks/day only about 10% and at three drinks/day, an increased risk. Sasaki's group[17] has also reported similar cortical cataract risk reduction associated with moderate alcohol consumption. Our investigations have revealed that alcohol itself is an antioxidant.[18] Furthermore, beers and wines, as well as whiskies, rum, etc., are aged in wood. Polyphenolic components extracted from the oak casks can contribute significantly to the antioxidant intake of individuals.

In studies related to heart disease risk, Gey[19] has suggested that inadequate dietary intake of antioxidants, in particular vitamin E, may contribute to the increased risk of heart disease in populations with low vitamin E plasma levels. Consistent with this, a study of risk of cardiac disease as a function of alcoholic beverage consumption revealed a J-shaped curve for heart disease risk reduction: a 50% risk reduction for one drink per day, rising from that level with increased alcoholic beverage intake.[20,21] This was almost identical to the J-shaped curve for cataract risk reduction. Although the effects of alcohol on heart disease had previously been attributed to an increased level of plasma high-density lipoprotein (HDL), this explanation is not valid for the eye lens, which obtains its nutrients from the aqueous humor, an ultrafiltrate of plasma which (to prevent light scattering) contains no HDL. A more likely explanation is that the concentration of plasma antioxidants is responsible for both effects, accounting for the similarity of the J-shaped curves for risk reduction. Antioxidants can protect against the formation of oxidized low-density lipoprotein,[22–25] which is preferentially incorporated into the atherosclerotic plaque. Antioxidants can also protect the eye lens cell membranes and mitochondrial membranes from damage, which in turn could lead to intracellular calcium elevation, calpain activation, spectrin/fodrin proteolysis, and blebbing of the cell membrane found in cortical cataract formation[26–28] (Figure 8.1).

Taken together, this suggests that antioxidants may be of more benefit in populations that have low antioxidant intakes, and it provides a rationale for the lack of cataract risk reduction for vitamin E in the Italian American Study[8]: the Mediterranean diet is known to be rich in antioxidants. With fruits, vegetables, olives and olive oil, and wine contributing significant amounts of antioxidants, perhaps a threshold for risk reduction or additional antioxidants is exceeded, above which no additional effect can be observed. By contrast, the northern European diet (e.g., Scotland) or the rice-based diet of the Orient may be lower in antioxidants, and so a significant benefit would be expected by adding antioxidants to the diet in the form of fruits, vegetables, or alcoholic beverages, reducing risk of cataract and atherosclerosis.

Vitamin A appears to be essential for lens epithelium. Deficiency in childhood, which is common in third world countries, may increase adult cataract risk.[29] Antioxidants could also protect lens fiber mitochondria, which are destroyed by brief periods of hyperglycemia, permanently damaging the ability of the cells to synthesize ATP[30] which is necessary for the synthesis of glutathione.[31]

MACULAR DEGENERATION

Rod and cone cells of the retina grow continuously, renewing themselves. The retinal pigment epithelium (RPE) is activated each morning by light to phagocytose the oldest segments of the rods and cones, digesting their cellular biochemical components. Oxidative stress, associated with light exposure of the retina, can result in cross-linked, oxidized proteins, which are not completely digested by the proteasomes and proteases of the RPE. Oxidized protein fragments and the metabolite A_2E (two retinaldehyde molecules linked by ethanolamine) accumulate as lipofuscin,[32,33] an insoluble fluorescent hydrophobic residue that accumulates, eventually forming small globules in the retina called drusen. High concentrations of A_2E are toxic to the retinal pigment epithelial cells.[32] This process is associated with retinal ischemia. As the degeneration progresses, the drusen grow and can separate the rods and cones from the RPE. Growth of new blood vessels stimulated by retinal ischemia (angiogenesis) signals the transition from "dry" macular degeneration to "wet" macular degeneration.

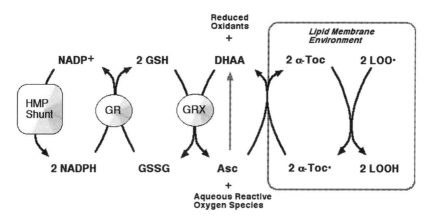

FIGURE 8.1 Antioxidant cycling in the cell. Protection of cell and mitochondrial biomembranes from peroxidation damage to both the protein and lipid components does not rely on large quantities of tocopherol (Toc). What is necessary are all of the intermediate components, including glutathione and ascorbate. Components of the cycle regenerate in reducing/oxidation reactions, and the system is supplied with reducing equivalents in the form of NADPH from an active hexose monophosphate (HMP) shunt. Lipid peroxy radicals (LOO¡) are kept reduced by the chain-breaking reaction with tocopherol. Tocopherol radicals (Toc¡) are more non-reactive due to localization of the free radical over the large ring system of tocopherol's structure. However, Toc¡ can be readily reduced by ascorbate (Asc) at the water/membrane interface in a non-enzymatic reaction, thus regenerating reduced Toc. Dehydroascorbate (DHAA) is in turn reduced to Asc by the non-enzymatic and enzymatic redox reaction with reduced glutathione (GSH). NADPH provides reducing power to glutathione reductase (GR) for reducing glutathione disulfide (GSSG) to GSH. Disease or environmental stresses that compromise the energy (ATP) required to maintain the synthetic supply of glutathione (gamma-glutamyl cysteinyl glycine), or that compromise the supply of NADPH, will lead to a failure of the entire system. Compromise of the antioxidant cycling system can then allow further oxidation damage to systems supplying ATP or NADPH, leading to an accelerating loss of viability termed an *oxidation spiral*[9] that ultimately leads to death of the cell.

Absorption of light in the eyes characteristically occurs as follows: the cornea absorbs all light < 295 nm wavelength, so the shortest wavelength reaching the retina is 400 nm, since all light < 400 nm is absorbed by the lens.[34,35] As we can see in the following sections, the retina has developed mechanisms to protect against damage from blue light.

Nutritional and lifestyle strategies for reducing the risk of macular degeneration include:

1. Vitamin E dietary supplementation (400 IU/day), which Sheila West[36] has shown reduces the risk of macular degeneration by approximately 57%.
2. In a randomized, placebo-controlled trial, the Age Related Eye Disease Study Group reported that a combined regimen of antioxidants (vitamin C: 500 mg, vitamin E: 400 IU, and beta-carotene: 15 mg.), and zinc (80 mg, as zinc oxide, with copper, 2 mg., as cupric oxide) reduced the risk of

progression of macular degeneration from the dry to the wet form. This group suggested that people "with extensive intermediate-size drusen, at least one large druse, noncentral geographic atrophy in one or both eyes, or advanced AMD or vision loss due to AMD in one eye and without contraindications such as smoking" should "consider taking a supplement of antioxidants plus zinc such as that used in this study."[15]

3. Dietary choices to increase intake of lutein and zeaxanthin, carotenoids which are specifically accumulated in the fovea or macular area.[37] The yellow color they impart to the fovea is the result of absorption of blue light. Blue light has the highest energy photons of visible light and can photoactivate absorbing molecules in the retina, which results in oxidative changes.[38] Cytochemistry has revealed higher concentrations of lutein and zeaxanthin at the ends of rod cells, as well as in the RPE and fovea.[39] High concentrations of these carotenoids are present in sweet corn, yellow peppers, broccoli, etc.[40] Consumption of these vegetables will help reduce risk of macular degeneration by enhancing foveal protective concentrations of lutein and zeaxanthin.[41,42]

4. Preventing cataracts by the dietary strategies above also protects the retina because of the absorption in the lens of blue light by 3-hydroxykynurenine, a metabolite of tryptophan, which imparts a yellow color to the eye lens. This color increases with age.

5. Wearing yellow sunglasses, which can be obtained at sporting goods departments, also reduces the intensity of blue light which could potentially damage the retina. These glasses also absorb UV light, which in extreme cases can lead to corneal inflammation in "Labrador keratitis."[43-45]

6. Avoiding horizons, such as in outdoor occupations, reduces UV and blue light, which is also reduced more than 90% by use of hats with visors.[46]

7. If people use photochemically activated drugs for psoriasis, such as psoralen derivatives or herbal remedies such as St. John's Wort, often used to treat depression, it is important for them to avoid sunlight or high-intensity light therapy.[47]

Combining the above preventive measures should effectively diminish the risk of macular degeneration resulting from blue light exposure.

8. Vitamin A, essential for vision, also appears to be necessary for the retina.[48] Consumption of foods rich in vitamin A or its carotene precursors would reduce retinal vitamin A deficiency. Such deficiency can cause "night blindness" in mild deficiency (nyctalopia), and in more severe deficiency, keratomalacia, or scarring of the cornea.[49]

RETINOPATHY OF PREMATURITY (ROP)

This condition, formerly known as retrolental fibroplasia, is found frequently in premature infants who have been given oxygen to help improve their blood oxygen levels. Because vitamin E transfer to the fetus occurs late in pregnancy, infants born prior to 21 weeks are usually vitamin E deficient. Hittner and Knetzer[50] showed that risk of this neovascularization of the retina could be reduced by vitamin E supplementation. A number of clinical trials using vitamin E therapy were performed but were stopped because of infant deaths from a particular parenteral preparation. Meta-analysis of all studies, and of a selected group, showed significant risk reduction of ROP.[51] Although expert opinion felt that vitamin E supplementation of infant parenteral or dietary formulas was not justified to reduce the risk of ROP, infant formulas currently contain supplemental concentrations of vitamin E (400 IU in a 70-kg adult). The justification for this supplementation appears to be the prevention of hemolytic disease in the newborn, which occurs because of erythrocyte fragility in vitamin E deficiency. Johnson has shown that if treatment is delayed until age 3 months, the risk of necrotizing enterocolitis and sepsis is very low and reversal of neovascularization of the retina is maximal.[52]

DIABETIC RETINOPATHY

King and Bursick have conducted several studies in experimental models and are currently involved in a clinical trial of vitamin E supplementation to investigate the possible reduction of risk of diabetic retinopathy.[53,54] The early stages of degenerative change in the retina, involving induction of the enzyme protein kinase C, appear to be prevented or reversed by vitamin E.

VITAMINS E, A, AND RETINITIS PIGMENTOSA (RP)

While the very act of retinal degeneration, loss of photoreceptors, such as occurs in RP, can lead to extra free radical stresses from the release of cellular debris, there are examples of a more direct involvement of oxidation and the requirement of vitamin E for the healthy functional retina. Quite recently, it has been shown that mutations of the liver alpha-tocopherol transport protein (alpha-TTP), which lead to lowered serum alpha-tocopherol levels, cause retinitis pigmentosa and ataxia.[55,56] Supplementation in these patients alleviates the progression of RP and ataxia. So far, the acetate ester of tocopherol has been used. One could speculate that the more water-soluble form of tocopherol, tocopherol succinate ester, would be a better treatment because the water-soluble form can be delivered independent of the lipid component.

Alpha-TTP-based retinal disease, such as that described above, is a good example of the importance of understanding the exact disease cause when assessing the use of nutritional treatments for disease. In this case, there is a clear and logical benefit of the use of tocopherol supplementation because there is a below-normal concentration in the first place. The benefits of vitamin E supplementation are not that clear for all cases of RP. In a frequently cited 6-year study, a daily supplement of 15,000 IU vitamin A decreased the average yearly decline in measuring the 30 Hz cone ERG amplitudes of test subjects. Unfortunately, the addition of 400 IU/d of vitamin E alone or in combination with vitamin A actually increased the decline for the higher-amplitude cohorts (but not randomized patients) compared to patients getting trace amounts of vitamins A and E (75 and 3 IU/d, respectively).[57] However, the actual percentage gains or losses will require more than a single study: when the same authors used visual acuity (letters lost per year), the average decline in visual acuity was not statistically significant. Furthermore, the negative effect of vitamin E in the study does not seem to be retinal based because the patients on the vitamin E supplement had greatly decreased serum vitamin A levels. Was this some effect of the base carrier for the E supplement? This confuses the issue greatly. It is clear that the use of antioxidants as a nutritional-based supplementation for general retinitis pigmentosa does not show a clear benefit according to our current state of knowledge. Furthermore, it is apparent that for each case of RP the exact mechanism of the defect should be taken into account. This will be easier as more mechanisms of various RPs are determined at the molecular level.

Nature and Nurture in Retinal Disease

Rod Visual Cycle and ABCR, a Lesson in Genetics and Oxidation Stress

Rhodopsin, a G-protein coupled receptor, is arguably the most central component of vision; it is the light-detecting protein, formed from the combination of the rod cell opsin protein and 11-cis-retinal (11-Ral). Absorption of a photon of light by 11-Ral leads to transition from its excited state to the all trans-retinal (t-Ral). This activated form of rhodopsin (Rh*) activates the heterotrimeric G-protein, transducin, which releases the photoreceptor-specific phosphodiesterase PDE6 from its inhibitory subunits. This results in the hydrolysis of cGMP, which leads to the closing of cGMP-gated calcium channels in the rod cell membrane, causing hyperpolarization and starting the first signal transduction events via synapses with rod bipolar cells. Thus begins the neurotransmission that will involve several other classes of neurons on its way to the visual cortex. It is important to note that Rh detects a photon once and then activates transducin. Phosphorylation of Rh* and the interaction with arrestin serve to turn off the Rh* and aid in release of t-Ral from opsin itself. Within the outer disk membranes, where Rh resides, opsin must bind new 11-Ral so the receptor can be used again. Unfortunately, this is not just a matter of storing large caches of 11-Ral, since the concentrations of 11-Ral would be too high in the disks. In that situation, free 11-Ral could still be converted to t-Ral by a photon of light and then there would also be a large accumulation of t-Ral. That would be similar to the invertebrate visual system, where 11-Ral is converted to t-Ral, and then another photon is required to convert back to 11-Ral again.

In contrast, in the vertebrate rod visual cycle, t-Ral is transported from the disk lumen across the disk membrane into the cytoplasm of the rod outer segment[58] (Figure 8.2). Current thinking holds that t-Ral is then reduced to all-trans-retinol (t-Rol) and is transported, bound to the intercellular retinal binding protein (IRBP), by a shuttle mechanism into the extracellular space and then into the cytoplasm of the RPE cells. It is in the RPE that t-Rol is converted back into 11-Ral for its transport back to the rod cell. T-Rol is transiently esterified by lecithin:retinol acetyl transferase (LRAT), isomerized, and hydrolyzed to 11-cis-retinol by isomerohydrolase, then oxidized to 11-cis-retinal by 11-cis-retinol dehydrogenase. Intermediates can be bound in the RPE cell by cellular retinol binding protein (CRBP) and cellular retinaldehyde binding protein (CRALBP). Regenerated 11-Ral can then be transported back into the rod cell. This two-cell-based visual cycle provides for recovery

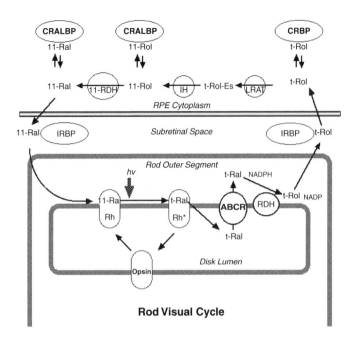

Rod Visual Cycle

FIGURE 8.2 The rod visual cycle and ABCR: Where nature meets nurture? Rhodopsin is continuously recycled in the vertebrate visual system in a two-cell compartment arrangement. Rhodopsin (Rh), a G-protein coupled receptor, is formed from the combination of the rod cell opsin protein with 11-cis-retinal (11-Ral). Adsorption of a photon of light by 11-Ral leads to transition from its excited state to the all trans-retinal (t-Ral). Rhodopsin (Rh*) activates the heterotrimeric G-protein, transducin, which releases the photoreceptor specific phosphodiesterase PDE 6 from its inhibitory subunits. Hydrolysis of cGMP leads to the closing of cGMP-gated calcium channels in the rod-cell membrane causing hyperpolarization and starting signal transduction via synapses with rod bipolar cells. Rh detects a photon once and then activates transducin. Phosphorylation of Rh*, and the interaction with arrestin, turn off the Rh* and aid in release of t-Ral from opsin. Within the outer disk membranes, where Rh resides, opsin must bind new 11-Ral so the receptor can be used again. The ABCR protein (ATP-binding cassette transporter) transports t-Ral from the disk lumen across the disk membrane into the cytoplasm of the rod outer segment. Current thinking holds that t-Ral is then reduced to all-trans-retinol (t-Rol), and transported into the extracellular space, shuttled by intercellular retinal binding protein (IRBP) and then into the cytoplasm of the RPA cells. In the RPE t-Rol is converted back into 11-Ral for transport back to the rod cell. T-Rol is transiently esterified by lecithin:retinol acetyl transferase (LRAT), isomerized and hydrolyzed to 11-cis-retinol by Isomerohydrolase, and oxidized to 11-cis-retinal by 11-cis-retinol dehydrogenase. Intermediates can be bound in the RPE cell by cellular retinol binding protein (CRBP) and cellular retinaldehyde binding protein (CRALBP). Regenerated 11-Ral can then be transported back into the rod cell. Mutations in the photoreceptor-specific ABCR cause Stargardt macular dystrophy, RP, and cone–rod dystrophy and is one example of mutations in a single protein leading to very different phenotypes and disease progression.[59–61] Accumulation of t-Ral in the disks is thought to increase susceptibility to light-induced damage, since t-Ral can act as a photosensitive agent.

of bleached photopigment back to 11-Ral even in the dark. Regeneration, therefore, is not dependent on light flux levels as in the invertebrate system.[58]

Mutations in the photoreceptor-specific ATP-binding cassette transporter, ABCR, cause Stargardt macular dystrophy, RP, and cone–rod dystrophy. This is one example of mutations in a single protein leading to very different phenotypes and disease progression.[59–61] While the differences in mutations themselves play a role in different disease manifestations, is there an environmental component influencing the outcome as well? It has been known for some time that genetic models of animal RP can be influenced, i.e., slowed in their progression, depending on the level of antioxidants such as tocopherol in the diet. What could possibly link this environmental effect (nurture) with the genetic predisposition (nature)? It is currently thought that the ABCR transporter moves t-Ral from the disk lumen into the rod outer segment cytoplasm, driven by the hydrolysis of ATP.[62] Recently, Sun and Nathans[63] have shown *in vitro* that all-trans-retinal itself is a potent photosensitizer that readily causes photo-dependent oxidation damage to normal ABCR protein, causing it to aggregate and lose its ATPase activity. Furthermore, two other structural proteins that co-localize with ABCR in the outer segment disk rims, Peripherin/RDS and ROM-1, are also susceptible to the same photo-dependent oxidation damage. Could this be one of many unknown examples of the direct interaction of genetic predisposition mixed with environmental stress (i.e., photo-dependent oxidation) to yield a varied phenotype and rate of disease progression?

OTHER FUNCTIONAL TARGETS SENSITIVE TO OXIDATION DAMAGE

There are already about 40 known disease genes with numerous mutations that cause retinitis pigmentosa and allied disease of the retina.[64] Any protein that can be damaged by mutation to its sequence, leading to loss of photoreceptors, is a potential disease-causing protein if this chemical alteration negates normal function. Such damage can also be caused by other stressors, such as oxidation damage. There are many more proteins that could be evaluated for susceptibility to oxidation damage. Most proteins of the visual transduction cascade itself, such as rhodopsin, may cause disease when mutations affect their function. A few examples of genes are rhodopsin (*RHO*), catalytic alpha and beta subunits of PDE6 (*PDE6A/B*), rod cyclic nucleotide gated channel (*CNGA1*), arrestin/S-antigen (*SAG*), alpha-transducin (GNAT1), rhodopsin kinase (*RHOK*), guanylate cyclase activator 1A (*GUCA1A*), and retina-specific guanylate cyclase (*GUCY2D*). Many more genes encoding proteins of the visual cycle, such as ABCR,[62,63] the tetraspanin disk structural proteins (i.e., *ROM1*), and some catabolic and mitochondrial proteins, are all potential places where nature meets nurture. Even transcription factors are functional examples of human retinal disease genes. Mutations in NRL (neural retina leucine zipper) and CRX (cone rod homeobox), two transcription factors that interact and activate transcription of several photoreceptor specific genes (i.e., rhodopsin), cause cone–rod dystrophy, Leber congenital amaurosis, and autosomal dominant retinitis pigmentosa.[65–67]

Any of these proteins are potential focal points for the effects of genetic predisposition combined with environmental stress, such as oxidation stress. In the era of the human genome, we will have the unprecedented capacity to begin evaluating the normal polymorphisms that create some variety in coding sequences of many genes. This information will allow for evaluation of these different polymorphisms and how they could influence the inherent susceptibility to oxidation stress.

GLAUCOMA

Although Russian workers have pioneered in this area,[68–72] with work indicating that antioxidant supplementation may reduce the risk of glaucoma, more recently the group of Azzi in Bern, Switzerland, has shown that tocopherol can inhibit the proliferation of human Tenon's capsule fibroblasts. This would be useful in maintaining holes in the trabecular meshwork after glaucoma filtering therapy.[73,74]

CORNEA

Like the retina, the cornea requires vitamin A. This requirement is partially spared by vitamin E[48,75,76] in the retina, but this has not been studied in the cornea. In more severe vitamin A deficiency than that causing night blindness, corneal damage can occur[49] in the condition known as xerophthalmia, which can lead to corneal scarring (keratomalacia) or even corneal lysis in extreme deficiency. This has occurred historically in children in third world countries who have not received supplemental vitamin A. Vitamin A deficiency is often associated with diarrhea, more prevalent in the tropics during the rainy season, and is an autocatalytic phenomenon, since vitamin A is required for all epithelial cells including those of the intestinal epithelium. Diarrhea decreases vitamin A absorption, which in turn reduces the ability to absorb vitamin A, so a vicious cycle ensues. For this reason, the World Health Organization (WHO) has initiated a vitamin A treatment program in the third world in which children receive about 1 million IU of vitamin A once every 6 months. Although this dose would be toxic if given repeatedly, it is stored in the liver and can be released slowly. This treatment has significantly reduced corneal blindness in third world countries.

Those experiencing vitamin A deficiency in childhood might have an increased risk of cataracts as adults. It is possible that this can be a factor in the higher incidence of cataracts in underdeveloped countries. Supporting this hypothesis, the population of the northern section of Nigeria, where white palm oil is commonly used for cooking, has an earlier age of cataract onset than the population of southern Nigeria, where red palm oil is used. The color of red palm oil indicates that it contains a higher concentration of beta carotene.[77]

KERATOCONUS

In this condition, which occurs infrequently, the softening of the cornea occurs, resulting in a cone-shaped cornea. Russian workers have reported efficacy of vitamin E supplementation for this condition.[78]

COMBINATIONS OF VITAMIN SUPPLEMENTS

Although it would appear to be intuitively obvious that combinations of vitamins, particularly those with antioxidant activity, should have beneficial effects, this may not be the case. At the Royal College of Surgeons, the treatment of retina degeneration associated cataracts in rats with dietary vitamin A, E, or C individually resulted in reduction of cataract and retinopathy.[79,80] Unexpectedly, the combination of vitamins A, C, and E at the same concentrations showed no significant reduction of cataract risk.

Studies of vitamins E and A on the retina indicated[75] that vitamin E can significantly protect vitamin A in vitamin A-deficient rats.

Taurine supplementation (*in vitro*) can protect against loss of other amino acids in the diabetic cataract.[81,82] Because taurine can have a strong antioxidant activity,[82] this suggests that rather complex interactions can occur as a result of protection by antioxidants.

ROLE OF DIET AND BEVERAGE CONSUMPTION IN HEALTH OF THE EYE AND VISION

Glutamate, although an antioxidant in its own right,[83] is toxic to the retina, causing retinal degeneration if concentrations are too high. Since monosodium glutamate is marketed as Accent®, a flavor-enhancing condiment, its use in food preparation should be carefully monitored for newborns and nursing or pregnant mothers, as well as those working in strong light (TV, film, and advertising, still photography, astronauts, pilots, farmers, seamen, construction workers, suntan parlor workers).

Antioxidant polyphenols and proanthocyanidins are found in significant quantities in most vegetables and fruits. Leske et al.[10] have found significant cataract risk reduction correlated with consumption of fruits and vegetables (as well as multivitamin supplements). Beverages such as beer, cider, wines, and distilled wines and spirits (if aged in wood) as well as a number of liqueurs (the best contain chocolate), coffees, and teas also contain significant levels of polyphenols. Teas and coffees are also available as specialized preparations, but these are all derived from plant sources, and so contain plant polyphenols, as do over-the-counter medications such as phycogenol and grape seed extract.

Honey[84,85] appears to have greater antioxidant activity depending on how dark the honey is colored. This color is related to the polyphenol content, as is the color of beverages aged in wood, such as bourbon, scotch, whisky, Canadian whisky, rum, etc. We have found that maple syrup is also an excellent source of antioxidant polyphenols (unpublished data).

Sweet corn is quite high in lutein. Broccoli, containing combined lutein and zeaxanthinal high concentrations, is a good source of these carotenoids. Foods which contain dark chocolate[86] are excellent sources of polyphenols, which are not present in white chocolate. Soy products, including soy milk, contain large quantities of phytates, flavones, and isoflavones. In fermented products such as soya sauce, miso, and tempeh, the phytates are significantly reduced. Long, slow cooking, which usually is a means to reduce phytates, is not effective with soy bean products. Phytates cause decreased calcium absorption from the diet. Low calcium (as well as elevated Ca^{++}) has been shown in model cataracts *in vitro* to cause cataracts.[87–91] Because of this, it would seem prudent to avoid making soy-derived foods (unless fermented) such as milk or meat substitutes a major component of diet.

Lycopene, available as nutritional supplements, but also a major component of tomato sauce and tomato paste, has been associated with reduction of risk of cataract as well as heart disease.[92]

POSSIBLY DAMAGING HERBAL MEDICATIONS

Some herbal medications have potentially dangerous side effects for the eye. As mentioned, light activation of endogenous molecules to a triplet state can make possible the regeneration of singlet oxygen. Such molecules can be found in nature (e.g., riboflavin) or may be present in medications (e.g., hypericin, found in the antidepressant St. John's Wort).[47,93]

HERBAL DIETARY SUPPLEMENTS USED IN TRADITIONAL MEDICINES

Several medications recommended in Chinese traditional medicine for treatment of cataract were found to contain rather high concentrations of antioxidants,[94] consistent with the generally accepted view that antioxidants can contribute to cataract risk reduction, perhaps by protecting cell membranes or mitochondria.

Several medications also were found to have strong aldose reductase activity, which would reduce the osmotic stress resulting from conversion to glucose to sorbitol in diabetes.[95–99]

REFERENCES

1. Hayes, B.P. and R.F. Fisher. Influence of a prolonged period of low-dosage x-rays on the optic and ultrastructural appearances of cataract of the human lens. *Br. J. Ophthalmol.* 63: 457–464, 1979.
2. Jacques, P.F., A. Taylor, S.E. Hankinson, W.C. Willett, B. Mahnken, Y. Lee, K. Vaid, and M. Lahav. Long-term vitamin C supplement use and prevalence of early age-related lens opacities. *Am. J. Clin. Nutr.* 66: 911–916, 1997.
3. Robertson, J.M., A.P. Donner, and J.R. Trevithick. Vitamin E intake and risk of cataracts in humans. *Ann. N.Y. Acad. Sci.* 570: 372–382, 1989.
4. Robertson, J.M., A.P. Donner, and J.R. Trevithick. A possible role for vitamins C and E in cataract prevention. *Am. J. Clin. Nutr.* 53: 346S-351S, 1991.
5. Brown, S.B. and K.J. Mellish. Verteporfin: a milestone in ophthalmology and photodynamic therapy. *Expert. Opin. Pharmacother.* 2: 351–361, 2001.
6. Harding, S. Photodynamic therapy in the treatment of subfoveal choroidal neovascularisation. *Eye.* 15: 407–412, 2001.
7. Shuler, M.F., J. L. Borrillo, and A.C. Ho. Photodynamic therapy update. *Curr. Opin. Ophthalmol.* 12: 202–206, 2001.
8. Belpoliti, M., G. Maraini, G. Alberti, R. Corona, and S. Crateri. Enzyme activities in human lens epithelium of age-related cataract. *Invest. Ophthalmol. Vis. Sci.* 34: 2843–2847, 1993.
9. Trevithick, J.R. and K.P. Mitton. Vitamins C and E in cataract risk reduction. *Int. Ophthalmol. Clin.* 40: 59–69, 2000.

10. Leske, M.C., S.Y. Wu, L. Hyman, R. Sperduto, B. Underwood, L.T. Chylack, R.C. Milton, S. Srivastava, and N. Ansari. Biochemical factors in the lens opacities. Case-control study. The Lens Opacities Case-Control Study Group. *Arch. Ophthalmol.* 113: 1113–1119, 1995.

11. Leske, M.C., L.T. Chylack, Jr., Q. He, S.Y. Wu, E. Schoenfeld, J. Friend, and J. Wolfe. Antioxidant vitamins and nuclear opacities: the longitudinal study of cataract. *Ophthalmology.* 105: 831–836, 1998.

12. Chylack L.T., Jr. The REACT Trial: antioxidant micronutrient mixture to slow progression of age-related cataract. *Ophthalmic. Res.* 32: 87, 2001.

13. Taylor H.R. Supplemental vitamin A and risk of cataract and AMD: Results from the VECAT Study. *Invest Ophthalmol. Vis. Sci.* 42: 8518, 2001.

14. A randomized, placebo-controlled, clinical trial of high-dose supplementation with vitamins C and E and beta carotene for age-related cataract and vision loss: AREDS report no.9. *Arch. Ophthalmol.* 119: 1439–1452, 2001.

15. A randomized, placebo-controlled, clinical trial of high-dose supplementation with vitamins C and E, beta carotene, and zinc for age- related macular degeneration and vision loss: AREDS report no.8. *Arch. Ophthalmol.* 119: 1417–1436, 2001.

16. Clayton, R.M., J. Cuthbert, J. Duffy, J. Seth, C.I. Phillips, R.S. Bartholomew, and J. M. Reid. Some risk factors associated with cataract in S.E. Scotland: a pilot study. *Trans. Ophthalmol. Soc. U.K.* 102 Pt 3: 331–336, 1982.

17. Katoh, N., F. Jonasson, H. Sasaki, M. Kojima, M. Ono, N. Takahashi, and K. Sasaki. Cortical lens opacification in Iceland. Risk factor analysis — Reykjavik Eye Study. *Acta Ophthalmol. Scand.* 79: 154–159, 2001.

18. Trevithick, C.C., M.M. Chartrand, J. Wahlman, F. Rahman, M. Hirst, and J.R. Trevithick. Shaken, not stirred: bioanalytical study of the antioxidant activities of martinis. *BMJ.* 319: 1600–1602, 1999.

19. Gey, K.F. Vitamins E plus C and interacting conutrients required for optimal health. A critical and constructive review of epidemiology and supplementation data regarding cardiovascular disease and cancer. *Biofactors.* 7: 113–174, 1998.

20. Kiechl, S., J. Willeit, G. Egger, M. Oberhollenzer, and F. Aichner. Alcohol consumption and carotid atherosclerosis: evidence of dose- dependent atherogenic and anti-atherogenic effects. Results from the Bruneck Study. *Stroke.* 25: 1593–1598, 1994.

21. Kiechl, S., J. Willeit, G. Rungger, G. Egger, F. Oberhollenzer, and E. Bonora. Alcohol consumption and atherosclerosis: what is the relation? Prospective results from the Bruneck Study. *Stroke.* 29: 900–907, 1998.

22. Trevithick, C.C., J. A. Vinson, J. Caulfield, F. Rahman, T. Derksen, L. Bocksch, S. Hong, A. Stefan, K. Teufel, N. Wu, M. Hirst, and J.R. Trevithick. Is ethanol an important antioxidant in alcoholic beverages associated with risk reduction of cataract and atherosclerosis? *Redox. Rep.* 4: 89–93, 1999.

23. Vinson, J.A., J. Proch, and P. Bose. Determination of quantity and quality of polyphenol antioxidants in foods and beverages. *Methods Enzymol.* 335: 103–114, 2001.

24. Vinson, J.A., K. Teufel, and N. Wu. Red wine, dealcoholized red wine, and especially grape juice, inhibit atherosclerosis in a hamster model. *Atherosclerosis.* 156: 67–72, 2001.

25. Vinson, J.A. Flavonoids in foods as *in vitro* and *in vivo* antioxidants. *Adv. Exp. Med. Biol.* 439: 151–164, 1998.

26. Bhatnagar, A., N.H. Ansari, L. Wang, P. Khanna, C. Wang, and S.K. Srivastava. Calcium-mediated disintegrative globulization of isolated ocular lens fibers mimics cataractogenesis. *Exp. Eye Res.* 61: 303–310, 1995.

27. Kilic, F. and J.R. Trevithick. Modelling cortical cataractogenesis. XXIX. Calpain proteolysis of lens fodrin in cataract. *Biochem. Mol. Biol. Int.* 45: 963–978, 1998.

28. Creighton, M.O., J.R. Trevithick, G.Y. Mousa, D.H. Percy, A.J. McKinna, C. Dyson, H. Maisel, and R. Bradley. Globular bodies: a primary cause of the opacity in senile and diabetic posterior cortical subcapsular cataracts? *Can. J. Ophthalmol.* 13: 166–181, 1978.

29. Linklater, H.A., T. Dzialoszynski, H.L. McLeod, S.E. Sanford, and J. R. Trevithick. Modelling cortical cataractogenesis. XII: supplemental vitamin A treatment reduces gamma-crystallin leakage from lenses in diabetic rats. *Lens Eye Toxic. Res.* 9: 115–126, 1992.

30. Linklater, H.A., T. Dzialoszynski, H.L. McLeod, S.E. Sanford, and J. R. Trevithick. Modelling cortical cataractogenesis VIII: effects of butylated hydroxytoluene (BHT) in reducing protein leakage from lenses in diabetic rats. *Exp. Eye Res.* 43: 305–314, 1986.

31. Bantseev, V.L., K.L. Herbert, J.R. Trevithick, and J.G. Sivak. Mitochondria of rat lenses: distribution near and at the sutures. *Curr. Eye Res.* 19: 506–516, 1999.

32. Cubeddu, R., P. Taroni, D.N. Hu, N. Sakai, K. Nakanishi, and J.E. Roberts. Photophysical studies of A2-E, putative precursor of lipofuscin, in human retinal pigment epithelial cells. *Photochem. Photobiol.* 70: 172–175, 1999.

33. Busch, E.M., T.G. Gorgels, J.E. Roberts, and D. van Norren. The effects of two stereoisomers of N-acetylcysteine on photochemical damage by UVA and blue light in rat retina. *Photochem. Photobiol.* 70: 353–358, 1999.

34. McDermott, M., R. Chiesa, J.E. Roberts, and J. Dillon. Photooxidation of specific residues in alpha-crystallin polypeptides. *Biochemistry.* 30: 8653–8660, 1991.

35. Dillon, J., R. Chiesa, R.H. Wang, and M. McDermott. Molecular changes during the photooxidation of alpha-crystallin in the presence of uroporphyrin. *Photochem. Photobiol.* 57: 526–530, 1993.

36. West, S., S. Vitale, J. Hallfrisch, B. Munoz, D. Muller, S. Bressler, and N.M. Bressler. Are antioxidants or supplements protective for age-related macular degeneration? *Arch. Ophthalmol.* 112: 222–227, 1994.

37. Leung, I., M. Tso, W. Li, and T. Lam. Absorption and tissue distribution of zeaxanthin and lutein in rhesus monkeys after taking Fructus lycii (Gou Qi Zi) extract. *Invest. Ophthalmol. Vis. Sci.* 42: 466–471, 2001.

38. Landrum, J.T., R.A. Bone, H. Joa, M.D. Kilburn, L.L. Moore, and K.E. Sprague. A one year study of the macular pigment: the effect of 140 days of a lutein supplement. *Exp. Eye Res.* 65: 57–62, 1997.

39. Van Kuijk, F.J. Role of lutein in prevention of age-related macular degeneration. *Abstracts, Oxygen Club of California.* 84, 2001.

40. Sommerburg, O., J. E. Keunen, A.C. Bird, and F.J. Van Kuijk. Fruits and vegetables that are sources for lutein and zeaxanthin: the macular pigment in human eyes. *Br. J. Ophthalmol.* 82: 907–910, 1998.

41. Mares-Perlman, J.A. Too soon for lutein supplements. *Am. J. Clin. Nutr.* 70: 431–432, 1999.

42. Lavine, J.B. Lutein supplementation or green vegetables. *Am. J. Clin. Nutr.* 71: 1210–1211, 2000.

43. Zigman, S. Ocular light damage. *Photochem. Photobiol.* 57: 1060–1068, 1993.

44. Zigman, S. Environmental near-UV radiation and cataracts. *Optom. Vis. Sci.* 72: 899–901, 1995.

45. Zigman, S. Lens UVA photobiology. *J. Ocul. Pharmacol. Ther.* 16: 161–165, 2000.

46. Sasaki, K., H. Sasaki, M. Kojima, Y.B. Shui, O. Hockwin, F. Jonasson, H.M. Cheng, M. Ono, and N. Katoh. Epidemiological studies on UV-related cataract in climatically different countries. *J. Epidemiol.* 9: S33-S38, 1999.

47. Schey, K.L., S. Patat, C.F. Chignell, M. Datillo, R.H. Wang, and J. E. Roberts. Photo-oxidation of lens alpha-crystallin by hypericin (active ingredient in St. John's Wort). *Photochem. Photobiol.* 72: 200–203, 2000.

48. Robison, W.G., Jr., T. Kuwabara, and J. G. Bieri. Deficiencies of vitamins E and A in the rat. Retinal damage and lipofuscin accumulation. *Invest. Ophthalmol. Vis. Sci.* 19: 1030–1037, 1980.

49. Carter-Dawson, L., M. Tanaka, T. Kuwabara, and J.G. Bieri. Early corneal changes in vitamin A deficient rats. *Exp. Eye Res.* 30: 261–269, 1980.

50. Kretzer, F.L. and H.M. Hittner. Retinopathy of prematurity: clinical implications of retinal development. *Arch. Dis. Child.* 63: 1151–1167, 1988.

51. Trevithick, J.R., K.P. Mitton, and J.M. Robertson. Vitamin E and the eye, in *Vitamin E in Health and Disease*. Packer, L. and Fuchs J. (Eds.) Marcel Dekker, New York, 1992, Chapter 61, 873–896.

52. Johnson, L.H., G.E. Quinn, S. Abbasi, and F.W. Bowen. Retinopathy of prematurity: prevalence and treatment over a 20 year period at Pennsylvania Hospital. *Doc. Ophthalmol.* 74: 213–222, 1990.

53. Bursell, S.E. and G.L. King. Can protein kinase C inhibition and vitamin E prevent the development of diabetic vascular complications? *Diabetes Res. Clin. Pract.* 45: 169–182, 1999.

54. Chasan-Taber, L., W.C. Willett, J.M. Seddon, M.J. Stampfer, B. Rosner, G.A. Colditz, F.E. Speizer, and S.E. Hankinson. A prospective study of carotenoid and vitamin A intakes and risk of cataract extraction in US women. *Am. J. Clin. Nutr.* 70: 509–516, 1999.

55. Yokota, T., T. Shiojiri, T. Gotoda, and H. Arai. Retinitis pigmentosa and ataxia caused by a mutation in the gene for the alpha-tocopherol-transfer protein. *N. Engl. J. Med.* 335: 1770–1771, 1996.

56. Yokota, T., T. Uchihara, J. Kumagai, T. Shiojiri, J. J. Pang, M. Arita, H. Arai, M. Hayashi, M. Kiyosawa, R. Okeda, and H. Mizusawa. Postmortem study of ataxia with retinitis pigmentosa by mutation of the alpha-tocopherol transfer protein gene. *J. Neurol. Neurosurg. Psychiatry.* 68: 521–525, 2000.

57. Berson, E.L., B. Rosner, M.A. Sandberg, K.C. Hayes, B.W. Nicholson, C. Weigel-DiFrano, and W. Willett. Vitamin A supplementation for retinitis pigmentosa. *Arch. Ophthalmol.* 111: 1456–1459, 1993.

58. Saari, J.C. Biochemistry of visual pigment regeneration: the Friedenwald lecture. *Invest Ophthalmol. Vis. Sci.* 41: 337–348, 2000.

59. Allikmets, R., N.F. Shroyer, N. Singh, J. M. Seddon, R.A. Lewis, P.S. Bernstein, A. Peiffer, N.A. Zabriskie, Y. Li, A. Hutchinson, M. Dean, J. R. Lupski, and M. Leppert. Mutation of the Stargardt disease gene (ABCR) in age-related macular degeneration. *Science.* 277: 1805–1807, 1997.

60. Martinez-Mir, A., E. Paloma, R. Allikmets, C. Ayuso, T. del Rio, M. Dean, L. Vilageliu, R. Gonzalez-Duarte, and S. Balcells. Retinitis pigmentosa caused by a homozygous mutation in the Stargardt disease gene ABCR. *Nat. Genet.* 18: 11–12, 1998.

61. Cremers, F.P., D.J. van de Pol, M. van Driel, A.I. den Hollander, F.J. van Haren, N.V. Knoers, N. Tijmes, A.A. Bergen, K. Rohrschneider, A. Blankenagel, A.J. Pinckers, A.F. Deutman, and C.B. Hoyng. Autosomal recessive retinitis pigmentosa and cone-rod dystrophy caused by splice site mutations in the Stargardt's disease gene ABCR. *Hum. Mol. Genet.* 7: 355–362, 1998.

62. Sun, H., R.S. Molday, and J. Nathans. Retinal stimulates ATP hydrolysis by purified and reconstituted ABCR, the photoreceptor-specific ATP-binding cassette transporter responsible for Stargardt disease. *J. Biol. Chem.* 274: 8269–8281, 1999.

63. Sun, H. and J. Nathans. ABCR, the ATP-binding cassette transporter responsible for Stargardt macular dystrophy, is an efficient target of all-trans-retinal-mediated photooxidative damage *in vitro*. Implications for retinal disease. *J. Biol. Chem.* 276: 11766–11774, 2001.

64. Phelan, J.K. and D. Bok. A brief review of retinitis pigmentosa and the identified retinitis pigmentosa genes. *Mol. Vis.* 6: 116–124, 2000.

65. Sohocki, M.M., L.S. Sullivan, H.A. Mintz-Hittner, D. Birch, J. R. Heckenlively, C.L. Freund, R.R. McInnes, and S.P. Daiger. A range of clinical phenotypes associated with mutations in CRX, a photoreceptor transcription-factor gene. *Am. J. Hum. Genet.* 63: 1307–1315, 1998.

66. Bessant, D.A., A.M. Payne, K.P. Mitton, Q.L. Wang, P.K. Swain, C. Plant, A.C. Bird, D.J. Zack, A. Swaroop, and S.S. Bhattacharya. A mutation in NRL is associated with autosomal dominant retinitis pigmentosa. *Nat. Genet.* 21: 355–356, 1999.

67. Mitton, K.P., P.K. Swain, S. Chen, S. Xu, D.J. Zack, and A. Swaroop. The leucine zipper of NRL interacts with the CRX homeodomain. A possible mechanism of transcriptional synergy in rhodopsin regulation. *J. Biol. Chem.* 275: 29794–29799, 2000.

68. Filina, A.A. Antioxidant therapy of primary glaucoma. *Vestn. Oftalmol.* 110: 33–35, 1994.

69. Hilsdorf, C. On the decrease of intraocular pressure by intravenous drop infusion of 20 per cent sodium ascorbinate. *Klin. Monatsbl. Augenheilkd.* 150: 352–358, 1967.

70. Fishbein, S.L. and S. Goodstein. The pressure lowering effect of ascorbic acid. *Ann. Ophthalmol.* 4: 487–491, 1972.

71. Kramorenko, I.S., T.A. Dobritsa, Z.A. Imanbaeva, and E.A. Egorov. Emoxipine in the treatment of primary glaucoma. *Vestn. Oftalmol.* 108: 14–15, 1992.

72. Kurysheva, N.I., M.I. Vinetskaia, V.P. Erichev, M.L. Demchuk, and S.I. Kuryshev. Contribution of free-radical reactions of chamber humor to the development of primary open-angle glaucoma. *Vestn. Oftalmol.* 112: 3–5, 1996.

73. Haas, A.L., D. Boscoboinik, D.S. Mojon, M. Bohnke, and A. Azzi. Vitamin E inhibits proliferation of human Tenon's capsule fibroblasts *in vitro*. *Ophthalmic Res.* 28: 171–175, 1996.

74. Pinilla, I., J. M. Larrosa, V. Polo, and F.M. Honrubia. Alpha-tocopherol derivatives in an experimental model of filtering surgery. *Ophthalmic Res.* 31: 440–445, 1999.

75. Robison, W.G., Jr., T. Kuwabara, and J. G. Bieri. Vitamin E deficiency and the retina: photoreceptor and pigment epithelial changes. *Invest. Ophthalmol. Vis. Sci.* 18: 683–690, 1979.

76. Bieri, J.G., T.J. Tolliver, W.G. Robison, Jr., and T. Kuwabara. Lipofuscin in vitamin E deficiency and the possible role of retinol. *Lipids.* 15: 10–13, 1980.

77. Imafidon C. personal communication. Cataract prevalence in different regions of Nigeria, 1995.

78. Ahlrot-Westerlund, B. and A. Norrby. Remarkable success of antioxidant treatment (selenomethionine and vitamin E) to a 34-year-old patient with posterior subcapsular cataract, keratoconus, severe atopic eczema and asthma. *Acta Ophthalmol. (Copenh).* 66: 237–238, 1988.

79. Clarke, I.S., T. Dzialoszynski, S.E. Sanford, and J.R. Trevithick. Dietary strategies to prevent retinopathy in RCS rats. *Lens Eye Toxic. Res.* 6: 253–254, 1989.

80. Clarke, I.S., T. Dzialoszynski, S.E. Sanford, and J.R. Trevithick. A possible relationship between cataract, increased levels of the major heat shock protein HSP 70 and decreased levels of S-antigen in the retina of the RCS rat. *Exp. Eye Res.* 53: 545–548, 1991.

81. Mitton, K.P., H.A. Linklater, T. Dzialoszynski, S.E. Sanford, K. Starkey, and J.R. Trevithick. Modelling cortical cataractogenesis 21: in diabetic rat lenses taurine supplementation partially reduces damage resulting from osmotic compensation leading to osmolyte loss and antioxidant depletion. *Exp. Eye Res.* 69: 279–289, 1999.

82. Kilic, F., R. Bhardwaj, J. Caulfield, and J. R. Trevithick. Modelling cortical cataractogenesis 22: is *in vitro* reduction of damage in model diabetic rat cataract by taurine due to its antioxidant activity? *Exp. Eye Res.* 69: 291–300, 1999.

83. Trevithick, J.R., H.A. Linklater, K.P. Mitton, T. Dzialoszynski, and S.E. Sanford. Modeling cortical cataractogenesis: IX. Activity of vitamin E and esters in preventing cataracts and gamma-crystallin leakage from lenses in diabetic rats. *Ann. N.Y. Acad. Sci.* 570: 358–371, 1989.

84. Wahdan, H.A. Causes of the antimicrobial activity of honey. *Infection.* 26: 26–31, 1998.

85. Martos, I., F. Ferreres, L. Yao, B. D'Arcy, N. Caffin, and F.A. Tomas-Barberan. Flavonoids in monospecific eucalyptus honeys from Australia. *J. Agric. Food Chem.* 48: 4744–4748, 2000.

86. Vinson, J.A., J. Proch, and L. Zubik. Phenol antioxidant quantity and quality in foods: cocoa, dark chocolate, and milk chocolate. *J. Agric. Food Chem.* 47: 4821–4824, 1999.

87. Delamere, N.A., C.A. Paterson, and D.L. Holmes. Hypocalcemic cataract. I. An animal model and cation distribution study. *Metab. Pediatr. Ophthalmol.* 5: 77–82, 1981.

88. Arora, R., P.S. Menon, S.K. Angra, S. Ghose, and A. Virmani. Hypocalcemic cataract secondary to idiopathic hypoparathyroidism. *Indian Pediatr.* 26: 1157–1159, 1989.

89. Gupta, M.M. Calcium imbalance in hypoparathyroidism. *J. Assoc. Physicians India.* 39: 616–618, 1991.

90. Takahashi, H. Ca(2+)-ATPase activity in the hypocalcemic cataract. *Nippon Ganka Gakkai Zasshi* 98: 142–149, 1994.

91. Kilic, F. and J.R. Trevithick. Modelling cortical cataractogenesis. XXIX. Calpain proteolysis of lens fodrin in cataract. *Biochem. Mol. Biol. Int.* 45: 963–978, 1998.

92. Pastori, M., H. Pfander, D. Boscoboinik, and A. Azzi. Lycopene in association with alpha-tocopherol inhibits at physiological concentrations proliferation of prostate carcinoma cells. *Biochem. Biophys. Res. Commun.* 250: 582–585, 1998.

93. Sgarbossa, A., N. Angelini, D. Gioffre, T. Youssef, F. Lenci, and J. E. Roberts. The uptake, location and fluorescence of hypericin in bovine intact lens. *Curr. Eye Res.* 21: 597–601, 2000.

94. Fujiwara, H., N. Tanaka, and T. Suzuki. The effects of anti-cataract drugs on free radicals formation in lenses. *Nippon Ganka Gakkai Zasshi* 95: 1071–1076, 1991.

95. Usuki, Y., S. Usuki, and S. Hommura. Successful treatment of a senile diabetic woman with cataract with goshajinkigan. *Am. J. Chin. Med.* 19: 259–263, 1991.

96. Chiou, G.C., N.J. Stolowich, Y.Q. Zheng, Z.F. Shen, M. Zhu, and Z.D. Min. Effects of some natural products on sugar cataract studied with nuclear magnetic resonance spectroscopy. *J. Ocul. Pharmacol.* 8: 115–120, 1992.

97. Kubo, M., H. Matsuda, K. Tokuoka, Y. Kobayashi, S. Ma, and T. Tanaka. Studies of anti-cataract drugs from natural sources. I. Effects of a methanolic extract and the alkaloidal components from Corydalis tuber on *in vitro* aldose reductase activity. *Biol. Pharm. Bull.* 17: 458–459, 1994.

98. Du, G.H., Y. Qiu, Y.E. Tian, and J.T. Zhang. Prevention of galactose-induced cata-
     ractogenesis in rats by salvianolic acid A. *Yao Xue. Xue. Bao.* 30: 561–566, 1995.
99. Matsuda, H., H. Cai, M. Kubo, H. Tosa, and M. Iinuma. Study on anti-cataract drugs
     from natural sources. II. Effects of buddlejae flos on *in vitro* aldose reductase activity.
     Biol. Pharm. Bull. 18: 463–466, 1995.

9 Nutritional Assessment and Management of the Cancer Patient

Joel B. Mason and Sang-Woon Choi

CONTENTS

INTRODUCTION

Protein-calorie malnutrition, typically referred to simply as "malnutrition," is a common but not invariable feature of cancer, the most frequent manifestation of which is weight loss. A large, multicenter survey of over 3000 patients awaiting the initiation of chemotherapy, for instance, observed weight losses exceeding 4% in one third of the patients and, in one half of those patients, the magnitude of loss exceeded 10%.[22] Weight losses exceeding 10% are of particular significance because there is considerable evidence that weight loss of this magnitude in the setting of illness leads to significant increases in morbidity and mortality.[48]

The likelihood that a particular individual will sustain substantial weight loss appears to be related to many factors. The type of cancer is an important factor in determining the likelihood of developing malnutrition, although the presence of metastases, whether the cancer physically impedes normal food intake, and intervening emotional issues such as depression are also important determinants. In the aforementioned survey,[22] only 4 to 7% of individuals with leukemias, sarcomas, and breast cancer had >10% weight loss, whereas individuals with gastrointestinal cancers had a much higher likelihood of this degree of weight loss: 14% of patients with colon cancer and 25 to 40% of individuals with pancreatic and gastric cancers. Not surprisingly, carcinoma of the oropharynx, a condition that frequently interferes with chewing and swallowing, results in >10% weight loss in over 40% of patients.[75]

Nevertheless, arriving at precise figures for the prevalence of significant protein-calorie malnutrition among cancer patients is a frustrating task because it depends entirely on the nature of the particular assessment tool that is used to define malnutrition. It is therefore not surprising that other surveys, using nutritional assessment criteria such as the creatinine–height index (a measure of muscle mass), have found that up to 90% of hospitalized cancer patients have evidence of significant malnutrition.[81] Ultimately, it is perhaps more constructive merely to be aware that malnutrition of a degree that worsens clinical outcomes is common among cancer patients and is particularly common in cancers of the gastrointestinal tract. Physicians need to be cognizant of this reality in their assessment and management of the patient.

Although it has not been unequivocally proven that malnutrition independently contributes to the morbidity, mortality, and diminished quality of life associated with cancer, this contention is almost certainly true. A plethora of case-control and prospective cohort studies in cancer patients indicate that a substantial degree of malnutrition diminishes tolerance and responsiveness to chemotherapy[3,22] and radiotherapy,[62] increases perioperative morbidity,[86] worsens the quality of life,[3] and diminishes the likelihood of survival.[3,22,53,61,115] Most important, clinicians must realize that prompt attention to meeting the nutritional needs of the malnourished patient improves the clinical outcome of many types of ill patients, as has been shown repeatedly in prospective, randomized trials (reviewed in Reference 73). Improvements in clinical outcome as a result of aggressive nutritional support have, however, been more difficult to demonstrate in the cancer patient, probably because other factors have such a major impact on the clinical course. Nevertheless, aggressive nutritional support has genuine benefits to offer the cancer patient in selected circumstances. A detailed discussion of these circumstances appears later in this chapter.

MALNUTRITION IN THE CANCER PATIENT: MECHANISMS

The development of malnutrition in the cancer patient is usually multifactorial. Table 9.1 outlines many of the factors that are often observed to contribute to this problem.

A BODY COMPARTMENT PERSPECTIVE

The type of tissue that is lost when an individual loses weight is critical in determining the pathologic ramifications of weight loss. Over 98% of the metabolic machinery that maintains normal homeostasis resides in the lean body mass (LBM), and it is the maintenance of this body compartment that is most critical for health. The lean body mass can be subdivided into skeletal muscle mass; visceral lean mass (which includes the major organs) and extracellular lean mass such as the interstitial fluid, blood serum, and the mineral matrix of the skeleton (Figure 9.1). These are useful distinctions because the body draws on each of these compartments

---

TABLE 9.1
Factors in the Cancer Patient that Contribute to the Development of Protein-Calorie Malnutrition

A. Insufficient dietary intake
   1. Suppression of appetite
      - Mediated by cytokines, other humoral factors
      - Mediated by emotional depression
      - Mediated by loss of taste sensation (neural destruction, drug effects, paraneoplastic syndrome)
      - Learned aversion to eating due to adverse symptoms
      - Nausea, vomiting, other symptoms due to surgery, radiation, or chemotherapy

   2. Physical impairment of deglutition
      - Effects on chewing or swallowing mechanisms
      - Reduction in saliva production (tumor invasion, effects due to surgery, radiation, or drugs)
      - Mass effect of tumor
      - Radiation- or chemotherapy-induced mucositis
      - Surgical interruption of swallowing mechanism

B. Alterations in physiology and metabolism
   1. Malabsorption/maldigestion due to tumor or to therapy
   2. Constipation/GI dysmotility due to surgical ablation of autonomic innervation of gut or to narcotics and sedatives
   3. Increases in net protein catabolism
   4. Inefficiency in energy consumption/increases in overall caloric expenditure

From Mason JB. Gastrointestinal cancer: nutritional support, in *Principles and Practice of Gastrointestinal Oncology* . Kelsen D, et al. (Eds.) Philadelphia: Lippincott Williams & Wilkins, 2002, 127–142. With permission.

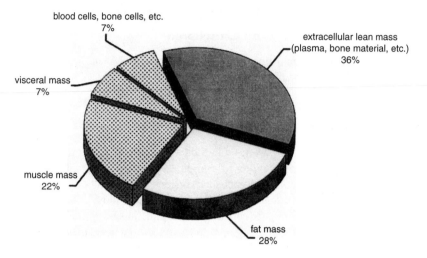

FIGURE 9.1   Body composition analysis by weight in a healthy adult. Speckled segments and the gray segment collectively represent lean body mass. The specked segments alone represent body cell mass. (From Mason JB. Gastrointestinal cancer: nutritional support, in *Principles and Practice of Gastrointestinal Oncology.* Kelsen D, et al. (Eds.) Philadelphia: Lippincott Williams & Wilkins, 2002, 127–142. With permission.)

differently in the course of weight loss. In the cancer patient, there is usually a disproportionately large contraction of the skeletal muscle and fat mass with relative sparing of the visceral mass. In this respect, the weight loss is similar to that seen in many acute illnesses of a nonmalignant nature.[123] For example, in one study of cancer patients who had lost, on the average, about one quarter of their pre-illness weight, fat mass and skeletal muscle mass each decreased by 75 to 80% of control values, whereas visceral lean mass was spared, not decreasing significantly.[31] By comparison, the weight loss in simple starvation is less detrimental because the body preferentially utilizes adipose tissue for energy needs and, therefore, the percentage loss in skeletal muscle mass is considerably less than the proportional loss of fat mass. For example, in healthy volunteers fed a calorically inadequate diet for 3 months, approximately one quarter of the initial body weight was lost; this was accompanied by a 70% decline in fat mass but by only a 24% drop in lean mass.[55] Table 9.2 summarizes the relative losses in body compartments in simple starvation vs. those observed in cancer.

The loss in weight and muscle mass (now referred to as "wasting"[91]) seen in cancer is generally regarded as a physiologic adaptation to stress: the body sacrifices large portions of the muscle mass to spare more immediately critical functions in the visceral mass. However, there are clear limitations to this adaptive response. In the first place, contraction of the skeletal muscle mass leads to muscle weakness, decreased work tolerance, and measurable decreases in functional status.[32] Furthermore, sparing the visceral mass is only relative, and sustained weight loss will eventually also affect this compartment.

TABLE 9.2
Body Compartment Losses in Starvation and Cancer

	Skeletal Muscle Wasting	Visceral Wasting	Loss of Fat Mass
Starvation	+	+/−[a]	+++
Wasting in cancer	+++	++/−[a]	+++

[a] Relatively spared early in the process; but can become pronounced with extended starvation or wasting

From Mason JB. Gastrointestinal cancer: nutritional support, in *Principles and Practice of Gastrointestinal Oncology* . Kelsen D, et al. (Eds.) Philadelphia: Lippincott Williams & Wilkins, 2002, 127–142. With permission.

MECHANISMS OF WASTING IN CANCER: ANOREXIA

Anorexia is a common contributor to wasting in cancer. Particularly with cancers of the gastrointestinal tract, the act of eating may incite a variety of adverse symptoms: pain, vomiting, diarrhea, etc., and therefore anorexia can become a means of avoiding such symptoms. The tumor mass alone may preclude adequate ingestion of food, a factor which is particularly prevalent in cancers of the oropharynx, esophagus, and stomach. In addition, therapeutic modalities involving drugs, radiation, and surgery often induce anorexia directly or deter the patient from eating so as to avoid the gastrointestinal side effects of therapy. A prime example of this is chemotherapy-induced mucositis. The emotional adjustment associated with dealing with a major cancer continues to be a common precipitant of depression and anxiety:[82] these emotional states are also important in producing a state of anorexia.

Nevertheless, anorexia is commonly present even in the absence of any of the above-mentioned factors and may even be the presenting symptom of the cancer. Anorexia in such a setting is thought to be largely due to the effects of particular cytokines (Table 9.3), originating in host cells that are descendants of the white blood cell lineage (macrophages and lymphocytes), and which are responding to the presence of the neoplasm. Fibroblasts also seem capable of secreting some of these cytokines. A highly reproducible and remarkable degree of anorexia is observed with administration of TNF-alpha,[104] IL-1,[84] and gamma-interferon.[60]

MECHANISMS OF WASTING IN CANCER: ALTERATIONS IN METABOLISM

A wide spectrum of alterations in protein, lipid, and carbohydrate metabolism is commonly observed in patients bearing cancers (Table 9.3). In concert with the other factors outlined in Table 9.1, these factors contribute to the development of malnutrition. Although much of the work pertaining to mechanisms has been performed in cell culture and animal models, studies in humans have been in general accord with findings in the nonhuman models.

Effects on Protein Metabolism and the Lean Body Mass

Skeletal muscle is the body compartment where most of the contraction of lean body mass occurs.[31] The overriding functional significance of this is underscored by the observation that the extent to which this compartment is diminished correlates inversely with the likelihood of survival.[81] The decrease in skeletal muscle mass appears to be due to both a reduction in muscle protein synthesis and an increase in muscle protein degradation.[70] Total body protein turnover tends to increase in this setting and is often present even before clinically evident wasting has occurred.[30,43]

TNF-alpha, IL-6, and probably IL-1 and gamma interferon, appear to play major roles in mediating the dissolution of skeletal muscle in the wasting associated with cancer. Exogenous administration of TNF-alpha and IL-6 induces muscle wasting[24,67] and is overcome by specific antibodies directed against TNF-alpha.[18] TNF-alpha and IL-6 may not be the direct effectors of the response; rather, they may act by stimulating the secretion or expression of "downstream" mediators. One candidate for such a downstream mediator is protein mobilizing factor (PMF), a proteoglycan that induces proteolysis in isolated muscle preparations and a reduction in lean mass in intact animals.[112] PMF has been found in the urine of cancer patients with weight loss, but not in those without weight loss,[111] nor in the urine of patients who have lost weight due to non-neoplastic illnesses. Thus, it appears to be highly specific. Nevertheless, observations regarding PMF have been difficult to reproduce, and its existence and function in cancer wasting remains a matter of debate.

Effects on Lipid Metabolism and the Adipose Tissue

In the wasting associated with cancer, adipose tissue constitutes the major source of energy (though less so than in simple starvation) and, therefore, a large decrease in fat mass is observed. The net efflux of glycerol and fatty acids from adipose tissue that is observed in cancer wasting appears to be due to at least three factors: 1) an increased rate of lipolysis in adipose tissue, apparently mediated by TNF-alpha, leukemia inhibitory factor (LIF) and lipid mobilizing factor(LMF)[8,42,72]; 2) a decrease in *de novo* lipogenesis in the adipose tissue, mediated largely by TNF-alpha and IL-1[114]; and 3) diminished activity of lipoprotein lipase. This enzyme is necessary for the uptake of fatty acids from circulating lipoproteins and the diminished activity in cancer appears to be mediated by TNF-alpha, IL-6, interferon-gamma, and LIF.[36,105] The decrease in lipoprotein lipase, in particular, explains why cancer patients have a diminished ability to clear an exogenous lipid load[78] and usually have elevated plasma glycerol and triglyceride levels.[90] Hypertriglyceridemia in cancer patients may also be related to increased rates of hepatic lipogenesis, since several of the cytokines implicated in cancer wasting, including TNF-alpha, IL-1, and interferon-alpha, each stimulate hepatic lipogenesis.[41] Table 9.3 summarizes the mediators of these alterations in fat metabolism.

TABLE 9.3
Major Cytokines Believed to Be Involved in Cancer Anorexia and Wasting

Cytokine	Production	Effects Related to Anorexia or Cachexia
Tumor necrosis factor (TNF)-α	Blood levels do not correlate with presence of malnutrition	Injection induces anorexia, weight loss, and cachexia
	Produced by macrophages	May increase resting energy expenditure
		Has hypothalamic effects in inducing anorexia
		May have local gastrointestinal effects such as delay of gastric emptying
		Inhibits lipoprotein lipase
		Causes hypertriglyceridemia
		Depletes total body fat stores
		Increases skeletal protein breakdown
		Increases synthesis of acute phase reactants
		Increases hepatic glucose output and gluconeogenesis
Interleukin-1 (IL-1)	Serum levels do not correlate with malnutrition	Injection induces anorexia, weight loss, and cachexia, more so than TNF-α
	Produced by macrophages	May increase resting energy expenditure
		Has hypothalamic effects in inducing anorexia
		Causes similar effects on fat metabolism as TNF-α
		Causes similar effects on protein metabolism as TNF-α
Interleukin-6 (IL-6)	Production induced by TNF-α and IL-1	Induces hepatic gluconeogenesis
	Levels correlate with extent of tumor burden in animal models	Increases production of acute phase reactants
	Produced by macrophages and fibroblasts	Increases lipolysis
	Produced by activated T lymphocytes	Augments effects of TNF-α on lipid metabolism
		Inhibits lipoprotein lipase
		Inhibition by antibodies in tumor bearing animals reduces weight loss
		Increases anorexia

Adapted from Mutlu EA and Mobarhan S. *Nutr. Clin. Care.* 3(1):6, 2000. With permission.

Effects on Carbohydrate and Energy Metabolism

Cancer frequently produces a state where the host expends more calories per kilogram of lean mass than is normal: this state of hypermetabolism is inherently less energy efficient and therefore predisposes to weight loss.

The Cori cycle, in which lactate produced by the cancer or by peripheral tissues is reconverted to glucose in the liver, is an inefficient means of salvaging glucose, as it consumes six molecules of ATP per cycle. If the cancer, or other tissue, is producing significant quantities of lactate by anaerobic glycolysis, which only yields two molecules of ATP per molecule of glucose substrate, substantial net loss of energy occurs (a so-called "futile cycle"). Increased activity of the Cori cycle has been reported to exist in individuals with cancer and, more specifically, in those cancer patients with weight loss.[50] Nevertheless, the quantitative contribution to cancer wasting made by excessive activity of the Cori cycle is not known.

Other commonly altered aspects of carbohydrate metabolism include increased rates of gluconeogenesis and glucose flux, and the development of impaired insulin secretion, as well as a modest degree of insulin insensitivity. The latter induces impaired glucose utilization in peripheral tissues and glucose intolerance.[108] Similar alterations in glucose metabolism are observed in any condition associated with a systemic inflammatory response and are thought to be due to TNF-alpha.[89] These changes contrast with weight loss unrelated to illness or cancer, where increased insulin sensitivity is maintained.[108]

ASSESSING NUTRITIONAL STATUS:
A BRIEF INTRODUCTION

Providing nutritional support in a rational manner requires that the clinician acquire an objective means of systematically categorizing patients into those who are either well-nourished or mildly malnourished vs. those who have a moderate-to-severe degree of malnutrition. It is the patients in the latter category who will benefit from an aggressive approach to nutritional support. Patients with moderate-to-severe degrees of malnutrition have demonstrable impairments in most physiologic processes associated with their malnutrition and suffer significantly greater morbidity and mortality as a result. Most important, the added morbidity and mortality can be attenuated or eliminated by diligent attention to their nutritional needs (reviewed in reference 73). Similar benefits of aggressive nutritional support cannot usually be demonstrated in well-nourished or mildly malnourished patients.

A comprehensive assessment of protein-energy status involves history taking (including a diet history), physical exam, anthropometric measures of nutritional status (e.g., weight, mid-arm muscle circumference, triceps fat fold), biochemical tests such as the measurement of serum albumin or prealbumin, and the evaluation of body compartment sizes with tools such as body impedance analysis or dual-photon absorptiometry. The means of performing this type of comprehensive assessment is reviewed elsewhere[80] but is beyond the scope of this chapter. Far simpler algorithms, which are surprisingly accurate, can be used by the clinician if the primary purpose is merely to categorize patients as either well- or mildly malnourished vs. moderately to severely malnourished.

Perhaps the most straightforward means is to determine the percent of unintentional weight loss the patient has suffered as a result of disease. Since disproportionately large degrees of protein catabolism accompany acute inflammatory illnesses and cancer wasting, an unintentional weight loss of $\geq 10\%$ of pre-morbid weight, due to disease, translates into a contraction of 15 to 20% of the critical protein-containing compartment of the body.[48] Beyond this threshold, physiologic functions are impaired, and increased morbidity and mortality are observed.[48] Clinical trials have shown patients who exceed this threshold will benefit from aggressive nutritional support.[6,113]

Body weight can, however, be misleading. A common example is the patient with cirrhosis and ascites, where the weight of the ascites masks the loss in lean body mass. Potential confusion surrounding the nutritional assessment of cirrhotics, however, has been clarified by studies utilizing sophisticated methods of assessing total body protein which have demonstrated that essentially all patients who are Childs–Pugh Class B and C have lost more than 20% of total body protein; more surprising is that half of the patients with Childs classification A fall into the same category.[88]

Two other commonly used means of assessing protein-calorie status are the creatinine–height index (CHI) and the Prognostic Nutritional Index (PNI). The creatinine–height index, i.e., the amount of creatinine excreted in the urine in 24 hours, corrected for the patient's height, is an accurate reflection of muscle mass, because a constant percentage (about 2%) of muscle creatine is converted to creatinine each day. However, incomplete urine collections, excessive meat ingestion, corticosteroid therapy, and abnormal or unstable renal function can each alter apparent or actual creatinine excretion independent of changes in muscle mass. Patients whose CHI lies 20% or more below the normative value may be considered as having a moderate-to-severe degree of malnutrition (Table 9.4).

The Prognostic Nutritional Index (PNI) is one of several indices that represent a weighted regression of nutritional and physiologic parameters. The PNI has been shown to be a valid predictor of postoperative complications and mortality among inpatient cancer patients who are about to undergo surgery.[125] The disadvantage of the PNI is that it requires the measurement of serum albumin and transferrin, triceps skinfold (the accurate measure of which is highly operator-dependent), and delayed skin hypersensitivity. Although the PNI indicates both nutritional status and severity of illness, a value greater than 40% suggests moderate-to-severe malnutrition.

EFFICACY OF NUTRITIONAL SUPPORT

"Aggressive" nutritional support, here defined as "using whatever means is necessary and practical to meet the nutritional needs of the patient," will not benefit every patient with cancer. A decision on whether aggressive nutritional support is warranted in a particular patient, therefore, requires that the clinician understand what can reasonably be expected from embarking on such a course.

TABLE 9.4
Normative Values for Creatinine Excretion Based on Height

Men[a]		Women[b]	
Height, cm	Ideal Creatinine, mg	Height, cm	Ideal Creatinine, mg
157.5	1288	147.3	830
160.0	1325	149.9	851
162.6	1359	152.4	875
165.1	1386	154.9	900
167.6	1426	157.9	925
170.2	1467	160.0	949
172.7	1513	162.6	977
175.3	1555	165.1	1006
177.8	1596	167.6	1044
180.3	1642	170.2	1076
182.9	1691	172.7	1109
185.4	1739	175.3	1141
188.0	1785	177.8	1174
190.5	1831	180.3	1206
193.0	1891	182.9	1240

Note: Creatinine-Height Index = Actual 24-h urinary creatinine excretion/normative value for height and sex

[a] Creatinine coefficient (men) n = 23 mg/kg of ideal body weight.
[b] Creatinine coefficient (women) = 18 mg/kg of ideal body weight.

From Blackburn GL et al. *J. Parent. Ent. Nutr.* 1:11, 1977. With permission.

It is important to recognize that the metabolic changes that accompany cancer wasting make it exceedingly difficult to correct the nutritional deficits that are present. In a patient with an untreated cancer, nutritional support will not generally lead to an increase in the protein compartment of the body; a gain in weight may not occur, and when it does, much of it is due to water and to an expanded fat mass.[99] Despite these limitations, however, *even in the absence of weight gain or demonstrable increases in serum proteins that reflect protein-energy status (e.g., albumin, prealbumin, retinol binding protein), a course of nutritional support to an appropriate patient can improve physiologic functions and clinical outcome.* Once a tumor mass has been eliminated or caused to shrink, many of these metabolic aberrations disappear,[92] and, therefore, it is reasonable to expect that aggressive nutritional support might effect an increase in lean body mass.

The next several subheadings review common clinical scenarios where there exists substantial evidence that aggressive nutritional support provides benefit to the patient. Contained in the text accompanying each scenario are qualifications that describe the particular circumstances under which an aggressive approach is indicated.

THE CANCER PATIENT UNDERGOING MAJOR SURGERY

Nutritional support is most beneficial in moderately to severely malnourished cancer patients who are scheduled to undergo major surgery. Aggressive nutritional support for 7 or more days prior to surgery reduces perioperative complications, and sometimes mortality, in malnourished patients.[9,28,77,113] The VA Cooperative Trial,[113] a multicenter trial encompassing nearly 500 subjects of whom two thirds had cancer, demonstrated an important qualification of this benefit: patients who were categorized as "severely" malnourished and who were randomized to receive preoperative total parenteral nutrition (TPN) realized nearly a 90% decline in noninfectious perioperative complications, whereas no benefits were observed in mildly malnourished or well-nourished individuals. Thus, in trials that confine themselves to moderately to severely malnourished patients, preoperative nutrition support conveys large benefits: one recent trial, which enrolled 90 patients with gastric or colorectal cancers undergoing surgery, demonstrated a 35% decline in overall complications as well as a significant reduction in mortality.[9] The fact that the benefits of preoperative nutritional support are confined to those with a substantial degree of malnutrition is not terribly surprising, and is the same conclusion reached by recent meta-analyses.[21,45] Deferring aggressive nutritional support until after surgery does not appear to have the same ability to diminish perioperative complications, as underscored by many trials.[9,14]

The benefits of aggressive nutritional support in the preoperative period are not confined to the use of total parenteral nutrition: provision of nutrients via an enteral approach is also of substantial benefit. The number of trials done with preoperative enteral support are far fewer and less compelling than those with TPN, but what studies exist indicate that preoperative enteral support conveys the same nutritional[100] and clinical[34] benefits as does TPN. As was the case with TPN, postoperative enteral nutrition in the absence of aggressive preoperative support is less likely to convey benefit to the patient,[103] although it has been reported to improve clinical outcomes in certain instances.[6] An additional circumstance where postoperative enteral nutrition alone appears to convey benefit is to those individuals who receive a specialized immunomodulatory enteral formula (see below).

In all patients, it is preferable to provide nutritional support via an enteral, versus a parenteral, route. Enteral feeding through a nasoenteric tube, gastrostomy, or jejunostomy leads to fewer metabolic difficulties, avoids the complications associated with placement of a central intravenous line, and is generally less expensive. Although still controversial,[66] individual prospective, randomized trials[57] generally indicate that serious infections (even disregarding infections of the central venous catheter) are significantly less frequent in tube-fed patients. A recent well-performed meta-analysis, encompassing 20 prospective, randomized, controlled trials comparing tube feeding vs. TPN has lent considerable support to the concept that tube feeding is preferable.[13] Even in the subanalysis where only those studies that exclusively contained cancer patients were examined, the benefit remained.

In times past, an enteral approach was frequently not considered in the setting of major abdominal surgery, but it is now evident that postoperative ileus is largely due to gastroparesis. Therefore, placement of an intraoperative nasojejunal tube or jejunostomy allows rapid institution of enteral feeds within 24 hours after surgery in most instances.

PATIENTS UNDERGOING CHEMOTHERAPY OR RADIATION THERAPY

Prospective cohort analyses certainly suggest that malnutrition is a risk factor for diminished responsiveness to chemotherapy, increased toxicity with the drugs, poorer quality of life, and shorter survival.[3] Nevertheless, early intervention trials that examined the efficacy of nutritional support in chemotherapy showed no benefits, culminating in a meta-analysis that concluded that routine nutritional support was not indicated.[74] Subsequent intervention trials, which have stratified the patients by nutritional status, have generally shown that nutritional gains can be achieved in malnourished patients.[20,52] However, the improvement in nutritional status has not led to a reduction in toxicity. If such patients subsequently proceed to surgery, the nutritional support appears to improve their perioperative course.[52]

Aggressive nutritional support in a cancer patient has frequently raised the concern whether the support might lead to an acceleration of tumor growth. In animal models of cancer, intravenous and enteral repletion of malnourished tumor-bearing animals stimulates tumor growth.[87] Human data regarding this issue are scarce, but it has been observed that nutritional repletion of malnourished cancer patients may stimulate DNA synthesis in the tumor.[52] There is, nevertheless, a conspicuous lack of reports indicating any clinically significant acceleration of tumor growth with aggressive nutrition support. In our opinion, therefore, this concern should not discourage the use of aggressive support in an otherwise appropriate setting. This issue should be of even less concern in those individuals who are embarking on chemotherapy, since aggressive nutrition support does not enhance DNA synthesis if chemotherapy is being administered concurrently.[52] Indeed, providing nutritional support and placing more tumor cells in the vulnerable DNA synthesis phase of the cell cycle may enhance sensitivity to cycle-specific chemotherapeutic agents;[52] this is an attractive hypothesis but one that has yet to be proven.

The utility of aggressive nutrition support in patients undergoing radiation therapy has been most extensively studied in individuals who have head and neck and esophageal cancers because: 1) they tend to have mechanical difficulties with deglutition, 2) they frequently are substantially malnourished, and 3) radiation therapy is a commonly used modality of treatment. There is now reasonable evidence in these patients that placement of a percutaneous endoscopic gastrostomy (PEG) tube and administration of supplemental tube feedings during and after the course of radiation therapy prevents further deterioration of nutritional status.[10,97] In head and neck cancers, supplemental PEG feedings have also been shown to significantly improve objective indicators of the quality of life.[97] Although improvements in survival or decreased morbidity have not yet been demonstrated, the improved quality of life alone may warrant its use in this setting.

PATIENTS UNDERGOING BONE MARROW TRANSPLANT

One situation where the routine, prophylactic use of TPN has been shown to be of benefit, even in well-nourished individuals, is in bone marrow transplantation.[119]

Well-nourished individuals presumably benefit in this setting because the cytoreductive chemotherapy and radiation tend to cause prolonged and severe GI dysfunction, limiting oral intake for weeks at a time when physiologic stresses have substantially increased energy and protein requirements. In the past, there was some concern that conventional doses of intravenous lipids might increase the risk of bacteremias and fungemias, but prospective, randomized trials have largely dispelled this concern about added infection risk.[63,79] The small increase in immunosuppression due to the use of conventional lipid emulsions that contain n-6 fatty acids may in fact diminish the likelihood of morbidity and mortality from graft-versus-host disease.[79]

There has been considerable reluctance to utilize enteral tube feeding regimens in bone marrow transplant because it was feared the feeding tube would injure a mucosal surface whose integrity is already impaired by mucositis. However, prospective, controlled studies have shown that tube feeding does not increase diarrhea, impair nutritional restitution, increase the duration of hospitalization, or impact the survival rate when compared to TPN.[76,107] Advantages of the enteral approach include fewer episodes of hyperglycemia and a decrease of more than 50% in the cost of nutritional support.[107]

The use of TPN supplemented with glutamine in bone marrow transplant patients is discussed in a subsequent section.

NUTRITIONAL SUPPORT WITH AGENTS THAT COMBAT ANOREXIA

Aggressive nutritional support does not invariably require tube feeding or TPN. If inadequate oral intake is due to anorexia, pharmacologic management of the anorexia is an excellent starting point for therapy. Resorting to enteric tube feeding or TPN can then be reserved for those who fail such an approach.

Loss of appetite can be a result of both tumor and host-derived substances. It can also be due to altered taste, to dysphagia and other gastrointestinal symptoms, to emotional depression, or to the therapeutic modalities used to treat cancer. Anorexia is very prevalent, encountered in 15 to 40% of patients at diagnosis and in 65% in advanced disease.[23]

A variety of pharmacologic agents have been evaluated for the treatment and symptomatic improvement of cancer anorexia and cancer wasting.

Progestational Agents

Both megestrol acetate (MA) and medroxyprogesterone acetate (MPA) are synthetic derivatives of progesterone intended for oral use; MPA is also available for injection as an intramuscular depot. Both MA and MPA are well tolerated but are relatively contraindicated in individuals who have sustained thromboembolic events, because these drugs are thought to enhance thrombotic potential.

Clinical trials using progesterone derivatives to treat hormone-responsive breast cancers reported appetite and weight gain as a side effect of MA therapy.[109] Several randomized, controlled trials of MA subsequently confirmed a positive effect on appetite and weight gain in a wide variety of cancers.[15,68,69] Typical weight gains over several months have been in the range of 3 to 6 kgs, compared to what are usually weight losses in the placebo group. The data have shown a very strong dose-responsive effect on appetite and food intake when the daily dose was increased from 160 to 800 mgs. Doses above 800 mg daily do not seem to provide additional gains.[69] Significant increases in subcutaneous fat accompany the weight gain, and it is this increase in adiposity that appears to be the predominant form of tissue accrual.[15] Accrual of muscle mass or lean body mass has not yet been convincingly shown in cancer patients; however, in a parallel trial in AIDS patients with wasting, a significant increase in lean mass was observed.[116] Although appetite increases promptly, the median response time to achieve maximal weight gain is 6 to 10 weeks.[116] A not insignificant benefit of MA therapy is that patients report an increase in "energy."[15] Similarly, in the trials in patients with AIDS, a very significant increase in the "sense of well being" has often been observed.[85,116]

Side effects of MA therapy include male impotence, vaginal spotting, mild pitting edema, and thromboembolic phenomena. Thromboembolic events tend to occur in about 5% of patients, but this figure, being similar to the incidence in the placebo group, suggests the increase in risk is quite minor.

MPA is a similar progestational agent used in Europe. Two recent randomized, placebo-controlled trials in cancer patients[101,102] demonstrated effects similar to those of MA: 500 mg twice daily for 12 weeks significantly increased food intake, arrested weight loss, and increased the fat mass. As was true of MA, no increase in fat-free mass was observed. MPA is well tolerated; contraindications are the same as for MA.

Cannabinoids

Cannabinoids are derivatives of the *Cannabis sativa* plant. Dronabinol is the most commonly used pharmacologic form, a synthetic and orally bioavailable form of tetrahydrocannabinol. It is FDA-approved for chemotherapy-induced nausea and vomiting, as well as AIDS- and cancer-associated anorexia. Dronabinol is an effective antiemetic agent for cancer patients suffering chemotherapy-induced nausea.[94] Its appetite stimulant effects have been known for some time. Significant increases in appetite are uniformly observed in trials with either AIDS[7] or cancer patients, but reductions in weight loss have not yet been convincingly demonstrated. In AIDS patients, dronabinol was as effective as megestrol in stimulating appetite but weight gain was only seen with MA.[110] Side effects of dronabinol are euphoria, dizziness, somnolence, and confusion: as a result, dose reduction or withdrawal occurs in approximately 25% of patients.

Prokinetic Agents

Some patients with cancer have symptoms related to gastroparesis or impaired motility of the gut, arising from tumor infiltration, or surgical or pharmacologic interruption of the autonomic innervation of the gastrointestinal tract. The prokinetic agents, metoclopramide, 10 mg orally QID, and domperidone, 10 mg orally QID, are beneficial in the relief of anorexia, nausea, and early satiety caused by poor GI motility.

A common limiting side effect of metoclopramide is sedation, although a reduction in dose can often correct this problem. Sedation is not as pronounced a problem with domperidone, but the drug is not currently available in the U.S. Another prokinetic drug, cisapride, is very effective but has been removed from the market due to rare, life-threatening prolongation of the QT interval and subsequent dysrhythmias.

TARGETED NUTRIENT THERAPY

Particular nutrients are sometimes administered in quantities that exceed basal metabolic requirements in order to elicit a beneficial physiologic response. This is sometimes referred to as "targeted nutrient therapy," and the nutrient in this circumstance is often referred to as a "nutraceutical" because it is being used as a pharmacologic agent rather than just to meet nutritional requirements. Examples used in cancer patients include omega-3 fatty acids, glutamine, arginine, vitamin E, and RNA.

OMEGA-3 POLYUNSATURATED FATTY ACIDS

The source of fat in conventional enteral and parenteral nutritional formulas is typically a vegetable oil rich in omega-6 polyunsaturated fatty acids. Replacing those fatty acids with one or both of the two major omega-3 polyunsaturated fatty acids, eicosapentaenoic acid (EPA) and docosahexaenoic acid (DHA), may improve patient outcome, including the attenuation or reversal of cancer wasting.

Exchanging dietary n-3 fatty acids for n-6 fatty acids, or even just supplementing a constant level of n-6 intake with n-3 fatty acids, is known to change the fatty acid composition of immune cell membranes and alter the immune response.[26] In healthy volunteers, supplemental n-3 fatty acids generally downregulate the systemic reaction to inflammatory mediators and diminish immunoresponsiveness in general, probably through their inhibition of TNF-alpha, IL-1, IL-2, and IL-6 release.[26,27,122] Interestingly, a paradoxical effect has sometimes been observed in malnourished cancer patients or those stressed by major surgery: cytokines such as TNF-alpha may increase with the administration of n-3 fatty acids and lymphocyte responsiveness has been shown to be enhanced.[37,40] Thus, how n-3 fatty acids alter the immune system may depend on the health of the individual taking the fatty acids.

n-3 fatty acids may also negatively affect the viability of an existing cancer in ways other than modulation of the immune response: in cell cultures of pancreatic or colon cancer cells, EPA induces apoptosis,[17,59] and in other studies, n-3 fatty acids seem to interfere with the microcirculation of a neoplasm, limiting its ability to expand.[5]

Regardless of the mechanism(s) by which they operate, administration of supplemental n-3 fatty acids may improve the clinical outcome of the patient, although definitive studies are not yet available. Improved survival among 60 cancer patients harboring a variety of solid neoplasms was observed with 18 grams of n-3 fatty acids per day.[40] Among malnourished cancer patients, improved performance scores were observed even though no apparent benefit in protein-calorie status had resulted from n-3 supplementation. Similarly, an uncontrolled study of patients with pancreatic cancer who were steadily losing weight observed that administration of n-3 fatty acids in conjunction with supplemental calories and protein brought about a weight gain and improved performance scores.[4] The observed decrease in the caloric expenditure per kg of lean mass in the latter study provides a plausible mechanism for the anabolic potential of n-3 fatty acids,[4] as do the reports that they can inhibit the release of the protein catabolic cytokines, TNF-alpha and IL-6.[26,27,122]

More compelling data are available on the possible clinical benefits of multi-modality supplements which contain, in addition to n-3 fatty acids, supraphysiologic quantities of arginine, RNA, and glutamine.

ARGININE, RNA, GLUTAMINE, AND MULTIMODALITY "IMMUNOMODULATORY" FORMULAS

Many studies in tumor-bearing, or otherwise stressed, animals support the concept that supplementation with arginine, RNA, or glutamine can enhance cell-mediated immunoresponsiveness and, in some cases, improve survival.[29,35,93] Demonstrating comparable results in human trials has been considerably less consistent when these nutrients have been used individually.

However, in a select type of patient — the bone marrow transplant recipient — two prospective, controlled, intervention trials have demonstrated significant improvements in clinical endpoints with the use of glutamine-supplemented TPN compared to TPN not containing the amino acid. It is important to note that this utility of glutamine has only been demonstrated in bone marrow transplant patients, specifically among those who are almost exclusively reliant on parenteral nutrition. In the first study,[124] transplant patients with malignances demonstrated significantly improved nitrogen balance, a diminished incidence of clinically significant infections, lower rates of microbial colonization, and a mean reduction in hospitalization of approximately 1 week. In the second study,[95] Schloerb et al. also observed a mean decrease in the duration of hospitalization of approximately 1 week but in the absence of any decrease in infections or other indicators of morbidity. A follow-up study by Schloerb et al.[96] examined whether any benefits of glutamine supplementation in bone marrow transplant patients could be demonstrated in those who were initially supported by enteral nutrition (and who therefore received oral glutamine) and who only then resorted to parenteral nutrition (containing glutamine) when necessary: no improvements in any clinical endpoints were seen in this setting. The reasonable conclusion, based on the available data, is that glutamine supplementation in the immediate post-transplant period may improve the clinical outcome for bone marrow transplant patients but only for those who are reliant solely on parenteral nutrition.

There is a growing consensus among clinical trials that specialized enteral formulas containing RNA, n-3 fatty acids, arginine, and glutamine improve clinical outcomes in patients undergoing surgery for gastrointestinal malignancies.[12,19,39,98] Two recent meta-analyses support this conclusion.[46,47] In patients receiving these specialized formulas, there occur significant decreases in the mediators of systemic inflammation, concurrent increases in cell-mediated immunity and, in some instances, an improvement in net protein balance.[11,38,39,49] In the majority of the trials, this translates into a remarkable decrease in perioperative infections,[12,19,39,98] and in the duration of hospitalization.[12,19,39] The most notable exception is a large trial in which the immunomodulatory formula provided no advantage over intravenous crystalloid to patients after surgery for gastrointestinal malignancies.[44] However, the patients in this study were particularly well nourished going into surgery: their mean weight loss was only 5% and the mean serum albumin was in the normal range. It is not surprising, therefore, that little benefit was realized with aggressive nutrition in a group of adequately nourished patients, even with a formula containing targeted nutrients. In patients who have sustained major trauma, administration of these formulas compared to conventional formulas has also been shown to result in improved outcomes.[58,118] It is important to recognize that the trials demonstrating the most clear-cut benefits are those in which administration of these formulas was begun either before surgery (as an oral supplement) or within 24 hours after surgery, typically through a surgically inserted jejunostomy tube. In conclusion, there is substantial evidence to indicate that the clinical course of malnourished patients with gastrointestinal malignancies can be significantly improved by multimodality, immunoenhancing formulas in situations involving a high degree of physiologic stress, such as major surgery. In instances where these formulas have also been compared to conventional TPN, they also convey a significant advantage.[39,49]

ORAL MUCOSITIS: TREATMENT WITH VITAMIN E OR GLUTAMINE

A prospective, randomized clinical trial, albeit of limited size, indicates that topical vitamin E[117] can provide significant healing and improvement in the oral mucositis associated with cytotoxic therapy. Similarly, glutamine suspensions that are swished in the mouth several times per day have been reported to improve radiation or cytotoxin-induced mucositis,[1,2,51] although this was not the case in one trial.[83] Although not definitively proven, these therapies offer essentially no risk of toxicity and are worth considering in patients suffering from this condition.

VITAMIN AND MINERAL DEFICIENCIES ASSOCIATED WITH SPECIFIC CIRCUMSTANCES

The focus of providing nutritional support should be the prevention or amelioration of protein-calorie malnutrition, but it is wise to anticipate vitamin or mineral deficiencies that may arise in patients with malignancies. The intent of this discussion is not to elaborate on the treatment of each of these deficiency states: this information can be found elsewhere. Rather, it is to point out often overlooked examples where these deficiencies may occur.

HIGH-GRADE BILIARY OBSTRUCTION OR BILIARY DIVERSION/ILEAL DAMAGE

Biliary obstruction of a high-grade nature, biliary diversion, or extensive loss of ileal function due to resection or radiation will cause fat malabsorption. In the case of biliary disease, insufficient bile to emulsify fat is reaching the lumen of the intestine and, in the case of ileal dysfunction, insufficient bile acids are recycled by the ileum, overwhelming the liver's ability to upregulate bile acid synthesis and secretion. Ileal disease or resection exceeding 100 cm in length is usually associated with steatorrhea; less extensive involvement can usually be compensated for by increased hepatic biliary synthesis.

Compensation for the calories lost in malabsorbed fat and the attendant diarrhea are usually the focus of treatment. However, fat-soluble vitamins are also malabsorbed, as are the divalent cationic minerals, calcium, magnesium, zinc, and copper. Deficiencies of vitamins A, D, and K are common in patients with longstanding cholestasis, and the likelihood of this occurring is proportional to the magnitude of the elevation in serum bilirubin.[54] Chronic total bilirubin elevations above 5 mg/dL should definitely raise suspicion. Many deficiency states in this setting are so subtle that diagnosis is only made when the index of suspicion is high and appropriate blood tests are obtained.[54] Deficiency of vitamin E is observed less commonly, but is certainly known to occur in other causes of chronic cholestasis, such as biliary atresia. Losses of calcium, magnesium, zinc, and copper presumably occur because these cations bind to the unabsorbed fatty acids present in the stool; losses are similarly proportional to the degree of cholestasis.[120] Symptomatic hypocalcemia, hypomagnesemia, and metabolic bone disease can occur.

B$_{12}$ DEFICIENCY DUE TO ILEAL OR GASTRIC INSULTS

Age is one of the major risk factors for the occurrence of cancers of the stomach, pancreas, and colorectum. Atrophic gastritis also increases with age, reaching a prevalence of nearly 40% by the 8th decade of life.[56] Elderly individuals with atrophic gastritis have lower vitamin B$_{12}$ levels than elder controls without the condition, due to diminished bioavailability of protein-bound B$_{12}$ in food. Furthermore, the clinical manifestations of B$_{12}$ deficiency can occur with only modest reductions in the plasma B$_{12}$ concentration, within what is considered the normal range. The condition is, therefore, best identified by the presence of elevated methylmalonic acid levels in the blood.[65] From 4 to 10% of elderly individuals whose plasma B$_{12}$ levels are in the "low-normal" range of 180 to 400 pg/mL probably have a cellular deficiency of the vitamin. Neurologic degeneration due to this subtle form of B$_{12}$ deficiency can arise without any hematologic manifestations, further obscuring the diagnosis.[64]

Further compromises in B$_{12}$ absorption can occur due to pharmacologic suppression of gastric acid secretion,[71] surgical ablation of a portion of the stomach or ileum, or radiation damage to those organs. The risk of B$_{12}$ deficiency goes up rapidly when more than 90 cm of ileum are lost or involved with disease.[33] The clinician, therefore, needs to remain cognizant that either neurologic or hematologic manifestations of B$_{12}$ deficiency may superimpose themselves on the myriad of other reasons that cancer patients develop abnormalities of the nervous system and bone marrow, and that a "normal" plasma B$_{12}$ level may not exclude the diagnosis.

"ALTERNATIVE" NUTRITION FOR THE TREATMENT OF CANCER

The major cancers are characterized by substantial morbidity and mortality. To date, conventional medicine can only provide modest opportunities for treatment and palliation, particularly for the latter stages of disease. It is therefore not surprising that many cancer patients take unproven natural remedies in conjunction with traditional treatment. For example, in a recent cohort study[16] which examined 480 patients with newly diagnosed breast cancer, 11% of the women had used alternative medicine before they were given a diagnosis of breast cancer and 28% started using some type of alternative medicine after the diagnosis was established.

The majority of patients using alternative therapy do not voluntarily discuss this therapy with their physicians. This lack of communication is not in the patient's best interests. Alternative therapies may contain harmful substances, or substances that interact with conventional therapies. Moreover, the physician can better manage the patient's care if he/she fully understands the patient's perspectives on treatment options. Therefore, patients should be asked about any unconventional therapy they are taking and should be reasonably informed about the types of therapies available.[25]

Table 9.5 shows several alternative nutrition remedies that are frequently used by cancer patients. None has been demonstrated to be efficacious in a compelling fashion, although several of them are presently being studied in scientifically rigorous trials.

PRACTICAL CONCLUSIONS

1. Wasting — a form of protein-calorie malnutrition that involves loss of weight and lean mass — is common in patients, particularly those with gastrointestinal and hepatobiliary cancers, and carries with it negative consequences in regard to morbid events, ability to withstand therapy, and survival. An unintentional loss of ≥10% of usual body weight constitutes a convenient and surprisingly accurate means of identifying those patients whose malnutrition is moderate to severe.

2. Routine identification of those cancer patients with moderate-to-severe malnutrition is important because it is these patients who will most clearly benefit from aggressive nutritional support preoperatively and during chemotherapy and radiation treatment. Aggressive nutritional support should also be provided to well- or mildly malnourished patients if they are likely to fall far short of meeting their nutritional needs for 7 or more days. An oral or enteral approach to aggressive nutritional support is more physiologic than a parenteral approach, less expensive, and fraught with fewer instances of serious morbidity.

3. All cancer patients undergoing allogenic bone marrow transplantation appear to benefit from prophylactic aggressive nutritional support. TPN has most often been used. With parenteral feeding, supplementing the TPN with glutamine may convey nutritional and immunologic benefits that translate into fewer infections and shorter hospitalizations.

TABLE 9.5
Alternative Nutrition Remedies Frequently Used by Cancer Patients

Treatment	Active agent	Purported activity	Potential side effects	References	Efficacy*
Bromelain	Protease, peroxidase, acid phosphatase	Reduces metastases	Decreases platelet activity, GI allergic reactions	Cancer Invest 1988;6:2441 J Cancer Res Clin Oncol 1988; 114:507	–
Green tea	Polyphenols	Reduces recurrence of breast cancer	Safe, but no long-term studies	Jpn J Cancer Res 1998;89:254 Jpn J Cancer Res 1993;23:186	+
Shark cartilage	Sphyrnastatin 1 and 2	Angiogenesis inhibitor	Nausea, vomiting, constipation, hepatitis	Anticancer Res 1998;18:4435 J Clin Oncol 1998;16:3649	–
Laetrile	Amygdalin (vitamin B_{17})	Tumoricidal	Cyanide toxicity	CA 1991;41:187 Ann Int Med 1978;89:389	–
Macrobiotic diet	Cereal grains, vegetables, seaweed, beans	Corrects cancer cachexia	Low in vitamin D, B_{12}, iron and calcium, relatively low calories	Nutr Rev 1992;50:106 Int J Cancer 1998;suppl 11:69	–
Gerson diet	Raw vegetarian diet, vegetable and fruit juices, calf's liver, coffee enemas	Detoxification of harmful substances	Very low fat content	Int J Cancer 1998;suppl 11:69 CA 1990;40:253	–
Livingstone therapy	Vegetarian whole-food diet, blood transfusion, supplemental vitamins, autologous vaccination, enema	Enhancement of immune function	Worse quality of life	N Engl J Med 1991;325:1180 CA 1990;40:103	–

* Efficacy here defined as evidence of a beneficial effect in at least one prospective, controlled trial in humans.

From Mason JB. Gastrointestinal cancer: nutritional support, in *Principles and Practice of Gastrointestinal Oncology*. Kelsen D, et al. (Eds.) Philadelphia: Lippincott Williams & Wilkins, 2002, 127–142. With permission.)

4. When aggressive nutritional support is provided to malnourished cancer patients undergoing major surgery, further gains may be realized by using "immunoenhancing" enteral formulas in the pre- and postoperative periods in the place of conventional tube feeds or TPN.

5. Certain circumstances are likely to produce select vitamin and mineral deficiencies. Diligent attention to these situations, with a proactive approach to prevention, will avert morbidity associated with these deficiencies.

6. A large proportion of patients with these cancers will use "alternative" medical treatments in conjunction with conventional treatments and often do not mention such treatments unless specifically requested to do so. Thorough and sensible management of the patient dictates an awareness of all treatments.

ACKNOWLEDGMENTS

Supported in part by the National Cancer Institute (3R01-CA59005–06A1S1), and by the U.S. Department of Agriculture, Agricultural Research Service contract 53–3K06–01. The contents of this publication do not necessarily reflect the policies or views of the U.S.D.A., nor does mention of trade names, commercial products, or organizations imply endorsement by the U.S. government.

REFERENCES

1. Anderson P, Ramsay N, Shu X, et al. Effect of low-dose oral glutamine on painful stomatitis during bone marrow transplantation. *Bone Marrow Transplant.* 22:339–344, 1998.
2. Anderson P, Schroeder G, and Skubitz K. Oral glutamine reduces the duration and severity of stomatitis after cytotoxic cancer chemotherapy. *Cancer.* 83: 1433–1439, 1998.
3. Andreyev H, Norman A, Oates J, and Cunningham D. Why do patients with weight loss have a worse outcome when undergoing chemotherapy for gastrointestinal malignancies? *Eur. J. Cancer.* 34:503–509, 1998.
4. Barber M, Ross J, Voss Tisdale M, and Fearon K. The effect of an oral nutritional supplement enriched with fish oil on weight loss in patients with pancreatic cancer. *Br. J. Cancer.* 81:80 86, 1999.
5. Baronzio G, Galante F, Gramaglia A, Barlocco A, de Grandi S, and Freitas I. Tumor microcirculation and its significance in therapy: possible role of omega-3 fatty acids as rheological modifiers. *Med. Hypotheses.* 50: 175–182, 1998.
6. Bastow M, Rawlings J, and Allison S. Benefits of supplementary tube feeding after fractured neck of femur. *Br. Med. J.* 287: 1589–1594, 1983.
7. Beal J, Olson R, Lefkowitz L, et al. Long-term efficacy and safety of dronabinol for acquired immunodeficiency syndrome-associated anorexia. *J. Pain Symptom Manage.* 14:7–14, 1997.
8. Beck S and Tisdale M. Production of lipolytic and proteolytic factors by a murine tumor-producing cachexia in the host. *Cancer Res.* 47:5919–5923, 1987.

9. Bozzetti F, Bavazzi C, Miceli R, et al. Perioperative tpn in malnourished, gi cancer patients: a randomized, clinical trial. *J. Parent. Ent. Nutr.* 24: 7–14, 2000.

10. Bozzetti F, Cozzaglio L, Gavazzi C, et al. Nutritional support in patients with cancer of the esophagus: impact on nutritional status, patient compliance, and survival. *Tumori.* 84: 681–686, 1998.

11. Braga M, Gianotti L, Cestari A, Vignali A, Pellegatta F, Dolci A, and Di Carlo V. Gut function and immune and inflammatory responses in patients perioperatively fed with supplemented enteral formulas. *Archiv. Surg.* 131: 1257–1265, 1996.

12. Braga M, Gianotti L, Radeelli G, Vignali A, Gilberto M, Gentilini O, and DiCarlo V. Perioperative immunonutrition in patients undergoing cancer surgery: results of a randomized double-blind phase 3 trial. *Archiv. Surg.* 134: 428–433, 1999.

13. Braunschweig CL, Levy P, Shean PM, et al. Enteral compared with parenteral nutrition: a meta-analysis. *Am. J. Clin. Nutr.* 74:534–542, 2001.

14. Brennan M, Pisters P, Posner M, Quesada O, and Shike M. A prospective random-ized trial of total parenteral nutrition after major pancreatic resection for malig-nancy. *Ann. Surg.* 220: 436–441, 1994.

15. Bruera E, Macmillan K, Kuehn N, Hanson J, and MacDonald R. A controlled trial of megestrol acetate on appetite, caloric intake, nutritional status, and other symptoms in patients with advanced cancer. *Cancer.* 66: 1279–1282, 1990.

16. Burstein HJ, Gelber S, Guadagnoli E, and Weeks JC. Use of alternative medicine by women with early-stage breast cancer. *N. Engl. J. Med.* 340:1733–1739, 1999.

17. Clarke R, Lund E, Latham P, Pinder A, and Johnson I. Effect of EPA on the proliferation and incidence of apoptosis in the colorectal cell one HT29. *Lipids.* 34: 1287–1295, 1999.

18. Costelli P, Carbo N, Tessitore L, Bagby C, Lopez-Soriano F, Argiles J, and Baccino F. Tumor necrosis factor-alpha mediates changes in tissue protein turnover in a rat cachexia model. *J. Clin. Invest.* 92: 2783–2789, 1993.

19. Daly J, Weintraub F, Shou J, Rosato E, and Lucia M. Enteral nutrition during multi-modality therapy in upper gastrointestinal cancer patients. *Ann. Surg.* 221: 327–338, 1995.

20. De Cicco M, Panarello G, Fantin D, et al. Parenteral nutrition in cancer patients receiv-ing chemotherapy: effects on toxicity and nutritional status. *J. Parent. Ent. Nutr.* 17: 513–518, 1993.

21. Detsky A, Baker J, O'Rourke K, and Goel V. Perioperative parenteral nutrition: a meta-analysis. *Ann. Int. Med.* 107: 195–203, 1987.

22. Dewys WD, Begg C, Lavin P, et al. Prognostic effect of weight loss prior to chemo-therapy in cancer patients. *Am. J. Med.* 69:491–497, 1980.

23. Donnelly S and Walsh D. The symptoms of advanced cancer. *Seminars Oncol.* 22(suppl 3): 67–72, 1995.

24. Ebisui C, Tsujinaka T, Morimoto T, et al. Interleukin-6 induces proteolysis by activating intracellular proteases in C2C12 myotubes. *Clin. Sci.* 89: 431–439, 1995.

25. Eisenberg DM, Kessler RC, Foster C, Norlock FE, Calkins DR, and Delbanco TL. Unconventional medicine in the United States — prevalence, costs, and patterns of use. *N. Engl. J. Med.* 328:246–252, 1993.

26. Endres S, Ghorbani R, Kelley V, et al. The effect of dietary supplementation with n-3 polyunsaturated fatty acids on the synthesis of IL-1, and TNF by mononuclear cells. *N. Engl. J. Med.* 320: 265–271, 1989.

27. Endres S, Meydani S, Ghorbani R, Schindler R, and Dinarello C. Dietary supplemen-tation with n-3 fatty acids suppresses IL-2 production and mononuclear cell prolifera-tion. *J. Leukoc. Biol.* 54: 599–603, 1993.

28. Fan S, Lo C, Lai E, Chu K, Liu C, and Wong J. Perioperative nutritional support in patient undergoing hepatectomy for hepatocellular carcinoma. *N. Engl. J. Med.* 331: 1547–1552, 1994.

29. Fanslow W, Kulkarni A, Van Buren C, and Rudolph F. Effect of nucleotide restriction and supplementation on resistance to experimental murine candidiasis. *J. Parent. Ent. Nutr.* 12:49–52, 1988.

30. Fearon K, Hansell D, Preston T, et al. Influence of whole body protein turnover rate on resting energy expenditure in patients with cancer. *Cancer Res.* 48: 2590–2595, 1988.

31. Fearon KCH and Preston T. Body composition in cancer cachexia. *Infusionstherapie.* 17(suppl. 3): 63–66, 1990.

32. Fiatarone MA, O'Neill EF, Ryan ND, et al. Exercise training and nutritional supplementation for physical frailty in very elderly people. *N. Engl. J. Med.* 23:1769–1775, 1994.

33. Filipsson S, Hulten L, and Lindstedt G. Malabsorption of fat and vitamin B_{12} before and after intestinal resection for Crohn's Disease. *Scand. J. Gastroenterol.* 13: 529–536, 1978.

34. Flynn M and Leightty F. Preoperative outpatient nutritional support of patient with squamous cancer of the upper aerodigestive tract. *Am. J. Surg.* 154: 359–362, 1987.

35. Fox A, Kripke S, DePaula J, Berman J, Settle R, and Rombeau J. Effect of glutamine-supplemented enteral diet on methotrexate induced enterocolitis. *J. Parent. Ent. Nutr.* 12:325–331, 1988.

36. Fried S and Zechner R. Cachectin/tumor necrosis factor decreases human adipose tissue lipoprotein lipase mRNA levels, synthesis, and activity. *J. Lipid Res.* 30:1917–1923, 1989.

37. Furukawa K, Tashiro T, Yamamori H, et al. Effects of soybean oil emulsion and eicossapentaenoic acid on stress response and immune function after a severely stressful operation. *Ann. Surg.* 229: 255–261, 1999.

38. Gianotti L, Braga M, Fortis C, et al. A prospective, randomized clinical trial on perioperative feeding with an arginine-, n-3 fatty acid-, and RNA-enriched enteral diet: effect on host response and nutritional status. *J. Parent. Ent. Nutr.* 23:314–320, 1999.

39. Gianotti L, Braga M, Vignali A, Balzano G, Zerbi A, Bisagni P, and Di Carlo V. Effect of route of delivery and formulation of postoperative nutrition support in patient undergoing major operations for malignant neoplasms. *Archiv. Surg.* 132: 1222–1230, 1997.

40. Gogos C, Ginopoulos P, Salsa B, Apostolidou E, Zoumbos N, and Kalfarentzos F. Dietary omega-3 polyunsaturated fatty acids plus vitamin E restore immunodeficiency and prolong survival for severely ill patients with generalized malignancy. *Cancer.* 82: 395–402, 1998.

41. Grunfeld C, Dinarello C, and Feingold K. Tumor necrosis factor-alpha, interleukin-1, and interferon-alpha stimulate triglyceride synthesis in HepG2 cells. *Metabolism.* 40:894–898, 1991.

42. Hauner H, Petruschke T, Russ M, Rohrig K, and Eckel J. Effects of TNF-alpha on glucose transport and lipid metabolism of newly-differentiated human fat cells in culture. *Diabetologica.* 38:764–771, 1995.

43. Heber D, Chlebowski R, Ishibashi D, Herrold J, and Block J. Abnormalities in glucose and protein metabolism in non-cachectic lung cancer patients. *Cancer Res.* 42: 4815–4819, 1982.

44. Heslin M, Latkany L, Leung D, et al. A prospective, randomized trial of early enteral feeding after resection of upper gastrointestinal malignancy. *Ann. Surg.* 226: 567–580, 1997.

45. Heyland DK, MacDonald S, Keefe L, and Drover J. Total parenteral nutrition in the critically ill patient: a meta-analysis. *JAMA.* 16: 2013–2019, 1998.

46. Heyland DK, Novak F, Drover JW, et al. Should immunonutrition become routine in critically ill patients? *JAMA.* 286:944–953, 2001.

47. Heys S, Walker L, Smith I, and Eremin O. Enteral nutritional supplementation with key nutrients in patients with critical illness and cancer: a meta-analysis of randomized controlled clinical trials. *Ann. Surg.* 229: 467–477, 1999.

48. Hill G. Body composition research: implications for the practice of clinical nutrition. *J. Parent. Ent. Nutr.* 16:197–218, 1992.

49. Hochwald S, Harrison L, Heslin M, Burt M, and Brennan M. Early postoperative enteral feeding improves whole body protein kinetics in upper gastrointestinal cancer patients. *Am. J. Surg.* 174: 325–330, 1997.

50. Holyrode C, Babuzda T, Putnam R, et al. *Cancer Res.* 35:3710, 1975.

51. Huang E, Leung S, Wang C, et al. Oral glutamine to alleviate radiation-induced oral mucositis: a pilot randomized trial. *Int. J. Radiat. Oncol. Biol. Phys.* 46: 535–539, 2000.

52. Jin D, Phillips M, and Byles J. Effects of parenteral nutrition support and chemotherapy on the phasic composition of tumor cells in gastrointestinal cancer. *J. Parent. Ent. Nutr.* 23: 237–241, 1999.

53. Kama NA, Coskun T, Yuksek YN, and Yazgan A. Factors affecting post-operative mortality in malignant biliary tract obstruction. *Hepatogastroenterology.* 46:103–107, 1999.

54. Kaplan M, Elta G, Furie B, Sadowski J, and Russell R. Fat-soluble vitamin nutriture in primary biliary cirrhosis. *Gastroenterology.* 95: 787–792, 1988.

55. Keys A, Brozek J, Henschel A, et al. *The Biology of Human Starvation*. Minneapolis: University of Minnesota Press, 1950.

56. Krasinski S, Russell R, Samloff I, et al. Fundic atrophic gastritis in an elderly population: effect on hemoglobin and several nutritional indicators. *J. Am. Geriatr. Soc.* 34:800–806, 1986.

57. Kudsk K, Croce M, Fabian T, et al. Enteral versus parenteral feeding: effects on septic morbidity after blunt and penetrating abdominal trauma. *Ann. Surg.* 215: 503–511, 1992.

58. Kudsk K, Minard G, Croce M, et al. A randomized trial of isonitrogenous enteral diets after severe trauma. *Ann. Surg.* 224: 531–543, 1996.

59. Lai P, Ross J, Fearon K, Anderson J, and Carter D. Cell cycle arrest and induction of apoptosis in pancreatic cancer cells exposed to EPA *in vitro. Br. J. Cancer.* 74: 1375–1383, 1996.

60. Langstein H, Doherty G, Fraker D, Buresh C, and Norton J. The roles of gamma-interferon and tumor necrosis factor alpha in an experimental rat model of cancer cachexia. *Cancer Res.* 51:2302–2306, 1991.

61. Lanzotti VJ, Thomas DR, Boyle LE, Smith TL, Gehan EA, and Samuels ML. Survival with inoperable lung cancer. *Cancer.* 39: 303–313, 1977.

62. Lee JH, Machtay M, Unger LD, Weinstein GS, Weber RS, Chalian AA, and Rosenthal DI. Prophylactic gastrostomy tubes in patients undergoing intensive irradiation for cancer of the head and neck. *Arch. Otolaryngol. Head Neck Surg.* 124: 871–875, 1998.

63. Lenssen P, Bruemmer B, Bowden R, Gooley T, Aker S, and Mattson D. Intravenous lipid dose and incidence of bacteremias and fungemia in patients undergoing bone marrow transplantation. *Am. J. Clin. Nutr.* 67: 927–933, 1998.

64. Lindenbaum J, Savage D, Stabler S, et al. Diagnosis of cobalamin deficiency: relative sensitivities of serum cobalamin, methylmalonic acid, and total homocysteine levels. *Am. J. Hematol.* 34: 99–107, 1990.

65. Lindenbaum J, Savage D, Stabler S, et al. Neuropsychiatric disorders caused by cobalamin deficiency in the absence of anemia or macrocytosis. *N. Engl. J. Med.* 318: 1720–1728, 1988.

66. Lipman T. Grains or veins: is enteral nutrition really better than parenteral nutrition? A look at the evidence. *J. Parent. Ent. Nutr.* 22: 167–182, 1998.

67. Llovera M, Lopez-Soriano F, and Argiles J. Effects of tumor necrosis factor-alpha on muscle protein turnover in female Wistar rats. *J. Natl. Cancer Inst.* 85: 1334–1339, 1993.

68. Loprinzi J, Ellison N, Schaid D, et al. Controlled trial of megestrol acetate for the treatment of cancer anorexia and cachexia. *J. Natl. Cancer Inst.* 82: 1127–1132, 1990.

69. Loprinzi J, Michalak J, Schaid D, et al. Phase III evaluation of four doses of megestrol acetate as therapy for patients with cancer anorexia and/or cachexia. *J. Clin. Oncol.* 11:762–767, 1993.

70. Lundholm K, Bylund A, Holm J, and Schersten T. Skeletal muscle metabolism in patients with malignant tumor. *Eur. J. Cancer.* 12: 465–473, 1976.

71. Marcuard S, Albernaz I, and Khazanie P. Omeprazole therapy causes malabsorption of cyanocobalamin. *Ann. Intern. Med.* 120: 211–215, 1994.

72. Marshall M, Doerrler W, Feingold K, and Grunfeld C. Leukemia inhibitory factor induces changes in lipid metabolism in cultured adipocytes. *Endocrinology.* 135:141–147, 1994.

73. Mason JB. A clinical nutritionist's search for meaning: why should we bother to feed the acutely ill, hospitalized patient? *Nutrition.* 12:279–281, 1996.

74. McGeer A, Detsky A, and O'Rourke K. Parenteral nutrition in cancer patients undergoing chemotherapy: a meta-analysis. *Nutrition.* 6: 233–240, 1990.

75. Mick R, Vokes E, Weichselbaum RR, and Panje WR. Prognostic factors in advanced head and neck cancer patient undergoing multimodality therapy. *Otolaryngol. Head Neck Surg.* 105:62–73, 1991.

76. Mulder P, Bouman J, Gietema J, Van Rijsbergen H, Mulder N, Van der Geest S, and De Vries E. Hyperalimentation in autologous bone marrow transplantation for solid tumors. *Cancer.* 64: 2045–2052, 1989.

77. Muller J, Brenner U, Dienst C, and Pichlmaier H. Preoperative parenteral feeding in patients with gastrointestinal carcinoma. *Lancet.* 1: 68–71, 1982.

78. Muscaritoli M, Cangiano C, Cascino A, et al. Plasma clearance of exogenous lipids in patients with malignant disease. *Nutrition.* 6:147–151, 1990.

79. Muscaritoli M, Conversano L, Torelli G, et al. Clinical and metabolic effects of different parenteral regimens in patients undergoing allogeneic bone marrow transplantation. *Transplantation.* 66: 610–616, 1998.

80. Newton J and Halsted C. Clinical and functional assessment of adults; Heymsfield S, Baumgartner R, and Pan S-F. Nutritional assessment of malnutrition by anthropometric methods; Alcock N. Laboratory test for assessing nutritional status; Dwyer J. Dietary assessment, in *Modern Nutrition in Health and Disease, 9th ed.* Shils M, et al. (Eds.) Williams and Wilkins, Baltimore, 1999, 895–962.

81. Nixon DW, Heymsfield SB, Cohen AE, et al. Protein-calorie undernutrition in hospitalized cancer patients. *Am. J. Med.* 68;683–690, 1980.

82. Nordin K and Glimelius B. Psychological reactions in newly diagnosed gastrointestinal cancer patients. *Acta Oncol.* 36: 803–810, 1997.

83. Okuno S, Woodhouse C, Loprinzi C, et al. Phase III controlled evaluation of glutamine for decreasing stomatitis in patients receiving 5-FU-based chemotherapy. *Am. J. Clin. Oncol.* 22: 258–261, 1999.

84. Opara E, Laviano A, Meguid M, and Yang Z. Correlation between food intake and CSF IL-1 in anorectic tumor bearing rats. *Neuroreport.* 6: 750–752, 1995.

85. Oster M, Enders S, Samuels S, Cone L, Hooton T, Browder H, and Flynn N. Megestrol acetate in patients with AIDS and cachexia. *Ann. Intern. Med.* 121: 400–408, 1994.

86. Patil PK, Patel SG, Mistry RC, Deshpande RK, and Desai PB. Cancer of the esophagus; esophagogastric anastomotic leak-a retrospective study of predisposing factors. *J. Surg. Oncol.* 49:163–167, 1992.

87. Popp M, Morrison S, and Brennan M. Total parenteral nutrition in a methylchloranthrene-induced rat sarcoma model. *Cancer Treat. Reports.* 65(suppl 5): 137–143, 1981.

88. Prijatmoko D, Strauss B, Lambert J, et al. Early detection of protein depletion in alcoholic cirrhosis: role of body composition analysis. *Gastroenterology.* 105: 1839–1845, 1993.

89. Qi C and Pekala P. Tumor necrosis factor-alpha-induced insulin resistance in adipocytes. *Proc. Soc. Exp. Biol. Med.* 223: 128–135, 2000.

90. Rofe, A, Bourgeois C, Coyle P, Taylor A, and Abdi E. Altered insulin response to glucose in weight-losing cancer patients. *Anticancer Res.* 14:647–650, 1994.

91. Roubenoff R, Heymsfield SB, Kehayias JJ, Cammon JG, and Rosenberg IH. Standardization of nomenclature of body composition in weight loss. *Am. J. Clin. Nutr.* 67:492–493, 1998.

92. Russell D, Shike M, Marliss E, et al. Effects of total parenteral nutrition and chemotherapy on the metabolic derangements in small cell lung cancer. *Cancer Res.* 44: 1706–1711, 1984.

93. Saito H, Trocki O, Wang S, Gonce S, Joffe S, and Alexander J. Metabolic and immune effects of dietary arginine supplementation after burn. *Arch. Surg.* 122: 784–789, 1987.

94. Sallan S, Cronin C, Zelen M, and Zinberg N. Antiemetics in patients receiving chemotherapy for cancer. *N. Engl. J. Med.* 302: 135–138, 1980.

95. Schloerb P and Amare M. Total parenteral nutrition with glutamine in bone marrow transplantation and other clinical applications. *J. Parent. Ent. Nutr.* 17: 407–413, 1993.

96. Schloerb P and Skikne B. Oral and parenteral glutamine in bone marrow transplantation: a randomized, double blind study. *J. Parent. Ent. Nutr.* 23: 117–122, 1999.

97. Senft M, Fietkau R, Iro H, Sailer D, and Sauer R. The influence of supportive nutritional therapy via percutaneous endoscopically guided gastrostomy on the quality of life of cancer patients. *Support Care Cancer.* 1:272–275, 1993.

98. Senkal M, Zumtobel V, Vauer K-H, et al. Outcome and cost-effectiveness of perioperative enteral immunonutrition in patients undergoing elective upper gastrointestinal tract surgery: a prospective, randomized trial. *Archiv. Surg.* 134: 1309–1316, 1999.

99. Shike M, Russell D, Detsky A, et al. Changes in body composition in patients with small-cell lung cancer. *Ann. Int. Med.* 101: 303–309, 1984.

100. Shirabe K, Matsumata T, Shimada M, et al. A comparison of parenteral hyperalimentation and early enteral feeding regarding systemic immunity after major hepatic resection-a randomized, prospective study. *Hepatogastroenterology.* 44: 205–209, 1997.

101. Simons J, Aaronson N, Vansteenkiste J, et al. Effects of medroxyprogesterone acetate on appetite, weight, and quality of life in advanced stage non-hormone sensitive cancer: a placebo controlled multicenter trial. *J. Clin. Oncol.* 14:1077–1084, 1996.

102. Simons J, Schols A, Hoefnagels J, Westerterp K, ten Velde G, and Wouters E. Effects of medroxyprogesterone acetate on food intake, body composition, and resting energy expenditure in patients with advanced, nonhormone-sensitive cancer. *Cancer.* 82:553–560, 1998.

103. Smale BF, Mullen JL, Buzby GP, et al. The efficacy of nutritional assessment and support in cancer surgery. *Cancer.* 47:2375–2381, 1981.

104. Smith R, Hartemink R, Hollinshead J, and Gillett D. Fine bore jejunostomy feeding following major abdominal surgery: a controlled randomized clinical trial. *Br. J. Surg.* 72: 458–461, 1985.

105. Stovroff M, Fraker D, Swedenborg J, and Norton J. Cachectin/tumor necrosis factor, a possible mediator of cancer anorexia in the rat. *Cancer Res.* 48: 920–925, 1988.

106. Strassman G and Kambayashi T. Inhibition of experimental cancer cachexia by anti-cytokine and anti-cytokine receptor therapy. *Cytokines Mol. Ther.* 1:107–113, 1995.

107. Swain C, Tavill A, and Neale G. Studies of tryptophan and albumin metabolism in a patient with carcinoid syndrome, pellagra, and hypoproteinemia. *Gastroenterology.* 71: 484–489, 1976.

108. Szeluga D, Stuart R, Brookmeyer R, Utermohlen V, and Santos G. Nutritional support of bone marrow transplant recipients: a prospective, randomized clinical trial comparing tpn to an enteral feeding program. *Cancer Res.* 47: 3309–3316, 1987.

109. Tayek J, Manglik S, and Abemayor E. Insulin secretion, glucose production and insulin sensitivity in underweight and normal-weight volunteers, and in underweight and normal-weight cancer patients: a clinical research center study. *Metabolism.* 46: 140–145, 1997.

110. Tchekmedyian NS, Tait N, Abrams J, and Aisner J. High-dose megestrol acetate in the treatment of advanced breast cancer. *Semin. Oncol.* 113: 37–43, 1986.

111. Timpone J, Wrigth D, Li N, et al. The safety and pharmacokinetics of single-agent and combination therapy with megestrol acetate and dronabinol for the treatment of HIV wasting syndrome. *AIDS Res. Hum. Retrovirus.* 13: 305–315, 1997.

112. Todorov P, Cariuk P, McDevitt T, Coles B, Fearon K, and Tisdale M. Characterization of a cancer cachectic factor. *Nature.* 379: 739–742, 1996.

113. Todorov P, McDevitt T, Cariuk P, Coles B, Deacon M, and Tisdale M. Induction of muscle protein degradation and weight loss by a tumor product. *Cancer Res.* 56: 1256–1261, 1996.

114. V.A. TPN Cooperative Study Group. Perioperative total parenteral nutrition in surgical patients. *N. Engl. J. Med.* 325: 525–532, 1991.

115. Valverde A, Teruel T, Navarro P, Benito M, and Lorenzo M. Tumor necrosis factor-alpha causes insulin receptor substrate-2-mediated insulin resistance and inhibits insulin-induced adipogenesis in fetal brown adipocytes. *Endocrinology.* 139:1229–1238, 1998.

116. Van Bokhorst-de van der Schuer, Van Leeuwen PA, Kuik DJ, Klop WM, Sauerwein HP, Snow GB, and Quak JJ. The impact of nutritional status on the prognoses of patients with advanced head and neck cancer. *Cancer.* 86:519–527, 1999.

117. Von Roenn J, Armstrong D, Kotler D, et al. Megestrol acetate in patients with AIDS-related cachexia. *Ann. Int. Med.* 121: 393–399, 1994.

118. Wadleigh R, Redman R, Graham M, Krasnow S, Anderson A, and Cohen M. Vitamin E in the treatment of chemotherapy-induced mucositis. *Am. J. Med.* 92: 481–484, 1992.

119. Weimann A, Bastian L, Bischoff W, et al. Influence of arginine, n-3 fatty acids and nucleotide-supplemented enteral support on systemic inflammatory response syndrome and multiple organ failure in patients after severe trauma. *Nutrition.* 14: 165–172, 1998.

120. Weisdorf S, Lysne J, Wind D, et al. Positive effect of prophylactic tpn on long-term outcome of bone marrow transplantation. *Transplantation.* 43: 833–838, 1987.

121. Whelton M, Kehayoglou A, Agnew J, Tumberg L, Sherlock S. Calcium absorption in parenchymatous and biliary liver disease. *Gut.* 12: 978–983, 1971.

122. Wicks C, Somasundaram S, Bjarnason I, et al. Comparison of enteral feeding and tpn after liver transplantation. *Lancet.* 344: 837–840, 1994.

123. Wigmore S, Fearon K, Maingay J, and Ross J. Down-regulation of the acute-phase response in patients with pancreatic cancer cachexia receiving oral EPA is mediated via suppression of interleukin-6. *Clin. Sci. (Colch.)* 92:215–221, 1997.

124. Wilmore DW. Catabolic illness. Strategies for enhancing recovery. *N. Engl. J. Med.* 325:695–702, 1991.

125. Ziegler T, Young L, Benfell K, et al. Clinical and metabolic efficacy of glutamine-supplemented parenteral nutrition after bone marrow transplantation. *Ann. Int. Med.* 116: 821–828, 1992.

10 Nutritional Aspects of Trauma and Postsurgical Care

J. Stanley Smith and Wiley W. Souba

CONTENTS

0-8493-0945-X/03/$0.00+$1.50
© 2003 by CRC Press LLC

225

Historically, nutrition has been the specialty of physicians treating chronic medical diseases. Now surgeons, since the introduction of total parenteral nutrition (TPN), have become more involved in the nutritional care of debilitated malnourished individuals.[14,20,23,24] Due to an increased awareness on the part of physicians, malnutrition has been found to be common in hospitalized surgical patients, and the typical diets provided to patients contain inadequate amounts or proportions of certain nutrients. Our understanding of the relationship between nutrition and metabolism has been increased by TPN, and it has become ever more obvious that the optimal nutrition for a given patient depends on the primary diagnosis and underlying metabolic status. The mediators regulating the body's metabolic and nutritional response to injury, sepsis, and cancer have now been well described, and, today, nearly all hospitalized patients can be fed safely and effectively via several routes. As a consequence, surgeons must be familiar with the changes in body metabolism that develop during catabolic illness, as well as with the indications for perioperative nutritional support and the techniques for its delivery.

Although the disease process is usually the major cause of malnutrition, many patients develop a nutritional deficit and further weight loss during their hospitalization as the result of withholding meals for diagnostic tests and procedures. In addition, critically ill trauma and postoperative patients are frequently anorexic secondary to illness and confinement. These patients can be fed, but controlled trials done in patients with normal body composition undergoing elective operation show that such nutritional support produces little improvement in outcome. Therefore, limited weight loss in selected hospitalized patients is acceptable because short-term undernutrition does not prolong a life-limiting illness, nor does it complicate convalescence following a major operation or other therapy. Other patients, such as those sustaining major injury or a life-threatening complication such as sepsis, require vigorous nutritional care because of their severe catabolic state.

Understanding the relationship between nutrition and metabolism requires some basic review.

BASIC NUTRITIONAL BIOCHEMISTRY

BODY COMPOSITION

Total body mass is comprised of an aqueous component and a nonaqueous component. The nonaqueous portion is made up of bone, tendons, and mineral mass as well as adipose tissue. The aqueous phase includes the intracellular mass made up of skeletal muscle, internal organs, skin, and circulating blood cells. Also contributing to the aqueous portion is the interstitial fluid and the intravascular volume. Total body water in the average-sized (70-kg) adult male makes up about 55 to 60% (~40 L) of total body mass. Of these 40 liters, about 22 liters are intracellular, 14 are interstitial fluid, and the plasma volume adds approximately 3 to 3.5 liters. Body composition varies as a function of age and sex and is altered following injury or surgery. These changes are characterized by a loss of lean body mass, a loss of body fat, and an expansion of the extracellular fluid compartment. Thus, the energy producing and utilizing body mass becomes diminished.

TABLE 10.1
Energy Reserves of a Healthy (70-kg) Adult Male

Energy Source	Kilograms	Kilocalorie Value
Protein		
Skeletal Muscle	6	24,000
Other	6	24,000
Fat	14	126,000
Glycogen		
Muscle	0.15	600
-Liver	0.075	300
-Free glucose	0.02	80

Bioelectrical impedance is one way to measure changes in lean body mass. This tool is based on the principle of electrical resistance being proportional to the fluid and electrolyte content of tissue. Since the lean body mass or body cell mass has most of the fluid and electrolyte content, passage of an electrical current through the body at the bedside can measure the lean body, or "fat free," mass. This appears to work well in both healthy and critically ill patients and can give the practitioner a guide to accurately gauge the needed nutritional support.[10]

For energy, the body contains fuel reserves that it can mobilize and utilize during times of starvation or stress (Table 10.1). By far the greatest energy component is fat, which is calorically dense providing about 9 kilocalories (kcal) per gram. Body protein comprises the next largest mass of utilizable energy, but amino acids yield only about 4 kcal/g. Unlike fat reserves, body protein does not constitute energy storage, but rather is a structural and functional component of the body. Therefore, loss of body protein, if severe, has functional consequences since, following injury, proteolysis is accelerated to generate amino acids to support gluconeogenesis and other key synthetic processes. In the long run, a chronic catabolic state can lead to erosion of body protein stores such that susceptibility to infection is increased, wound healing is impaired, and outcome is unfavorably impacted.[23]

ENERGY METABOLISM

From a simplistic standpoint, the human body is an engine burning fuel to generate energy that, in turn, is used to perform work, such as mechanical work (e.g., movement and breathing), transport work (e.g., carrier-mediated uptake of nutrients into cells), and synthetic work (biosynthesis of proteins and other complex molecules). Indeed, all of these are essential for life. The energy comes from the chemical bonds of the nutrients we consume. The human body has the capacity to "oxidize" several types of fuels; among these are glucose (carbohydrates), amino acids (proteins), fatty acids (lipids), ketone bodies, and alcohol. Thus, the human body converts energy stored in nutrients into internal (i.e., enzymatic catalysis) and external work (i.e., muscular contraction for locomotion). During starvation, the body must oxidize stored energy sources to generate this work. In humans, this process is relatively inefficient, since about half of this potential energy is lost as heat used to help maintain body temperature via sensitive regulatory mechanisms in the hypothalamus. Excess heat is released primarily through the skin via evaporation, radiation, convection, and conduction. After surgery, these central regulatory mechanisms often become "reset," leading to the development of fever, which, in turn, causes an increase in enzymatic reactions necessary to support the inflammatory process.

Amino acids, glucose, and fatty acids are the major sources of energy the body uses to perform work. Amino acids come from endogenous proteins through proteins in the diet, or as crystalline L-amino acids when administered intravenously. Glucose is produced when carbohydrates are broken down in the gut lumen or may be generated in the liver from other sugars. Fatty acids are derived from the hydrolysis of triglycerides. Glucose comes in the form of dextrose, which is a hydrated glucose molecule providing 3.4 kcal/g. One liter of D_5W contains 50 g of dextrose or 170 kcal. Therefore, the usual postoperative surgical patient given an intravenous glucose solution at 125 cc/hour receives about 500 kcal/day, far less than the actual number of kcal needed to meet energy requirements. However, this is enough glucose to stimulate pancreatic release of insulin, the primary anabolic hormone stimulating amino acid uptake and protein synthesis as well as mobilizing fat to the liver.

Triglycerides are made up of three fatty acids bound to a glycerol molecule. Naturally occurring fatty acids may be saturated (no double bonds) or unsaturated (one or more double bonds). In most tissues, fatty acids are readily oxidized for energy and are especially important energy sources for the heart, liver, and skeletal muscle. In adipose tissue, fatty acids may be re-esterified with glycerol and stored as triacylglycerols. Nearly the entire volume of an adipocyte is comprised of a large fat droplet. This stored fat is mobilized during starvation and stress, whereas structural lipid is generally preserved. When circulating in plasma, the major lipids are not in a free form. Free fatty acids are bound to albumin, whereas cholesterol, triglycerides, and phospholipids are transported as lipoprotein complexes. Lipoproteins are cleared from the circulation by the action of lipoprotein lipase, an enzyme located on the surface of the capillary endothelium. This enzyme catalyzes the breakdown of triglyceride to free fatty acids and glycerol. A second lipase, which regulates the supply of free fatty acids to tissues, is hormone sensitive. Present only in adipose tissue, it catalyzes the breakdown of stored triglycerides into glycerol and fatty acids, which are then released into the circulation.

The hormones, epinephrine, norepinephrine, and glucagons, bind to cell membrane receptors and thereby rapidly activate the hormone-sensitive lipase. Growth hormone and glucocorticoids also increase the activity of hormone-sensitive lipase, but they do so more slowly, inasmuch as their action involves *de novo* protein synthesis.[16] During stress, the activity of hormone-sensitive lipase mobilizes fat, but the decreased activity of lipoprotein lipase on endothelial cells impairs the clearance of fat from the bloodstream. Nonetheless, fat is an important fuel source for critically ill patients, and, as a general rule, the amount of fat administered to patients receiving total parenteral nutrition should amount to about 15 to 30% of total nonprotein caloric needs.

Before free fatty acids can be oxidized, they must be activated by condensation with coenzyme A in the cytoplasm. The resulting fatty acyl Co-A molecules are transported into the mitochondria by means of a shuttle system in which L-carnitine acts as an acyl carrier. This process may be rate limiting in severe stress states. Carnitine depletion has been shown to be characteristic of critical illness and, therefore, supplementation of TPN solutions with carnitine has been proposed to enhance endogenous utilization of fats as a fuel source. Although carnitine addition has not proven to be effective, it is an example of the many attempts to improve the metabolic care of critically ill patients by a nutritional approach.

Energy is measured in calories. A calorie is the amount of heat required to raise the temperature of 1 gram of water from 14.5°C to 15.5°C at a pressure of one standard atmosphere. A kilocalorie (1000 calories) is the unit of energy measurement used in the U.S. when referring to body metabolism and nutrition. Basal energy requirements are those obtained on normal resting subjects when no external work is done; basal energy is used mainly for cellular transport work and synthesis. When energy is measured for patients by indirect calorimetry, this is referred to as *resting energy expenditure* (REE).[11,13] In normal subjects, a surprisingly small percentage (<5%) of this energy is spent on cardiac output and the work of breathing. In contrast, in individuals with chronic obstructive lung disease or patients who are being weaned from ventilators, the work of breathing may account for 15 to 20% of caloric expenditure. Thus, the REE measures the basal energy expenditure plus any extra required by the patient's disease process.

Caloric requirements (metabolic rate) are related to oxygen consumption by the formula:

$$\text{Metabolic Rate} = 4.83 \times O_2 \text{ consumption}$$

The metabolic rate is expressed in kcal/unit time, and O_2 consumption (VO_2) is expressed in liters of oxygen consumed/unit time. For example, the average resting 70-kg male consumes about 200 ml of oxygen/min or 288 liters of oxygen/d. This is equal to ~1450 kcal/d.

In the majority of adult surgical patients, energy requirements can be estimated without the need for complicated formulas. Basal metabolic rate can be estimated from body weight alone (Table 10.2) except in extremely obese individuals. Although metabolic rate varies with age and sex, these factors are not major determinants of caloric needs. When caloric requirements for surgical patients are determined, they provide the physician with an estimation of how many kilocalories should be provided to the patient. Resting energy expenditure (REE) can be measured at the bedside

TABLE 10. 2
Basal Metabolic Rates (BMR) in Adults

Weight in kg	BMR (kcal/day)
40	1150
50	1300
60	1450
70	1600
80	1750
90	1900
100	2050
110	2200
120	2350
130	2500

TABLE 10.3
Estimate of Increased Caloric Requirements
for the Average Adult (70-kg) Male

Condition	kcal/day
Basal state	1450
Uncomplicated post-op surgery	1500–1700
Sepsis syndrome	2000–2400
Multiple trauma patient on a ventilator	2200–2600
Major burn victim	2500–3000

by indirect calorimetry.[11,13] Patients must be either intubated or able to tolerate a hood placed over their face. Patients should not be receiving O_2 in excess of 70%. Protocols have been developed to measure the REE with just a 5-minute steady state because most patients are rarely quiet for longer than that.[11] In general, energy needs increase as the severity of the illness increases (Table 10.3). Therefore, the expenditure of kilocalories only minimally increases after elective surgery. The largest increase in energy expenditure occurs in patients with severe multiple trauma or major thermal injury. Yet the average-sized adult who sustains a major burn rarely requires more than 3500 kcal/d for maintenance.

While the provision of enough calories is beneficial, giving too many calories can lead to fatty infiltration of the liver. Excess calories, especially in the form of sugars, stimulate the secretion of insulin. This leads to mobilization and deposition of fat in the liver and interferes with hepatic function. This is important because critically ill patients do not fit the factors used to estimate their energy requirements. Rather indirect calorimetry must be used to measure energy expenditure accurately.[13]

The only significant source of usable energy in mammalian tissues is the carbon–hydrogen bond in carbohydrates, fats, and amino acids. Intracellular catalysis breaks this bond, liberating energy, forming carbon dioxide and water, and generating adenosine triphosphate (ATP) by adding a phosphate to adenosine diphosphate (ADP). This requires a pool of available phosphate. In chronically malnourished or previously starved patients, the phosphate pool may be depleted rapidly. As a result, serum phosphate levels may drop to dangerously low levels when dextrose is reintroduced to the metabolism. The resulting hypophosphatemia can lead to cardiac arrest and is known as the "refeeding syndrome." Although not usually seen in elective surgical patients, the refeeding syndrome may be encountered in cancer or in critically injured patients who receive only borderline nutrition.

The general formula for glucose oxidation is:

$$C_6H_{12}O_6 + 6\ O_2 \longrightarrow 6\ CO_2 + 6\ H_2O$$

The respiratory quotient (RQ) is the ratio between the volume of CO_2 produced to the volume of O_2 utilized. $RQ = VCO_2/VO_2$. The RQ for the above reaction is the ratio of 6 CO_2 to 6 O_2, which equals 1.00.

For the oxidation of fats, more oxygen is required to oxidize the multiple carbon–carbon bonds and, thus, the RQ is less than one. The oxidation of triglyceride is as follows:

$$2C_{51}H_{98}O_6 + 145\ O_2 \longrightarrow 102\ CO_2 + 98\ H_2O$$

The RQ is 102/145, which equals 0.703.

The RQ for proteins varies because the composition of the various amino acids will vary from protein to protein. However, on the average, the RQ for protein is 0.8. The RQ is greater than 1 in overfeeding, where the process of lipogenesis synthesizes fatty acids from glucose. Very high carbohydrate loads may increase the RQ to 1.3 or even 1.4. This indicates that a portion of the infused glucose is converted to fat. As a result, considerable quantities of carbon dioxide are produced that must be removed from the body via breathing. This can be harmful to patients with pulmonary insufficiency, who may develop a high pCO_2 and respiratory failure or who cannot be weaned from the ventilator during high glucose feedings. In that situation, the glucose supply should be decreased to prevent acid/base abnormalities. This is another reason why overfeeding can be dangerous.

PROTEIN/AMINO ACID METABOLISM

Protein is the building block of life, and it is the most important nutrient. About one sixth of total body weight is composed of proteins, about half of which are intracellular and half extracellular. Extracellular proteins include those that circulate in the bloodstream (e.g., albumin, transferrin, hemoglobin) and those that comprise the intracellular matrix, such as collagen and other fibrous proteins. In man and other mammals, dietary protein is the source of most amino acids. Intestinal absorption is the only pathway by which the body obtains exogenous amino acids, unless they are provided therapeutically by the intravenous route. Enterocytes are responsible for amino acid absorption. Small peptides are absorbed passively, but free amino acids are transported transcellularly via specific amino acid transport systems. Each transport system relates to an integral cell-membrane-associated transporter protein and functions to translocate the amino acid from the intraluminal environment into the cytoplasm.

Many factors, including nutritional status, have been shown to alter amino acid transport activities. The first is a reversible hyperplasia of the intestinal mucosa following a prolonged period of oral hyperalimentation. The hyperplasia results from an increase in the number of epithelial cells and their absorptive surface area and can increase amino acid uptake up to fivefold. The second mechanism is a reversible several-fold increase of specific amino acid transport activities. Surgery, infection, narcotics, and cancer also influence intestinal transport activity. Ileus, which may preclude enteral nutrition altogether, changes the brush border transport activity and may also alter nutrition and absorption in surgical patients.

Intraluminal digestion of dietary protein provides free amino acids, di-, tri-, and polypeptides that are absorbed by the small intestine and delivered so they can be incorporated into new proteins or other biosynthetic products. Excess amino acids

TABLE 10.4
Amino Acid Requirements for Adults

Condition	g of Amino Acids/kg/day
Uncomplicated post-op course	1–1.5
Sepsis syndrome	1.5–2.0
Multiple trauma patient on a ventilator	1.5–2.0
Major burn victim	2.0–3.0

are degraded, and their carbon skeleton is oxidized to produce energy or is incorporated into glycogen or free fatty acids. In addition to the metabolism of dietary amino acids, the existing proteins in the cell are continuously recycled, such that total protein turnover in the body is about 300 g/d. Vertebrates cannot reutilize nitrogen with 100% efficiency; therefore, obligatory nitrogen losses occur, mainly in the urine. The major form of nitrogen lost in the urine is urea (85%), with lesser amounts excreted as creatine and ammonia. Urinary nitrogen losses will diminish in individuals fed a protein-free diet but can never become zero because of the body's inability to completely reutilize nitrogen. In stressed patients, the ability to adapt to starvation is compromised to the extent that breakdown of body proteins continues at a substantial rate. Although skeletal muscle bears the brunt of this protein wasting, net proteolysis also occurs in the gut and liver, thereby compromising the ability to synthesize functional proteins. This may show clinically in patients as impaired wound healing, infection, immunoincompetence, or breakdown of the gut mucosal barrier.

Because of these obligatory nitrogen losses accentuated by catabolic disease states, stressed surgical patients have increased nitrogen and energy requirements. The recommended daily allowance for protein in the US. is 0.8 g/kg/d; this requirement may triple in critically ill patients because of their profound catabolism and inefficient protein economy (Table 10.4). One gram of nitrogen corresponds to 6.25 g of protein.

Nitrogen (N) balance can be calculated from the difference between the amount of nitrogen taken in by the patient and the amount of nitrogen lost in the urine, stool, skin, wounds, and fistula drainage. The nitrogen balance equals the amount of nitrogen given minus the amount excreted in the urine in a 24-h sample. Nitrogen loss from a high output fistula may also be measured, but stool, skin, and wound losses usually cannot be measured and are estimated (i.e., diarrhea) or not calculated.

$$N_{balance} = gN_{intake} - gN_{out}$$

Positive $N_{balance}$ indicates $N_{in} > N_{out}$

Examples: growth (child), anabolism after surgery

Negative $N_{balance}$ indicates $N_{in} < N_{out}$

Examples: starvation, injury, severe infection

Surgical patients develop a negative nitrogen balance because of their underlying disease process. This shows as wasting of skeletal muscle secondary to protein breakdown, a rate that exceeds protein synthesis.

Metabolism of amino acids (from enteral or parenteral feedings) generates ammonia, which is one of the most toxic and reactive compounds in physiologic fluids. Blood ammonia levels are generally kept at nontoxic concentrations (20 to 40 μM) primarily because the liver converts ammonia to urea, a nontoxic soluble compound. Much of the ammonia used for urea synthesis arises from nitrogen catabolism in extrahepatic tissues and requires excretory or transport amino acids, primarily glutamine and alanine. These molecules serve as vehicles for transporting ammonia in a nontoxic form from peripheral tissues to the visceral organs. In these organs, ammonia is reformed and then either excreted (kidneys) or detoxified (liver). From the intestinal tract, the large ammonia load escapes the systemic circulation because of the biochemical pathways in the liver that detoxify it. Only the liver has all the enzymes of urea synthesis, and these enzymes are located only in periportal hepatocytes. Thus, diseases affecting the periportal areas (cirrhosis) will cause an increase in the amount of ammonia that reaches the systemic circulation.

Urea is a highly soluble (2 mol/l), nontoxic compound with a high nitrogen content (47%). Its biosynthesis requires little energy. Healthy human subjects who ingest a western diet that generally contains red meat excrete about 30 g of urea daily. This may increase to over 60 g/d in catabolic surgical patients. Urea accounts for 85% of the total urinary nitrogen, with the remaining 15% contributed by ammonia and creatinine.

THERMOREGULATION

Alterations in the body's central thermostat, located in the hypothalamus, almost always occur in patients with systemic infection. Core temperature reflects the balance between heat production and heat loss, both of which may be altered in surgical patients. Most of the heat produced in the fasting basal state has its origin in the brain and in the abdominal organs. Increase in heat production follows an infectious challenge and is due to resetting the hypothalamic setpoint by pro-inflammatory cytokines such as interleukin-1. Such an increase in body temperature is thought to be an adaptive one, since rates of cellular reactions increase as a function of temperature. The metabolic rate increases about 10% for each degree (C°) increase in temperature.

HOMEOSTATIC RESPONSES AND ADJUSTMENTS TO STRESS

The body's defense mechanisms initiate a complex set of responses within moments of injury or insult. These responses are essential for survival and maintain body stability when key physiologic processes are threatened. They are rapidly set into motion by various aspects of the injury response, such as volume loss, tissue damage, fear, and pain. Factors that later re-initiate or perpetuate these responses include invasive infection and starvation. The stress response is related to the severity of

TABLE 10.5
Differences between Elective Surgery and Accidental Injury

Type of Insult	Elective Operation	Trauma
Tissue damage	Minimal; tissues are dissected and reapproximated with care	Substantial; tissues usually torn or ripped; debridement often necessary
Shock	Uncommon: preoperative hydration employed and fluid status is carefully monitored	Fluid resuscitation often delayed; blood loss can be substantial leading to shock
Pain/Fear/anxiety	Generally can be alleviated with pre-op medication or epidural analgesia	Generally present, unable to prevent ahead of event
Infection	Uncommon; prophylactic antibiotics often administered	Common with contamination, hypotension, and tissue devitalization
Stress response	Controlled, lesser magnitude; starvation better tolerated	Uncontrolled; occurs proportional to the magnitude of the injury; malnutrition is poorly tolerated

injury; that is, the greater the insult, the more pronounced the specific event. The human genome incorporates genes encoding the synthesis of key hormones and peptides, allowing the body's response to such insults to be remarkably robust. By evolution, these biologic responses result from a process favoring survival of the fittest in the struggle to preserve the species. From a teleological standpoint, these mechanisms are designed to benefit the organism, enhance recovery, and assure a relatively speedy return to health.

Whether caused by accidental or surgical injury, the body's responses are similar. However, a given setting will influence the extent and magnitude of the stress response (Table 10.5). Accidental injury is sudden and uncontrolled; tissues are torn, ripped, bruised, and contaminated. The associated volume and blood losses may be substantial, leading to tissue ischemia that, if prolonged, will result in damage and cellular death even in tissues that were not initially traumatized. Pain, excitement, and fear are generally heightened and uncontrolled. As a consequence, the magnitude of the physiologic response to major accidental injury can be considerable.

In contrast, the "selective" tissue trauma that occurs within an operating room is calculated, planned, and monitored. Although elective surgery causes pain, often interrupts food intake, and is generally associated with tissue removal, the periop-erative management of elective surgical patients is designed to attenuate such changes. Patients are seen preoperatively and evaluated to determine the need for nutritional support or additional medical therapy. Commonly, patients are hydrated and given prophylactic antibiotics and anxiolytics prior to surgery. In the operating room, the surgical site is prepared in a sterile fashion to minimize contamination, and a variety of physiologic parameters, i.e., blood pressure, pulse, and urine output, are continually monitored. Blood and blood products are available for replacement. During the actual operation, tissues are carefully dissected and incised to minimize tissue trauma, then reapproximated with care. Drugs are selected to block undesirable cardiovascular effects, and specific techniques, such as local, regional, or epidural anesthesia, are used to minimize postoperative pain. Accordingly, the responses to elective surgery are generally lesser in magnitude than those following major trauma.

Specific Components of the Stress Response

Volume Loss and Tissue Hypoperfusion

Following reduction of the circulating blood volume, the body immediately attempts to compensate in order to maintain adequate organ perfusion. Pressure receptors (aortic arch and carotid artery) and volume (stretch) receptors in the wall of the left atrium detect the fall in blood volume and start signaling the brain. Heart rate, stroke volume, and, therefore, cardiac output increase. Afferent nerve signals stimulate the release of both antidiuretic hormone (ADH) and aldosterone. ADH is produced by the posterior pituitary gland in response to hypotonicity and functions to increase water reabsorption in the kidney. Aldosterone is produced via the renin–angiotensin system that is activated when the juxtaglomerular apparatus in the kidney notices a fall in pulse pressure. Aldosterone increases renal sodium reabsorption and thereby conserves intravascular water. However, these mechanisms are only partially effective and severe hemorrhage without adequate resuscitation leads to a prolonged low-flow state. Under these circumstances, oxygen delivery is inadequate to meet tissue demands, and the cell switches to anaerobic metabolism, leading to lactic acidosis.

Tissue Damage

Body tissue injury appears to be the most important factor initiating the stress response. Hypovolemia and malnutrition may act synergistically with tissue destruction but, in and of themselves, do not initiate the hypermetabolic/hypercatabolic response unless the conditions result in infection or tissue injury. However, prolonged underperfusion may lead to ischemia, cellular death, and the release of toxic products that can start the "stress" response. Afferent neural pathways from the wound signal the hypothalamus that injury has occurred; tissue destruction is generally sensed in the conscious patient as pain. Efferent pathways from the brain are immediately triggered and stimulate a number of responses designed to maintain homeostasis.

Pain and Fear

Pain and fear are established components of the stress response. Both lead to excessive production of the catecholamines preparing the body for the "fight or flight" response.

Lack of Nutrient Intake and the Consequences of Malnutrition

The metabolic response to injury and surgery requires increased energy expenditure. In many patients undergoing surgery, nutrient intake is inadequate for a period of time (1 to 5 days) following the operation. If energy intake is less than energy expenditure, oxidation of body fat stores and erosion of lean body mass will occur, with a resultant loss of weight. Body glycogen stores are limited and are depleted within 24 to 36 hours. Consequently, glucose, which is essential for the central nervous system and for white blood cells, must be synthesized *de novo*. Amino acids, released principally by skeletal muscle, are the major gluconeogenic precursors. Most injured patients can tolerate a loss of 15% of their preinjury weight without a significant increase in the risk of surgery. But when weight loss exceeds 15% of body weight, the complications of malnutrition interact with the stress process, impairing the body's ability to respond appropriately to the injury, and may weaken subsequent responses to added stresses such as infection.

A major goal of nutritional support in the trauma patient is to aid host defense by balancing the energy and nitrogen expenditures that occur following injury. In contrast to what happens in injured patients, the catabolic and hypermetabolic responses occurring after elective operations are of a lesser magnitude. There is less tissue destruction, and the neurohormonal/inflammatory response is less intense. Consequently, well-nourished patients undergoing major elective operations do not require nutritional support after surgery unless there will be no food intake for more than 7 days.

Invasive Infection

A major complication observed in surgical patients is infection. Most patients, particularly those in intensive care units, are exposed to a variety of infectious agents in the hospital. Normal barrier defense mechanisms are disrupted by multiple indwelling pieces of plastic, such as intravenous catheters and nasotracheal and nasogastric tubes. These plastic intruders may be immunosuppressive by themselves. Breakdown of skin and mucous membrane opens portals of entry for bacteria. Infection alone may initiate catabolic responses that are similar to, but not the same as, those described following injury in noninfected patients. Both processes cause fever, hyperventilation, tachycardia, accelerated gluconeogenesis, increased proteolysis, and lipolysis, with fat utilized as the principal fuel source. Inflammatory cells release a variety of soluble mediators that aid host resistance and wound repair. Undernutrition may, however, compromise the available host defense mechanisms and thereby increase the likelihood of invasive sepsis, multiple organ system failure, and death.

DETERMINANTS OF HOST RESPONSES TO SURGICAL STRESS

The host response to surgical stress comes from the interaction of the individual patient with the stressful stimulus. The host receives and integrates neural and humoral signals from the injury and mounts an appropriate response from the interaction of multiple organ systems. The nature, intensity, and duration of the stress determine what mediators are activated and the resulting functional changes. The response following a minor elective operation is similar to a comparable, brief period of fasting and bed rest. On the other hand, a major burn results in a long-lasting, severe drain on the body's energy and protein stores, a drain that is only ended with wound closure and resolution of sepsis. The differences between the body's response to simple starvation and major stress are shown in Table 10.6.

TABLE 10.6
Metabolic Differences between the Response to Simple Starvation and to Injury

Simple Starvation		Severe Injury
—	Basal metabolic rate	++
—	Presence of mediators	+++
Fat	Fuel used	Mixed
+++	Production of ketones	±
+	Urea formation	+++
_	Nitrogen balance	_ _ _
+	Gluconeogenesis	+++
+	Muscle breakdown	+++
+	Hepatic protein synthesis	+++

Body Composition

Posttraumatic nitrogen excretion is directly related to the mass of body protein.[12] The balance of nitrogen intake vs. output serves as a marker of protein metabolism. The net loss of a certain amount of nitrogen from the body implies the net breakdown of the corresponding amount of protein. In women, the size of the skeletal muscle mass is about one half that of age-matched men. Thus, in a muscular young man, nitrogen losses are most marked after injury, but they represent a smaller fraction of the body nitrogen than in women, in whom the absolute loss is least but the fractional loss is greatest.

Nutritional Status

A strong relationship between protein breakdown and postoperative complications has been demonstrated in nonseptic, nonimmunocompromised patients undergoing elective major gastrointestinal surgery.[5,7,18] Protein-depleted patients had significantly lower preoperative respiratory muscle strength and vital capacity, an increased incidence of postoperative pneumonia, and longer postoperative hospital stays. Impaired wound healing as well as decreased respiratory, hepatic, and muscle function in protein-depleted patients awaiting surgery have also been reported.

Age

Many of the changes in the metabolic responses to surgical illness that occur with aging can be attributed to alterations in body composition and to longstanding patterns of physical activity. Although weight remains more or less stable, fat mass tends to increase with age, whereas muscle mass tends to decrease. The loss of strength that accompanies immobility, starvation, and acute surgical illness may have marked functional consequences. The capacity of muscle to serve as a substrate source may be limited during prolonged illness in the elderly patient, and muscle strength may rapidly become inadequate for respiratory and other vital muscle functions that lead to a patient's inability to wean from the ventilator or to ambulate.

The change in resting energy expenditure that occurs with aging comes from, in large part, the change in body composition, specifically the decrease in muscle mass.[10,13] After the limited stress of an elective operation, increases in energy expenditure are independent of age. Endocrine responses to an elective operation or to trauma appear unaffected in older patients in terms of plasma cortisol levels and the urinary excretion of adrenaline, noradrenaline, and 17-hydroxycorticosteroids.

The prevalence of cardiovascular and pulmonary disease increases with age. Diminished arterial compliance, impaired vasoconstriction, altered autonomic function, sensitivity to catecholamines, and decreased baroreflex sensitivity may all impair the maintenance of cardiovascular homeostasis during acute surgical illness. Thus, the delivery of oxygen to the tissues may be impaired in the elderly at every step of the oxygen transport pathway and may be inadequate when oxygen demands are highest. The physiology of aging, in general terms, is characterized by a diminished response to a change in homeostasis and diminished ability to restore homeostasis.

Gender

Observed differences between the metabolic responses of men and women in general reflect differences in body composition. Lean body mass, expressed as a proportion of body weight, is lower in women than in men, and this difference is thought to account for the lower net loss of nitrogen in women after major elective abdominal surgery.

MEDIATORS OF THE STRESS RESPONSE

The operative stresses (elective injury) or trauma (accidental injury) mediate a response comprised of two components: a neurohormonal response and an inflammatory response. These work together to determine the magnitude of the response. The principal counterregulatory hormones involved are the catecholamines, the corticosteroids, and glucagon. The inflammatory component of injury involves the local elaboration of cytokines and the systemic activation of humoral cascades, involving complement, eicosanoids, and platelet-activating factor. These mediators stimulate angiogenesis, white cell migration, and ingrowth of fibroblasts. During elective surgical procedures, the inflammatory response remains localized within the wound, and only insignificant amounts of mediators find their way into the systemic circulation. Following accidental injury where there is massive tissue destruction or prolonged hypotension with cell injury, these mediators may be produced locally in the wound in excessive amounts, resulting in "spillover" into the systemic circulation. In addition, cells in other tissues (such as Kupffer cells in the liver) may become activated to produce these mediators.[8,21] Such responses can lead to a systemic response in which these mediators cause detrimental effects, such as hypotension and organ dysfunction.

Counterregulatory Hormones

Following moderate to severe injury, one sees a marked rise in the elaboration of the hormones glucagon, glucocorticoids, and epinephrine. During the ebb phase of injury, the sympathoadrenal axis helps maintain the pressure–flow relationships necessary for an intact cardiovascular system. With the onset of hypermetabolism, characteristic of the flow phase, these and other hormones exert a variety of metabolic effects. Glucagon signals the hepatocytes to produce glucose from hepatic glycogen stores and gluconeogenic precursors. Cortisol mobilizes amino acids from skeletal muscle and increases hepatic gluconeogenesis. The catecholamines stimulate hepatic glycolysis and gluconeogenesis and increase lactate production from skeletal muscle. Catecholamines also increase metabolic rate and stimulate lipolysis. The level of growth hormone is elevated, even in the presence of hyperglycemia, and thyroid hormone levels are reduced to low-normal concentrations. Infusion of these hormones into normal subjects reproduces many of the metabolic alterations characteristic of injury.[3,16]

Cytokines

Cytokines are produced at the site of injury by endothelial cells and by diverse immune cells throughout the body, and, therefore, occupy a pivotal position in the stress response.[8,21] Cytokines differ from systemic hormones in that they are produced not by one but by a variety of cell types and can exert their effects locally via direct cell-to-cell communication ("networking") in a paracrine or autocrine fashion. Cytokines can also stimulate the production of other cytokines in cascade-like fashion, thereby amplifying and diversifying the effects of the original cytokine. Occasionally, when produced in excess, cytokines may "spill over" into the systemic circulation and become detectable in the bloodstream. Under these circumstances, cytokines produce systemic responses via endocrine mechanisms.

The cytokines that seem to play the most important roles in regulating the metabolic response to injury are tumor necrosis factor-a (TNF, cachectin), interleukin-1 (IL-1), interleukin-2 (IL-2), interleukin-6 (IL-6), and interferon-g (IFN). These polypeptide signalers, produced by the organism in response to tissue injury, necrosis, bacteremia, or endotoxemia, induce both adaptive (e.g., stimulation of the acute phase response) and adverse (organ dysfunction) responses following severe injury. The production of cytokines is likely to be at maximum when the injury is most severe. Under these circumstances, local cytokine production may gain access to the systemic circulation and trigger detrimental responses, such as hypotension and organ dysfunction. Cytokine effects on the microvasculature produce an intense inflammatory reaction leading to ischemic and hemorrhagic necrosis.

TNF is the primary mediator of the systemic effects of endotoxin, producing anorexia, fever, tachypnea, and tachycardia at low doses, and hypotension, organ failure, and death at higher doses. TNF is produced primarily by macrophages, but lymphocytes, Kupffer cells, and other cell types have also been identified as sources of TNF. In healthy humans, plasma levels of TNF are quite low, generally ranging from 0 to 35 pg./ml. Concentrations in tissues are likely to be higher. In animal models, stimulation of TNF-producing cells with endotoxin induces both transcription and translation of the protein within minutes, with detection in serum after 20 minutes. In humans and animals, TNF levels peak between 90 minutes and 2 hours after injection of endotoxin. TNF then leads to IL-1 release.

IL-1, like TNF, has a variety of pro-inflammatory activities. A single low *in vivo* dose of IL-1 results in fever, neutrophilia, hypozincemia, increased hepatic acute phase protein synthesis, decreased albumin synthesis, anorexia, sleep, and release of ACTH, glucocorticoids, and insulin. At higher doses, IL-1 induces hypotension, leukopenia, tissue injury, and death in a manner characteristic of septic shock. IL-1 induces many of the same biologic effects as TNF, and the combined effects of these two cytokines is often greater than that of either alone.[21]

IL-6 is now recognized as the primary mediator of altered hepatic protein synthesis known as the acute phase protein synthetic response. Glucocorticoid hormones augment the cytokine effects on acute phase protein synthesis. Elevated levels of IL-6 are found in the circulation of patients with infections, trauma, and cancer. Interferons are a family of proteins originally identified for their ability to inhibit viral replication in infected cells. IFN-g is a Type II interferon, totally unrelated in structure and function to the type I interferons, which have antiviral properties. IFN-g has the ability to upregulate the number of TNF receptors on various cell types.[8]

So if Cytokines Have Detrimental Effects, Why Do They Exist?

The genes that regulate cytokine biosynthesis are highly conserved; this indicates that these peptides confer a survival advantage following injury. Although excess production can be dangerous to the host, cytokines exert a number of beneficial effects that appear to outweigh the detrimental effects seen in extreme pathophysiologic states.[8]

Mobilization of Amino Acids and Stimulation of Acute Phase Protein Synthesis. Cytokines act in concert with other mediators to promote mobilization of amino acids from skeletal muscle. This response provides key nutrients to support cellular metabolism at a time when the organism generally cannot acquire food because of associated immobility and anorexia. The primary metabolic component of the acute phase response affected by IL-6 is a qualitative alteration in hepatic protein synthesis with a resulting change in plasma protein composition. Characteristically, the number of proteins acting as serum transport and binding molecules (albumin, transferrin) is reduced, and the quantity of acute phase proteins (fibrinogen, C-reactive protein) is increased. Acute phase proteins are produced for the purpose of reducing the systemic effects of tissue damage. While the true function of many acute phase proteins remains unclear, many act as anti-proteases, opsonins, or coagulation and wound-healing factors, and are likely to inhibit the generalized tissue destruction that is associated with the local initiation of inflammation. For example, increases in fibrinogen enhance thrombus formation, while anti-proteases reduce tissue damage caused by proteases that are released by dead or dying cells. C-reactive protein has been hypothesized to have a scavenger function. This acute phase response confers a significant survival advantage following injury and infection.

Elevation in White Blood Cell Count. Leukocytosis is recognized clinically by an elevation in the circulating white blood cell count, with an increase in the proportion of immature cells. This has been attributed, in part, to the release of neutrophils and their precursors into the circulation from the bone marrow. Both TNF and IL-1 produce an increase in the number and immaturity of circulating neutrophils due to a direct action on the bone marrow. Locally produced TNF and IL-1 are also chemotactic for neutrophils.

Hypoferremia and hypozincemia. Serum iron and zinc levels are reduced in septic patients, an event that is cytokine-mediated. The decrease in serum iron is probably important in protecting the host against various bacteria. The reduction of iron can inhibit the growth rate of microorganisms that have a strict requirement for iron as a growth factor. Both TNF and IL-1 have been shown to produce hypoferremia, hypozincemia, and other alterations in trace element metabolism.

Localization of the Wound/Inflammatory Site. Isolation or containment of tissue injury may be an important response designed to minimize systemic effects resulting from the inflammation at the trauma site. This is accomplished by vasodilation, migration of neutrophils and monocytes to the wound, initiation of the coagulation cascade, and proliferation of endothelial cells and fibroblasts in later stages of wound healing. These effects function to confine the insult as much as possible and activate defense mechanisms to minimize adverse systemic effects, such as cardiovascular collapse and subsequent organ failure. Cytokines are involved in all of these functions. The wound becomes an organ of cytokine production in which local metabolism is controlled in part by cytokines.

Fever/Subjective Discomfort. Fever is the systemic response to invading microorganisms and their toxins elicited by changes in the microenvironment of the anterior hypothalamus. These febrile responses are cytokine-mediated. Fever has both beneficial and detrimental effects. It is generally believed that the generation of fever by endogenously produced substances has adaptive value and imparts a survival advantage for the organism. Fever induced by the injection of cytokines in humans is associated with symptoms of malaise, myalgia, headache, and joint pain. These constitutional symptoms are likely to be beneficial because they encourage the "sick" organism to seek shelter, safety, and rest, and to avoid additional stressful situations.

The Gut Mucosal Barrier Dysfunction as a Mediator of the Stress Response

Under certain circumstances, the gut may become a source of sepsis and serve as the motor of the systemic inflammatory response syndrome. The maintenance of an intact brush border and intercellular tight junctions prevents the movement of toxic substances into the intestinal lymphatics and circulation; these functions may become altered in critically ill patients. Maintenance of a gut mucosal barrier that effectively excludes luminal bacteria and toxins requires normal perfusion, an intact epithelium, and normal mucosal immune mechanisms.

Microbial translocation is the process by which microorganisms migrate across the mucosal barrier and invade the host. Translocation can be promoted in three general ways: 1) by altered permeability of the intestinal mucosa, 2) by decreased host defense, and 3) by an increased number of bacteria within the intestine.

Hemorrhagic shock, hypoxemia, sepsis, distant injury, or administration of cell toxins all lead to altered permeability. Glucocorticoid administration, immunosuppression, or protein depletion decrease host defense. Bacterial overgrowth, intestinal stasis, and even feeding bacteria to experimental animals increase the luminal bacterial count and may cause translocation. Even if actual migration of whole bacteria does not occur, endotoxins may be absorbed by the gut and enter the portal venous system. A number of retrospective and epidemiological studies have associated infection in specific patient populations with bacterial invasion from the gut.[2,9,25–27] These studies have demonstrated an increase in mucosal permeability in normal volunteers who received endotoxin and in infected burn patients. Because many of the factors that facilitate bacterial translocation occur simultaneously in surgical patients and because their effects may be additive or cumulative, patients in an intensive care unit may be extremely vulnerable to the invasion of enteric bacteria or to the absorption of their toxins. Such patients do not generally receive enteral feedings, and current parenteral therapy may result in gut atrophy. Methods currently used to support critically ill patients neither facilitate repair of the intestinal mucosa, nor maintain gut barrier function.

ELECTIVE OPERATIONS

PHYSIOLOGIC RESPONSES TO SURGERY

The physiologic responses to surgical stress are multiple and complex. One of the earliest consequences of a surgical procedure is the rise in circulating cortisol that occurs in response to a sudden outpouring of adrenocorticotropic hormone (ACTH) from the anterior pituitary gland. The rise in ACTH stimulates the adrenal cortex to elaborate cortisol that remains elevated for 24 to 48 hours after the operation.[3] Cortisol has generalized effects on tissue catabolism and mobilizes amino acids from skeletal muscle that provide substrates for wound healing and serve as precursors for the hepatic synthesis of acute phase proteins. The adrenal medulla is stimulated through the sympathetic nervous system, with elaboration of epinephrine and norepinephrine. These circulating neurotransmitters play important roles in circulatory stability and elicit metabolic responses if the augmented secretion rate continues over a prolonged period of time.

The neuroendocrine responses to surgery also modify the mechanisms that regulate salt and water excretion. Alterations in serum osmolarity and tonicity of body fluids secondary to anesthesia and operative stress stimulate the secretion of aldosterone and antidiuretic hormone (ADH). Aldosterone is a potent stimulator of renal sodium retention, whereas ADH stimulates renal tubular water reabsorption. Thus, the ability to excrete a water load after elective surgical procedures is lost, and weight gain secondary to salt and water retention is usual following an operation. Edema occurs in all surgical wounds and accumulates in proportion to the extent of tissue dissection and local trauma. This "third-spaced" fluid eventually returns to the circulation and is eliminated by diuresis 2 to 4 days after the operation.[6]

The endocrine pancreas also responds following an elective operation. In general, insulin elaboration is diminished, and glucagon concentrations rise. This response may be related to increased sympathetic activity or to the rise in circulating epinephrine that suppresses insulin release. The rise in glucagon and the corresponding drop in insulin are potent signals to accelerate hepatic glucose production, and, with the help of epinephrine and glucocorticoids, gluconeogenesis goes up.

The period of catabolism initiated by surgery and characterized by a combination of inadequate nutrition and alteration of the hormonal environment has been termed the "adrenergic–corticoid phase." This phase generally lasts 1 to 3 days and is followed by the "adrenergic–corticoid withdrawal phase," also lasting 1 to 3 days. The next period is the onset of anabolism, which occurs at a variable time in the patient's convalescence. In general, with no postoperative complications, this phase starts 3 to 6 days after an abdominal operation of the magnitude of a colectomy or gastrectomy; it is often signaled by the appearance of appetite, gastrointestinal activity, and commencement of oral feedings. The patient then enters a long period of positive nitrogen balance and weight gain. Protein synthesis is increased as a result of sustained enteral feedings, and lean body mass and muscular strength return.

NUTRITIONAL SUPPORT OF ELECTIVE SURGICAL PATIENTS

Most patients undergoing elective operations are adequately nourished. Unless the patient has suffered significant preoperative weight loss (>10 to 15%), or has a major intraoperative or postoperative complication, solutions containing 5% dextrose may be administered for 5 to 7 days before the return of feeding with no detrimental effect on outcome. Therefore, the increased cost of feedings and the potential complications associated with intravenous nutrition cannot be justified in these patients. The use of jejunal feedings in the postoperative period may be useful, especially for those undergoing extensive upper gastrointestinal surgery.

Nutritional Assessment

The two major objectives of nutritional assessment are: 1) determining the patient's nutritional status and 2) determining energy, protein, and other nutrient requirements. Part of the medical history and physical examination should be focused to inquire about associated diseases, history of weight loss, and to establish the diagnosis of cachexia, protein-energy malnutrition, or specific nutrient deficiencies. Measurements of skin fold thickness are helpful to determine fat mass, and a 24-hour urine collection with measurement of creatinine allows determination of the creatine–height index (CHI), a factor proportional to the size of muscle mass.[21]

Immunologic status has been used to evaluate nutritional status. Total peripheral lymphocyte counts, delayed hypersensitivity using a skin-test response to common antigens, and lymphocyte transformation have all been used as indicators of immunocompetence in the critically ill patient. The depressed immune function often returns to normal with nutritional repletion, but altered immunologic responses are not specific for nutritional deficiencies and can be observed in patients with advanced malignant disease or in those who have had a severe injury. Serum albumin, prealbumin, and transferrin are the most common serum proteins measured and correlate well with body protein deficiency.

Determining Nutritional Requirements

Total energy requirements are based on several factors: 1) the basal metabolic rate; 2) the degree of stress imposed by the disease process; and 3) the amount of energy expended with activity. Standard nomograms relate normal metabolic requirements to a person's age, sex, height, and weight. The principal determinants of nitrogen requirement in surgical patients are total energy intake, nitrogen intake, and the metabolic state of the patient. Malnourished patients with intact protein-conserving mechanisms can achieve nitrogen equilibrium when 7 to 8% of the total caloric needs are provided as protein. This translates into a calorie-to-nitrogen ratio of approximately 350 kcal/g nitrogen. Hypermetabolic, catabolic patients, on the other hand, have a diminished protein economy and require much more protein.

Routes of Feeding

For patients who can eat and who have a functioning gastrointestinal tract, adequate nutrition can best be provided by a regular hospital diet. The enteral route of feeding is always preferred, and a variety of enteral diets are available for patients who have a functioning intestinal tract and who will not or cannot eat. In these patients, nasogastric or nasojejunal feedings may be indicated.[17,18] For patients with a diseased or nonfunctioning gastrointestinal tract, parenteral nutrition may be necessary. Peripheral venous feedings provide dilute nutrients in a large fluid volume and rely on fat emulsions as a principal calorie source. Central venous feedings consist of hypertonic glucose and amino acid solutions infused through a catheter placed in the superior vena cava. Adequate calories can be administered in a small fluid volume, but this method of feeding requires placement and care of a central venous catheter and its possible infectious complications (Table 10.8).

TRAUMA

GENERAL OVERVIEW AND TIME COURSE OF THE INJURY RESPONSE

In the 1930s, Cuthbertson[6] described the time course for many of the post-traumatic responses and two distinct periods were identified. The early "ebb" or shock phase was usually brief (12 to 24 hours) and occurred immediately after injury. Blood pressure, cardiac output, body temperature, and oxygen consumption were reduced. These events were often associated with hemorrhage and resulted in hypoperfusion and lactic acidosis. With restoration of blood volume, ebb-phase alterations gave way to more accelerated responses. The "flow" phase was then characterized by hypermetabolism, increased cardiac output, increased urinary nitrogen losses, altered glucose metabolism, and accelerated tissue catabolism. The flow-phase responses to accidental injury are similar to those seen following elective operation. The response to injury, however, is usually much more intensive and extends over a long period of time.

CHARACTERISTICS OF THE FLOW PHASE OF THE INJURY RESPONSE

Hypermetabolism

Hypermetabolism is defined as an increase in basal metabolic rate (BMR) above that predicted on the basis of age, sex, and body size. Metabolic rate is usually determined by measuring the exchange of respiratory gases and by calculating heat production from oxygen consumption and carbon dioxide production (indirect calorimetry). The degree of hypermetabolism (increased oxygen consumption) is generally related to the severity of the injury. Patients with long-bone fractures have a 15 to 25% increase in metabolic rate, whereas the metabolic needs of patients with multiple injuries rise by 50%. Patients with severe burn injury (greater than 50% of BSA) have resting metabolic rates that may reach twice basal levels. These metabolic rates in trauma patients are contrasted with those in elective postoperative patients, who rarely raise their BMR by more than 10 to 15%.

Altered Glucose Metabolism

Hyperglycemia commonly occurs following injury, and the elevation of fasting blood sugar levels generally parallels the severity of stress in the ebb phase. During the ebb phase, insulin levels are low and glucose production is only slightly elevated. Later, during the flow phase, insulin concentrations are normal or elevated; yet, hyperglycemia persists. This phenomenon reflects insulin resistance. Hepatic glucose production is increased, and studies show that much of the new glucose generated by the liver arises from 3-carbon precursors released from peripheral tissues.

Measurements of substrate exchange across injured and uninjured extremities of severely burned patients indicate that glucose is used in small amounts by the uninjured extremity. In contrast, the injured extremity extracts large amounts of glucose.[2] The wound converts most of the glucose to lactate, which is recycled to the liver in the Cori cycle. Uninjured volunteers (controls) are able to dispose of an exogenous glucose load much more readily than injured patients.[21] Moreover, the quantity of insulin elaborated by the patients was greater than in control subjects; nonetheless, these rising insulin concentrations failed to increase glucose clearance in these patients. Other studies have demonstrated a failure to suppress hepatic glucose production in trauma patients during glucose loading or insulin infusion. Glucose loading or insulin infusion usually inhibits hepatic glucose production in normal subjects. Thus, profound insulin resistance occurs in injured patients.

Alterations in Protein Metabolism

Extensive urinary nitrogen loss occurs following major injury; its magnitude is related to the extent of the trauma, but also depends on the patient's previous nutritional status, as well as the age and sex of the patient, because these factors determine the size of the muscle mass. In unfed patients, protein breakdown rates exceed synthesis, with negative nitrogen balance resulting. Providing exogenous calories and nitrogen increases protein synthesis, but even with a slightly positive energy balance, the nitrogen loss is not overcome.[12] Only providing a higher than normal nitrogen intake can overcome the nitrogen loss.

Skeletal muscle is the major source of the nitrogen lost in the urine following extensive injury. Although increased quantities of amino acids are released by muscle following injury, the composition of amino acid efflux is skewed toward glutamine and alanine, each of which comprises about one third of the total amino acids released by skeletal muscle. The kidney extracts glutamine, where it contributes ammonium groups for ammonia generation, a process that excretes acid loads. The gastrointestinal tract also takes up glutamine to serve as an oxidative fuel. The gut enterocytes convert glutamine primarily to ammonia and alanine, which are then released into the portal venous blood. The ammonia is taken up by the liver and converted to urea. Alanine may serve as a gluconeogenic precursor. Following elective surgical stress, glutamine consumption by the bowel and the kidney is accelerated in a reaction that appears to be regulated by the increased elaboration of the glucocorticoids. Although skeletal muscle releases alanine at an accelerated rate, the gastrointestinal tract and kidney also contribute increased amounts of alanine. The alanine is then synthesized by the liver into glucose and acute phase proteins. Hence, glutamine and alanine are important participants in the transfer of nitrogen from skeletal muscle to visceral organs; however, their metabolic pathways favor the production of urea and ammonia, both of which are lost as nitrogen.

Alterations in Fat Metabolism

To support the increased gluconeogenesis and inter-organ substrate flux, stored triglyceride is mobilized and oxidized at an accelerated rate. The resulting lipolysis, due to continuous stimulation of the sympathetic nervous system, is not readily overcome by glucose administration. Although mobilization and use of free fatty acids are accelerated in injured subjects, ketosis during brief starvation is blunted, and the accelerated protein catabolism remains unchecked. If unfed, severely injured patients rapidly deplete their fat and protein stores. This increases their susceptibility to added stresses of hemorrhage, surgery, and infection, and contributes to organ failure, sepsis, and death.

Consequences of Malnutrition

The metabolic response to injury increases energy expenditure. If energy expenditure exceeds intake, oxidation of body fat stores and erosion of lean body mass will occur, with resultant loss of weight. When body weight loss exceeds 10 to 15%, the complications of undernutrition interact with the disease process, increasing morbidity and mortality rates.

The major impact of nutritional support in the trauma patient is to aid host defense. Trauma patients are exposed to a variety of infectious agents in the hospital, and their injuries and requirements of care increase their risk of infection. Undernutrition may compromise the available host defense mechanisms and increase the likelihood of invasive sepsis, multiple organ system failure, and death. Additional consequences of malnutrition include poor wound healing, decreased mobility and activity, the occurrence of pressure sores and decubitus ulcers, altered gastrointestinal function, and the occurrence of edema secondary to reduced colloid osmotic pressure.

Priorities of Care

Resuscitation, oxygenation, and arrest of hemorrhage are immediate priorities for survival. Wounds should then be repaired or stabilized as expeditiously as possible. Nutritional support is an essential part of the metabolic care of the critically ill trauma patient and should be instituted before significant weight loss occurs. Adequate nutrition supports normal responses that optimize wound healing and recovery.

Goals of Nutritional Support

The majority of injured persons are normally nourished at the time of injury, but the increased metabolic demands following injury will quickly lead to a malnourished state if the patient is not nutritionally supported. With stabilization of the patient's condition and development of a care plan, nutritional support can be gradually initiated. The goal of nutritional support is the maintenance of lean body cell mass and the limitation of weight loss to less than 10% of preinjury weight. The nutritional requirements of the trauma patient can be determined as follows:

1. Determine basal metabolic rate (BMR) for age, sex, and BSA (BMR in kcal/d)
2. Determine the percentage of increased metabolic rate due to the injury
3. If active, add 25% x BMR for hospital activity (walking, physical therapy, sitting, treatment)
4. The sum of steps 1 to 3 is an estimated daily caloric requirement for maintenance of body weight
5. Multiply body weight (kg) by 2 to determine amount of protein needed (in grams); if the subject is grossly overweight, use 125% of ideal body weight as an estimate of metabolically active mass
6. Give approximately 70% of caloric requirement as glucose; give remaining caloric requirement as fat; reassess energy and nitrogen needs at least twice weekly
7. If nutritional support seems inadequate because of progressive weight loss, consider direct measurement of oxygen consumption or measurement of nitrogen balance

SEPSIS

GENERAL OVERVIEW AND TIME COURSE OF THE METABOLIC RESPONSE TO SEPSIS

The response patterns following major infection are less predictable than those following elective operations and trauma. The invasion of the body by microorganisms initiates many host responses, including mobilization of phagocytes, an inflammatory response at the local site, fever, tachycardia, and other systemic responses. Systemic events during the hyperdynamic phase of sepsis can be categorized into two general types of response: 1) those related to the host's immunologic defenses and 2) those related to the body's general metabolic and circulatory adjustments to the infection. The changes in metabolism relate to alterations in glucose, nitrogen, and fat metabolism, as well as the redistribution of trace metals.

Systemic Metabolic Responses

Severe infection is characterized by fever, hypermetabolism, diminished protein economy, altered glucose dynamics, and accelerated lipolysis much like what occurs in the injured patient. Anorexia is commonly associated with systemic infection and contributes to the loss of lean body tissue. These effects are compounded in the patient with sepsis by multiorgan system failure, which includes the gastrointestinal tract, liver, heart, and lungs.

Hypermetabolism

Oxygen consumption is usually elevated in the infected patient. The extent of this increase is related to the severity of infection, with peak elevations reaching 50 to 60% above normal. Such responses often occur in the postoperative and post-injury periods secondary to severe pneumonia, intra-abdominal infection, or invasive wound infection. If the patient's metabolic rate is already elevated to a maximal extent because of severe injury, no further increase will be observed. In patients with only slightly accelerated rates of oxygen consumption, the presence of infection causes a rise in metabolic rate that appears additive to the pre-existing state. A portion of the increase in metabolism may be ascribed to the increase in reaction rate associated with fever (Q10 effect: calculations suggest that the metabolic rate increases 10 to 13% for each elevation of 1°C in central temperature). On resolution of the infection, the metabolic rate returns to normal.

Altered Glucose Dynamics

The increase in glucose production seen in infected patients appears to be additive to the augmented gluconeogenesis that occurs following injury. For example, uninfected burn patients have an accelerated glucose production rate approximately 50% above normal; with the onset of bacteremia in similar individuals, hepatic glucose production increases to twice basal levels. Glucose dynamics following infection are complex; profound hypoglycemia and diminished hepatic glucose production have also been described in animals and human patients. Studies in animals and in human patients show that deterioration in gluconeogenesis is associated with more progressive stages of infection and may be related to alterations in splanchnic blood flow.

Alterations in Protein Metabolism

Accelerated proteolysis, increased nitrogen excretion, and prolonged negative nitrogen balance occur following infection, and the response pattern is similar to that described for injury. Amino acid efflux from skeletal muscle is accelerated in patients with sepsis, and this flux is matched by accelerated visceral amino acid uptake. In infected burn patients, splanchnic uptake of amino acids is increased 50% above rates in uninfected burn patients with injuries of comparable size. These amino acids serve as glucose precursors and are used for the synthesis of acute phase proteins. Studies in animals have demonstrated that an increase in hepatic amino acid uptake during systemic infection is due to an increase in the activities of specific amino acid transporters that reside in the hepatocyte plasma membrane.

Severe infection is often associated with a hypercatabolic state that initiates marked changes in interorgan glutamine metabolism. The cycle may begin with a breakdown in the gut mucosal barrier, resulting in microbial translocation. Bacteria and their endotoxins stimulate macrophages to release cytokines that activate the pituitary/adrenal axis. The release of cortisol stimulates glutamine synthesis and release by tissues, such as the lungs and skeletal muscle. The bulk of the glutamine is taken up by the liver at the expense of the gut. Acidosis frequently occurs in the patient with sepsis and serves as a signal for accelerated glutamine uptake by the kidney. Glutamine liberates an ammonia ion that combines with a hydrogen ion and is excreted in the urine, thus participating in acid–base homeostasis. Because the glutamine arises from skeletal muscle proteolysis, this complication of sepsis is yet another stimulus of heightened skeletal muscle breakdown.

Alterations in Fat Metabolism

Fat is a major fuel oxidized in infected patients, and the increased metabolism of lipids from peripheral fat stores is especially prominent during a period of inadequate nutritional support. Lipolysis is most probably mediated by the heightened sympathetic activity, itself a potent stimulus for fat mobilization and accelerated oxidation. Serum triglyceride levels reflect the balance between rates of triglyceride production by the liver and use and storage by peripheral tissues. Marked hypertriglyceridemia has been associated with Gram-negative infection on occasion, but plasma triglyceride concentrations in patients with sepsis are usually normal or low. This indicates enhanced clearance by other organs. Infected patients cannot convert fatty acids to ketones efficiently in the liver, and hence do not adapt to starvation like fasted, unstressed individuals. The low ketone levels of infection may be a consequence of the hyperinsulinemia associated with catabolic states.

Changes in Trace Mineral Metabolism

Changes in the balances of magnesium, inorganic phosphate, zinc, and potassium generally follow alterations in nitrogen balance. In early infection, the iron-binding capacity of transferrin is usually unchanged, but during severe pyrogenic infections, serum iron and zinc levels tend to go down. These decreases cannot be accounted for merely by losses of the minerals from the body. Rather, both iron and zinc accumulate within the liver, and this accumulation appears to be another host defense mechanism. The administration of iron to the infected host, especially early in the disease, is contraindicated because increased serum iron concentrations may impair resistance. Unlike iron and zinc, copper levels generally rise, and the increased plasma concentrations can be ascribed almost entirely to the increase in ceruloplasmin produced by the liver.

NUTRITIONAL REQUIREMENTS AND SPECIAL FEEDING PROBLEMS

As with all patients, the primary objectives of nutritional assessment are to evaluate the septic patient's present nutritional status and to determine energy, protein, and macro- and micronutrient requirements. Weight gain and anabolism are generally difficult to achieve during the septic process, but they do occur once the disease process has abated. Total energy requirements can be calculated using the stress equation; mild-to-moderate infections increase energy requirements by 20 to 30%, and severe infection increases caloric needs ~50% above basal levels. The most severe complication of sepsis is the multiple organ dysfunction syndrome, which may result in death. The current treatment of systemic infection consists of: 1) removal or drainage of the septic source; 2) use of appropriate antibiotics; 3) supportive therapy of specific organ failure, whether cardiac, pulmonary, hepatic, renal, or gastrointestinal; and 4) vigorous support of the host through nutritional means.

Respiratory Insufficiency

A common problem associated with systemic infection is oxygenation and elimination of carbon dioxide. Patients often require intubation and vigorous ventilatory support. Most of the enteral and parenteral formulas used to provide nutritional support for critically ill patients contain large amounts of carbohydrate, which generate large quantities of carbon dioxide following oxidation. Such a large carbon dioxide load may worsen pulmonary function or may delay weaning from the ventilator. Should this become a problem, the carbohydrate load should be reduced to 50% of metabolic requirements, with a fat emulsion administered to provide additional calories.

Renal Failure

When renal failure becomes progressive, the early use of hemodialysis minimizes the effects of uremia superimposed on the metabolism of sepsis. Metabolic studies in patients with acute and chronic renal failure have limited the intake of nonessential amino acids in an attempt to lower urea production. Proteins of high biologic value, but in much smaller quantities (<0.5 g/kg/d) than usually given, are administered along with adequate calories, usually in the form of glucose. When enteral feedings are not feasible, a central venous infusion of an essential amino acid solution and hypertonic dextrose provides calories and a small quantity of nitrogen. This regimen will reduce protein catabolism while simultaneously controlling the rise in BUN. During dialysis, protein intake may be liberalized, but the BUN should be maintained below 100 mg/dl.

Gut Dysfunction

Sepsis causes marked changes in gastrointestinal function. Ileus, the most common abnormality, can result from intra-abdominal disease or from the effects of bacteria elsewhere. Breakdown of the gut mucosal barrier, with translocation of luminal bacteria and their toxins, can initiate a prolonged hypermetabolic state.

Hepatic Failure

Hepatic dysfunction is a common manifestation of septicemia. The degree of dysfunction varies and may appear early as a slight elevation of liver enzymes, or it may cause severe jaundice and hyperbilirubinemia. Hepatic dysfunction generally resolves on resolution of the sepsis, but if the inflammatory process persists, adjustments in the feeding formula become necessary. The carbohydrate load is reduced to no more than 50% of metabolic requirements, and additional calories can be provided in the form of a fat emulsion. The patient should be observed for the presence of encephalopathy; if this complication occurs, then the protein load should also be reduced.

Cardiac Dysfunction

The myocardial dysfunction that occurs in sepsis may be secondary to the elaboration of cytokines such as TNF or IL-1 that depress myocardial activity directly. Alternatively, the heart failure may be secondary to pulmonary insufficiency, resulting in an increased pulmonary vascular resistance and right ventricular overload. Malnourished patients with sepsis may be sensitive to volume overload, and use of a concentrated solution of hypertonic dextrose mixed with amino acids may be indicated to maximize calories and to minimize volume. In addition, 20% fat emulsion instead of 10% can be administered to provide additional energy.

CHOICE OF NUTRITION IN SURGICAL PATIENTS: ENTERAL OR PARENTERAL?

There is a physiologic advantage to enteral nutrition, but preoperative nutritional repletion via the enteral route has not been studied as extensively as preoperative TPN. Although its use can be associated with the development of nausea, diarrhea, and distention, we recommend the use of enteral nutrition (via a feeding tube or as between-meal supplements) in malnourished patients. Candidates must have a functional GI tract and must be able to receive adequate amounts of calories and nitrogen. In many critically ill patients, gastrointestinal complications related to enteral nutrition are high, resulting in, at least in part, a lower nutrient intake than ordered.[11,18]

TOTAL PARENTERAL NUTRITION AS PRIMARY THERAPY

Patients with Enterocutaneous Fistulae

Patients with gastrointestinal–cutaneous fistulas represent the classical indication for TPN. In such patients, oral intake of food almost invariably results in increased fistula output that can lead to metabolic disturbances, dehydration, and death. Several comprehensive reviews have concluded that total parenteral nutrition clearly impacts the treatment and course of disease for patients with GI fistulas. The following conclusions can be drawn from studies evaluating the use of TPN in patients with enterocutaneous fistulae: 1) TPN increases the spontaneous closure rate of enterocutaneous fistulas, but does not markedly decrease the mortality rate in patients with fistulas (improvements in mortality are mainly due to improved surgical and metabolic care); 2) if spontaneous closure of the fistula does not occur, patients are better prepared for operative intervention because of the nutritional support they have received; and 3) certain fistulas (radiated bowel) are associated with a higher failure rate of closure than others and should be treated surgically after a defined period of nutritional support (unless closure occurs).

Patients with Short Bowel Syndrome

Prospective, randomized trials designed specifically to examine the impact of TPN on patients with short bowel syndrome have not been initiated, mainly because such patients have no choice but to receive TPN. Most of these patients, who would have certainly died prior to the availability of TPN, now survive for long periods on home parenteral nutrition. In selected patients with residual small intestine (at least 18 inches), post-resectional hyperplasia may develop to the point where they can tolerate enteral feedings. Recent studies by Wilmore and colleagues[26] have demonstrated that the requirement for TPN could be decreased or even eliminated in patients with short gut syndrome by providing a nutritional regimen consisting of supplemental glutamine, growth hormone, and a modified high-carbohydrate, low-fat diet. With such a regimen, absorption of nutrients was increased and stool output was decreased. In addition, TPN requirements were reduced by 50%, as were the costs associated with care of these individuals. Discontinuation of the growth hormone did not increase TPN needs in these patients, once they had undergone successful gut rehabilitation.

Patients with Hepatic Failure

Individuals with liver disease are generally malnourished because of excessive alcohol intake and diminished food intake. These individuals are protein depleted, yet are intolerant of protein because of their tendency to become encephalopathic with a high nitrogen intake. Because of liver damage and portasystemic shunting, these patients develop derangements in circulating levels of amino acids. The plasma aromatic/BCAA ratio is increased, thus favoring the transport of aromatic amino acids across the blood–brain barrier. These amino acids are precursors of neurotransmitters that contribute to lethargy and encephalopathy. Treatment of individuals who have liver failure with solutions enriched in branched-chain amino acids and deficient in aromatic amino acids results in improved tolerance to the administered protein and in clinical improvement in the encephalopathic state.

Patients with Major Thermal Injury

Aggressive nutritional support in patients with major burns seems to be associated with improved survival, particularly when increased amounts of dietary protein are provided.[2] Burned patients often require ventilator support or suffer from ileus that prevents using the GI tract for feeding. Even if the GI tract is functional, burn patients are often unable to eat enough because of frequent trips to the operating room combined with the anorexia of severe injury. Most burn authorities believe that aggressive nutritional support in patients with major thermal injury has influenced outcome in a positive manner.

Patients with Acute Renal Failure

TPN with amino acids of high biologic value may decrease the mortality in patients with acute renal failure.[1] The use of solutions containing high-quality amino acids can improve nitrogen balance and diminish urea production. The thinking is that provision of only essential amino acids allows maximal reutilization of nitrogen for the synthesis of nonessential amino acids and, thereby, helps prevent rapid rises in blood urea nitrogen. This translates into a decreased frequency of dialysis. However, there is no advantage and, in fact, there is a cost disadvantage to using essential amino acid solutions if the patient is already being dialyzed every other day. Therefore, a balanced standard amino acid formulation is recommended for patients already on dialysis.

TOTAL PARENTERAL NUTRITION AS SECONDARY THERAPY

Prolonged Ileus after Operative Procedure

Occasionally, patients will develop a prolonged ileus after an abdominal procedure that precludes the use of the intestinal tract as a route of feeding. Such an occurrence is generally unpredictable and its cause is uncertain, but it may be due to narcotic use. If the patient is unable to eat by the 7th postoperative day, TPN should be started. The ileus may persist for several weeks, especially in cancer patients. Although provision of TPN does not influence the disease process *per se*, it is beneficial because it prevents further erosion of lean body mass.

Acute Radiation and Chemotherapy Enteritis

Malnourished patients who receive abdominal or pelvic radiation or chemotherapy may develop mucositis and enterocolitis that preclude using the GI tract for prolonged periods of time. In such individuals, TPN should be provided until the enteritis resolves and oral feeding can be resumed.

Preoperative TPN in the Perioperative Setting

In general, one cannot justify the use of preoperative nutrition (particularly parenteral feedings) unless a clear benefit to the patient can be demonstrated. The results of prospective, randomized trials evaluating the efficacy of preoperative TPN are conflicting. Variations in the nutritional status of patients studied, differences in types of diseases, and in the type and length of nutritional support that has been administered, and failure to accrue enough patients keep these studies from avoiding type II statistical error. The following questions are important to consider. Does preoperative nutritional support decrease the morbidity and mortality associated with major operative procedures? How long should nutritional support be administered? What type of nutritional repletion should be administered?

One of the best studies to date evaluating the efficacy of preoperative TPN was published by the Veterans Affairs Total Parenteral Nutrition Cooperative Study Group.[5] More than 3500 patients requiring mainly elective abdominal surgery were entered into this prospective, randomized trial. They were initially screened for evidence of malnutrition using subjective criteria or by determining their Nutritional Risk Index Score, which included objective criteria such as percentage of weight loss and serum albumin level. The patients were further divided into categories assigning them to one of four groups: well-nourished, borderline malnourished, moderately malnourished, or severely malnourished. Patients in each malnourished category were randomized to receive at least 7 days of preoperative TPN or to proceed with surgery without preoperative TPN. Patients randomized to receive TPN received 1000 kcal/d in excess of calculated caloric requirements (overfeeding). Lipid was provided on a daily basis. One criticism of this study was that patients were allowed to eat in addition to receiving parenteral feedings, providing even more overfeeding.

Analysis of the data from this study indicated that there was no difference in short-term or long-term complications between groups. Infectious complications, including pneumonia, abscess, and line sepsis, were significantly higher for patients receiving TPN. (Is this because they were overfed and deposited fat in their livers?) Noninfectious complications (impaired wound healing) were significantly lower only in those patients receiving TPN who were in the severely malnourished group (>15% weight loss and serum albumin <2.9 mg%). Consensus now dictates preoperative TPN should be provided only to severely malnourished patients who cannot be nourished by the enteral route.

Contraindications to the use of preoperative TPN include patients needing emergency operations and those who are only mildly or moderately malnourished. In such patients, the GI tract should be utilized for postoperative feeding if at all possible. Nasojejunal or jejunostomy tubes should also be considered in patients undergoing major upper abdominal procedures if oral feedings are unlikely to be resumed for 7 to 10 days after surgery.

COMPOSITION OF TPN FORMULATIONS

Total parenteral nutrition solutions are administered through a central venous catheter that is generally inserted into the subclavian vein. By this procedure, the hyperosmolar solution is diluted quickly by a high flow system. The composition of a standard TPN solution is shown in Table 10.7. Patients receiving TPN should be monitored regularly by measuring blood sugar, serum electrolytes, and doing liver function tests. Elevations in serum glucose are common in surgical patients receiving TPN, because of stress and insulin resistance. Hyperglycemia can generally be controlled by adding insulin to the TPN formulation or by decreasing the amount of glucose in the solution. An injured man can maximally oxidize about 5 to 6 mg of glucose/kg/min, equal to approximately 500 to 600 g/d (2000 to 2400 kcal/d) for a 70-kg man. Glucose calories in excess of this amount lead to hyperglycemia and glycosuria and are converted to fat. The amounts of the various electrolytes that are provided to patients receiving TPN will vary depending on factors such as previous

TABLE 10.7
Standard Central Venous Solution Components

Composition

Amino acids	100 g
Dextrose	300 g
Total N	100/6.25 = 16 g
Dextrose kcal	300 g ×3.4 kcal/g = 1020 kcal
Fat	20 g × 9 kcal/g = 180 kcal
mOsms/liter	~ 1700–2000

Volume

10% amino acid solution	1000 ml
50% dextrose solution	600 ml
Fat emulsion 10%	400 ml
Electrolytes + vitamins + minerals	~75 ml
Total volume	~ 2075 ml

Electrolytes Added to TPN Solutions	Usual Concentration	Range of Concentrations
Sodium (mEq/L)	60	0–150
Potassium (mEq/L)	40	0–80
Acetate (mEq/L)	50	50–150
Chloride (mEq/L)	50	0–150
Phosphate (mEq/L)	15	0–30
Calcium (mEq/L)[a]	4.5	0–20
Magnesium (mEq/L)	5	5–15

[a] Added as calcium gluconate or calcium chloride; one ampoule of calcium gluconate = 1 gm of calcium = 4.5 mEq

nutritional and hydration status. Careful monitoring is critical because as new cell mass is accrued, the intracellular ions potassium and phosphate may accumulate, leading to severe hypokalemia or hypophosphatemia. This drop in potassium and phosphate is called the "refeeding syndrome." Hypophosphatemia can develop rapidly and is much more life threatening than hyponatremia. Critically low levels of phosphate can lead to irreversible cardiac arrest. Vitamin and trace mineral requirements must also be taken into consideration.

The use of lipids in TPN was developed to meet the requirement for essential fatty acids, such as linoleic acid, and to meet the full caloric needs in hypermetabolic patients, in recognition of the complications associated with infusing large amounts of dextrose (hepatic steatosis). Intravenous fat emulsions are comprised of soy or safflower oils (vegetable fat emulsions) that contain mostly long-chain triglycerides (LCT) comprised of fatty acids with 16- and 18-carbon chain lengths. Adding fat provides essential linoleic acid, inhibits lipogenesis from carbohydrates, and lowers the RQ, which may benefit patients with respiratory compromise. However, the high content of ω-6 polyunsaturated fatty acids (in particular linoleic acid) in these emulsions may have harmful effects. Linoleic acid is a precursor for the synthesis of various prostaglandins and leukotrienes, which can cause immunosuppression and suppress cytokine activity. Standard intravenous fat emulsions may alter the phospholipid composition of cell membranes of the reticuloendothelial system. This may lead to changes in membrane fluidity that in turn may impair clearance of bacteria and toxins. In addition, ω-6 polyunsaturated fatty acids may alter the local production of cytokines impacting chemotaxis. Newer nutritional methods of modifying the catabolic response to injury and infection propose the use of the ω-3 fatty acids, which may decrease eicosanoid biosynthesis and thereby diminish the vasoconstriction, platelet aggregation, and immunosuppression that can occur when ω-6 derivatives are administered. Omega 3 fatty acids may benefit the critically ill patient and have been added to enteral but not to parenteral formulae.

It is important to realize that fat is an important fuel source for critically ill patients. Septic and injured patients seem to utilize endogenous fat as a preferred energy source. This may be related in part to the fact that counter-regulatory hormones stimulate fat mobilization. These patients show a relative unresponsiveness to the carbohydrate administration inasmuch as the free fatty acid (FFA) mobilization is only marginally decreased and FFA oxidation is not suppressed, as is the case in "pure starvation." Notwithstanding glucose infusion to levels above energy expenditure, the hormonal milieu favors fat mobilization and oxidation.

POTENTIAL COMPLICATIONS OF TPN

Advances in technology, monitoring, and catheter care have greatly reduced the incidence of complications associated with the use of TPN. The establishment of a nutrition support team (physician, dietitian, nurse, pharmacist) and the recognition that such a team constitutes an important part of overall patient care have been key factors in reducing complications.

Complications of TPN do occur and can be divided into three types: mechanical, metabolic, and infectious (Table 10.8).

TABLE 10.8A
Mechanical or Technical TPN Complications

Type of Complication	Cause	Treatment
Pneumothorax	Puncture/laceration of lung pleura	Serial CXR's; chest tube if indicated
Subclavian artery injury	Penetration of subclavian artery during needle stick	CXR; serial monitoring of vital signs
Air embolism	Aspiration of air into the subclavian vein and R heart	Place pt in Trendelenberg and left lateral decubitus; aspirate air
Catheter embolization	Shearing off the tip when withdrawing catheter	Retrieve catheter transvenously under fluoroscopic guidance
Venous thrombosis	Clot formation in great vein secondary to catheter	Heparinization if clinically significant
Catheter malposition	Tip of catheter directed into IJ or opposite subclavian	Reposition under fluoroscopy

TABLE 10.8B
Metabolic TPN Complications

Type	Cause	Treatment
Hyperglycemia	Excessive glucose calories or glucose intolerance	Decrease glucose calories; administer insulin
Hypoglycemia	Sudden cessation of TPN	Bolus 50% glucose solution Monitor blood glucose
Carbon dioxide retention	Infusion of glucose calories in excess of energy needs	Decrease glucose calories and replace with fat
Hyperglycemic, hyperosmolar nonketotic coma (HHNC)	Dehydration from excessive diuresis	Discontinue TPN immediately; give insulin; monitor glucose/electrolytes
Hyperchloremic metabolic acidosis	Excessive chloride administration	Give Na and K as acetate salts
Azotemia	Excessive amino acid administration with inadequate calories	Decrease amino acids; increase glucose calories
Essential fatty acid deficiency	Inadequate essential fatty acid administration	Administer fat solution
Hypertriglyceridemia	Rapid fat infusion of decreased fat clearance	Slow rate of fat infusion
Hypophosphatemia Hypocalcemia Hypomagnesemia Hypokalemia	Inadequate administration of electrolyte in question	Increase administration
Bleeding	Vitamin K deficiency	Administer Vitamin K

TABLE 10.8C
Septic Complications of TPN

Type	Cause	Treatment
Line sepsis	Catheter tip infected	Remove catheter; antibiotics
Infection at skin site	Bacteria at site of catheter entry into skin	Remove catheter; local wound care

THE EFFECTS OF TPN ON THE GASTROINTESTINAL TRACT

The majority of studies that have examined the effects of TPN on intestinal function and immunity have been done in animals. These studies clearly demonstrate that TPN has a disadvantage that is related to intestinal disuse. A unique feature of TPN is that patients can remain on bowel rest for prolonged periods of time without concomitant malnutrition, thereby facilitating the study of intestinal disuse as an independent variable. In rats, TPN results in significant disruption of the intestinal microflora and in bacterial translocation from the gut lumen to the mesenteric lymph nodes. In addition, when stresses such as burn injury, chemotherapy, or radiation are added, animals on TPN have a much higher mortality. This implicates bacterial translocation as the cause. This suggests that TPN may, under certain circumstances, predispose patients to complications resulting from an increase in gut-derived infections.[25]

A provocative study in human volunteers demonstrated that, compared to enterally fed volunteers, individuals receiving TPN had an accentuated systemic response to endotoxin challenge.[9] This is consistent with an impairment in gut barrier function during parenteral feedings and an increased release of bacteria or cytokines, leading to pronounced systemic responses and possible multiple organ failures.

NUTRITIONAL SUPPORT OF THE GUT IN CRITICALLY ILL PATIENTS

The intestinal tract has long been considered an organ of inactivity following operation or injury. Ileus is generally present, nasogastric decompression is often necessary, and, typically, the gut is not used functionally in the immediate postoperative period. In the past, digestion and absorption were thought to be the only physiologic role of the gut. Disuse of the GI tract, because of starvation or nutritional support by TPN, may lead to numerous physiologic derangements, as well as changes in gut microflora, impaired gut immune function, and disruption of the integrity of the mucosal barrier. Thus, maintaining gut function in the perioperative period may be essential. Treatment strategies designed to support the gut during critical illness should be directed toward providing appropriate nutrition aimed at the preservation of mucosal structure and function. Presumably, such efforts will assist the gut in its role as a metabolic processing station and as a barrier.

NUTRITION SUPPORT: ENTERAL FEEDINGS

Enteral feedings are probably the best single method of maintaining gut function. The trophic effects of luminal nutrition are key and well documented even if relatively small amounts of nutrients are provided.

Gut-Specific Nutrients

It is now clear that the composition of the diet as well as the route of delivery plays an important role in maintaining gut structure and function. Several gut-specific nutrients have been studied, but glutamine has received the most attention. Glutamine has been classified as a nonessential, or nutritionally dispensable, amino acid. This categorization implies that glutamine can be synthesized in adequate quantities from other amino acids and precursors and, therefore, does not have to be included in nutritional formulas. With few exceptions, glutamine is present in oral and enteral diets at the relatively low levels characteristic of its concentration in most animal and plant proteins, i.e., constituting about 7% of total amino acids.

Several recent studies have, however, demonstrated that glutamine may be a conditionally essential amino acid during critical illness, particularly in supporting the metabolic requirements of the intestinal mucosa.[19,25–27] In general, these studies demonstrate that dietary glutamine is not required under ordinary circumstances, but supplementation is beneficial when glutamine depletion is severe or when the intestinal mucosa is damaged by insults, such as chemotherapy and radiation therapy. Addition of glutamine to enteral diets reduces the incidence of gut translocation, but the degree of reduction depends on the amount of supplemental glutamine and the type of insult studied. Glutamine-enriched TPN partially attenuates the villous atrophy that develops during parenteral nutrition. The use of intravenous glutamine in humans appears to be safe and effective, to diminish complications, and to reduce hospital stay.[19,26]

In contrast to glutamine, short-chain fatty acids (SCFAs) are the primary energy source for colonocytes. Diets enriched in SCFAs increase colonic DNA content and mucosal appearance as well as strengthen colonic anastomoses in rats.[25,26]

THE USE OF GROWTH FACTORS TO SUPPORT THE GUT MUCOSA

Epidermal growth factor (EGF) is one of several growth factors that promote mucosal growth, tissue repair, and regeneration. Epidermal growth factor is a polypeptide secreted by submaxillary glands and by Brunner's glands of the small intestine. EGF acts on the gastrointestinal mucosa by stimulating DNA synthesis, as evidenced by increased thymidine uptake.[26]

OTHER METHODS OF MODIFYING THE CATABOLIC RESPONSE TO SURGERY AND CRITICAL ILLNESS

Besides nutritional intervention, other methods of modifying the physiologic and biochemical responses to an elective operative procedure have been used to reduce stress and to provide insight into response mechanisms. Anesthetic techniques, both local and regional, block afferent signals from the wound and interrupt sympathetic nervous efferent signals to the adrenal gland and possibly the liver. The overall effect of such blockade is to reduce the magnitude of the stress response. The central nervous system blockade brought about by blocking the efferent limb of the neuroendocrine response in sympathectomized animals is another means to interrupt afferent signals stimulated by operative procedures.

More recent studies have documented the safety and efficacy of long-term exogenous recombinant growth hormone (GH) administration.[16] GH stimulates protein synthesis during hypocaloric feedings and increases renal retention of sodium and potassium. Whether there is synergism between specialized nutrition and GH administration requires further study. Cyclooxygenase inhibitors, such as aspirin and ibuprofen, attenuate the symptoms and endocrine responses that occur with critical illness but do not alter cytokine elaboration. Ultimately, it should be possible to selectively block the deleterious effects of excessive cytokines, yet preserve their beneficial effects.

TECHNIQUES OF NUTRITIONAL SUPPORT

TRANSNASAL (NASOGASTRIC AND NASODUODENAL) FEEDING CATHETERS

The use of transnasal feeding catheters for intragastric feeding or for duodenal intubation is a popular method for providing nutritional support by the enteral route. The stomach is easily accessed by the passage of a soft flexible feeding tube. Intragastric feedings provide several advantages for the patient. The stomach has the capacity and reservoir for bolus feedings. Feeding into the stomach results in the stimulation of the biliary/pancreatic axis that is probably trophic for the small bowel, and gastric secretions will have a dilutional effect on the osmolarity of the feedings, thereby reducing the risk of diarrhea. The major risk of intragastric feeding is the regurgitation of gastric contents with consequent aspiration into the tracheobronchial tree. This risk is highest for patients who have an altered mental sensorium or who are paralyzed.

The placement of the feeding tube through the pylorus into the fourth portion of the duodenum reduces the risk of regurgitation and aspiration of feeding formulas. For placement of a transnasal intraduodenal feeding catheter, the patient should be in the sitting position with the neck slightly flexed. This will allow for the passage of a lubricated 8 French polyurethane feeding catheter (with a stylette in place) through the patient's nose in a posterior and inferior direction, bringing the catheter to the level of the pharynx. The head is brought back to a neutral position, and the patient is instructed to swallow while the feeding catheter is passed into the esophageal lumen. The advancement of the catheter is continued for a distance of about 45 to 50 cm. The stylette is removed, and the position of the catheter is confirmed either audibly or radiographically before feedings are initiated. Tubes can be positioned fluoroscopically if necessary. The patient who frequently removes the feeding catheter may be a candidate for a feeding tube "bridle" placed around the palate.

A variety of enteral formulas are commercially available and can be adjusted for caloric, nitrogen, or fiber content.

TECHNIQUE FOR GASTROSTOMY PLACEMENT

A feeding gastrostomy should be considered for patients who require long-term enteral nutrition and who have an obstructed esophagus or whose head and neck cancers preclude eating. A temporary Stamm gastrostomy is popular for accessing the gastric lumen and can be performed at the time of any major abdominal procedure. A square purse-string suture (0 polypropylene) with a 1.5-cm diameter is placed in the body of the anterior wall of the stomach. The ideal location is in the midportion of the stomach closer to the greater curvature in a relatively avascular site. A second hexagonal purse-string suture placed 1 cm outside the first one follows this. The feeding catheter (often a Foley catheter) is then brought through the abdominal wall (left upper quadrant). Using the electrocautery, a stab wound is made in the anterior portion of the stomach directly in the center of the concentric purse-string sutures. The feeding catheter is introduced into the gastric lumen, and the inner concentric suture is secured in place. The second purse-string is then secured, inverting the gastric mucosa completely. The Foley catheter balloon is inflated, and the stomach is drawn upward toward the anterior abdominal wall. Placement of Lembert silk sutures in all four quadrants around the Foley catheter will secure the stomach to the anterior abdominal wall. The Foley is then secured to the skin by the placement of sutures at the base of the catheter exit site.

TECHNIQUE FOR PERCUTANEOUS ENDOSCOPIC GASTROSTOMY

Percutaneous endoscopic gastrostomy (PEG) to provide access for gastric feedings can be performed without the need for a celiotomy or general anesthesia. With intravenous sedation and a topical anesthetic, a gastroscope is passed through the mouth into the esophagus and stomach. The light from the gastroscope is then transilluminated through the anterior abdominal wall in the epigastrium or left subcostal area. The abdomen is prepped and draped, and a wheal of local anesthetic is injected right at the area of transillumination. A 1-cm vertical incision is made, and an angiocath or needle is inserted through the abdominal wall into the stomach under the direct vision of the gastroscope. A wire loop is passed through the catheter into the stomach and grasped with a snare passed through the gastroscope. The scope and wire loop are withdrawn through the mouth and the wire is looped through a corresponding wire on the end of the PEG tube. By pulling the wire out through the abdominal wall, the PEG tube is pulled in through the mouth and out of the anterior abdomen with the path through the esophagus followed by the gastroscope. With the aid of the gastroscope, one can then ensure that the inner part of the PEG tube is seated properly against the stomach wall. The tube protruding through the abdominal wall may be attached securely at the exit site to prevent accidental dislodgment. The tube can later be removed with just a hard pull.

FEEDING CATHETER JEJUNOSTOMY PLACEMENT AND WITZEL JEJUNOSTOMY

A feeding catheter jejunostomy should be considered following any major upper abdominal procedure, if enteral nutrition support is likely to be prolonged. The simplest method is a needle jejunostomy performed fairly quickly at the end of the definitive operation. A 14- or 16-gauge needle is used to create a tunnel subserosally approximately 30 to 40 cm distal to the ligament of Treitz, and then the needle tip is introduced into the jejunal lumen. A feeding catheter is inserted through the needle and advanced 30 to 40 cm distally into the bowel lumen to the desired location, and the needle is then withdrawn. The loop of the jejunum is anchored to the parietal peritoneum with permanent sutures. The catheter is then secured to the skin with nylon sutures.

The Witzel technique can also be used to perform a feeding jejunostomy. A loop of proximal jejunum, 20 to 30 cm from the ligament of Treitz is delivered into the wound. A purse-string suture is placed on the antimesenteric border of the bowel, and the intestinal wall is then punctured with the aid of electrocautery in the center of the purse-string suture. A red rubber catheter (14F) is inserted into the lumen of the jejunum and advanced distally. The purse-string suture is then secured in place.

A serosal tunnel is then constructed with 000-silk sutures from the catheter's exit site, extending 5 to 6 cm proximally. With the aid of a separate stab incision, the catheter is then delivered through the abdominal wall. The adjacent loop of intestine is then anchored with two 000-silk sutures spread over 2 to 3 cm to prevent twisting of the loop and possible obstruction. The catheter is secured to the skin with a 3–0 nylon suture. Jejunal feeding catheters can be used immediately for feeding purposes following the operation.

PERIPHERAL INTRAVENOUS FEEDINGS

Peripheral veins may be used for infusion of glucose, amino acid solutions, and fat emulsions. The solutions must be nearly isotonic to avoid peripheral vein sclerosis. Solutions of 10% glucose may be used to increase the efficacy of amino acid utilization. Fat emulsions can be administered simultaneously with glucose and amino acid solutions, because they provide an efficient fuel source and are isotonic. The major disadvantage of peripherally administered solutions is that only limited amounts of calories can be delivered within tolerated fluid volumes. Indications for peripheral vein feeding are: 1) use as a supplement when enteral feedings can only be partially tolerated because of gastrointestinal dysfunction; 2) use as a method of nutritional support when the gastrointestinal tract must be kept relatively empty for short periods during diagnostic work-up; and 3) to provide preliminary feedings prior to subclavian catheter insertion in patients requiring TPN.

CONCLUSION

Nutrition is extremely important to the care of surgical patients because of the changes in their metabolism that surgery, trauma, or sepsis induce. Changes in body composition occur with the loss of lean body mass, an increase in energy requirements, and a decreased resistance to infection. Although we can minimize these changes, we cannot stop them. Thus, knowledge of the changes and ways we can minimize them is essential.

REFERENCES

1. Abel RM, Beck CH, Abbott WM, Ryan JA, Barnett GO, and Fischer JE. Improved survival from acute renal failure following treatment with intravenous essential L-amino acids and glucose: results of a prospective, double-blind study. *N. Engl. J. Med.* 288:695–9, 1973.
2. Alexander JW, MacMillan BG, Stinnert JD, Ogle CK, Bozian RC, Fischer JE, Oakes JB, Morris MJ, and Krummel R. Beneficial effects of aggressive protein feeding in severely burned children. *Ann. Surg.* 192:505–17, 1980.
3. Bessey PQ, Watters JM, Aoki TT, and Wilmore DW. Combined hormonal infusion simulates the metabolic response to injury. *Ann. Surg.* 200:264–81, 1984.
4. Brennan MF, Pisters PWT, Posner M, Quesada O, and Shike M. A prospective randomized trial of total parenteral nutrition after major pancreatic resection for malignancy. *Ann. Surg.* 220:436–444, 1994.

5. Buzby GP and The Veterans Affairs Total Parenteral Nutrition Cooperative Study Group. Perioperative total parenteral nutrition in surgical patients. *N. Engl. J. Med.* 325:525–32, 1991.

6. Cuthbertson DP. The metablolic response to injury and its nutritional implications: retrospect and prospect. *J. Parenter. Enteral Nutr.* May/June, 3(3), 108–129, 1979.

7. Fan ST, Lo CM, Lai ECS, Chu KM, Liu CL, and Wong J.. Perioperative nutritional support in patients undergoing hepatectomy for hepatocellular carcinoma. *N. Engl. J. Med.* 331: 1547–1552, 1994.

8. Fong Y and Lowry. Cytokines and the cellular response to injury and infection, in *Care of the Surgical Patient (Critical Care).* Wilmore, DW et al. (Eds.) Scientific American, 1992, Trauma Section. Chapter 7.

9. Fong Y, Marano MA, Barber A, He W, Moldawer LL, Bushman ED, Coyle SM, Shires GT, and Lowry SF. Total parenteral nutrition and bowel rest modify the metabolic response to endotoxin in humans. *Ann. Surg.* 210:449–457, 1989.

10. Frankenfield DC, Cooney RN, Smith JS, and Rowe WA. Bioelectrical impedance plethysmographic analysis of body composition in critically injured and healthy subjects. *Am. J. Clin. Nutr.* 69:426–31, 1999.

11. Frankenfield DC, Sarson GY, Blosser SA, Cooney RN, Smith JS. Validation of a 5-minute steady state indirect calorimetry protocol for resting energy expenditure in critically ill patients. *J. Am. Coll. Nutr.* 15:397–402, 1996.

12. Frankenfield DC, Smith JS, and Cooney RN. Accelerated nitrogen loss after traumatic injury is not attenuated by achievement of energy balance. *JPEN.* 21:324–9, 1997.

13. Frankenfield DC, Wiles CE, Bagley S, and Siegel JH. Relationships between resting and total energy expenditure in injured and septic patients. *Crit. Care Med.* 22:1796–1804, 1994.

14. Hasselgren P-O. Catabolic response to stress and injury: implications to regulation. *World J. Surg.* 24:1452–1459, 2000.

15. Hautamaki RD and Souba WW. Principles and techniques of nutritional support in the cancer patient, in *Atlas of Surgical Oncology.* Karakousis CP, Copeland EM, and Bland KI (Eds.). Philadelphia, WB Saunders, 1995, 741–748.

16. Jiang ZM, He GZ, Zhang SY, Wang XR, Yang NF, Zhu Y, and Wilmore DW. Low-dose growth hormone and hypocaloric nutrition attenuates the protein-catabolic response after major operation. *Ann. Surg.* 210:513–24, 1989.

17. McClave SA, Sexton LK, Spain DA, Adams JL, Owens NA, Sullins MB, Blandford BS, and Snider HL. Enteral tube feeding in the intensive care unit: factors impeding adequate delivery. *Crit. Care Med.* 27:1252–6, 1999.

18. Montejo JC. Enteral nutrition-related gastrointestinal complications in critically ill patients: a multicenter study. *Crit. Care Med.* 27:1447–53, 1999.

19. Scheltinga MR, Young LS, Benfell K, Bye RL, Zeigler TR, Santos AA, Antin JH, Schloerb PR, and Wilmore DW. Glutamine-enriched intravenous feedings attenuate extracellular fluid expansion after standard stress. *Ann. Surg.* 214:385–395, 1991.

20. Smith JS, Austen GA, and Souba WW. Nutrition and metabolism, in *Surgery, Scientific Principles and Practice. 3rd edition,* Greenfield L (Ed.). Philadelphia: Lippincott Williams and Wilkins, 2001, 43–68.

21. Souba WW and Wilmore DW. Diet and nutrition in the care of the patient with surgery, trauma, and sepsis, in *Modern Nutrition in Health and Disease. 8th edition.* Shils M and Young V (Eds.). Philadelphia, Lea and Febiger, 1994, 1202 – 1240.

22. Souba WW. Cytokines: key regulators of the nutritional/metabolic response to critical illness. *Curr. Prob. Surg.* 31 (7):577–652, 1994.
23. Souba WW. Homeostasis: bodily changes in trauma and surgery, in *Sabiston's Essentials of Surgery, 2nd edition.* Sabiston D and Lyerly K (Eds.). Philadelphia, WB Saunders 1994, 10–23.
24. Souba WW. Total parenteral nutrition, in *Current Practice of Surgery — CD-ROM,* Copeland EM III et al. (Eds.). New York, Churchill Livingstone, 1995.
25. van der Hulst RRWJ, van Kreel BK, von Meyenfeldt MF, et al. Glutamine and the preservation of gut integrity. *Lancet.* 341:1363–1365, 1993.
26. Wilmore DW, Byrne TA, Young LS, et al. A new treatment for patients with the short bowel syndrome: growth hormone, glutamine, and a modified diet. *Ann. Surg.* 222:243–54, 1995.
27. Ziegler TR, Young LS, Benfell K, Scheltinga M, Hortos K, Bye R, Morrow FD, Jacobs DO, Smith RJ, and Antin JH. Glutamine-supplemented parenteral nutrition improves nitrogen retention and reduces hospital mortality versus standard parenteral nutrition following bone marrow transplantation: a randomized human trial. *Ann. Intern. Med.* 116:821–28, 1992.

11 Nutritional Management of Immunocompromised Patients: Emphasis on HIV and AIDS Patients

*Sarah S. Richter, Suzanne S. Teuber,
and M. Eric Gershwin*

CONTENTS

INTRODUCTION

Nutritional status has a definite impact on immune system function. Thus, when considering the nutritional status of an immunocompromised patient, one of the physician's primary concerns should be that the patient's essential nutritional requirements are being met. This is true in cases where the immune system needs to be enhanced, as in HIV/AIDS, in patients undergoing radiation therapy, chemotherapy, severe burns, and in the aging patient, and also in situations where the immune system needs to be suppressed, as in transplant patients. By addressing the patient's nutritional status, improvements can be made in the patient's clinical health and well-being. The approach is a three-step process: treat the patient's nutritional deficiencies, and underlying causes of deficiency if there are any, determine if the patient is at risk for infection and adjust intake to minimize risk, and, finally, treat the disease state with nutrients known to be effective in ameliorating the symptoms or cause of the disease.

The primary concern in treating any patient is that his or her nutritional needs for basic life processes are met. Because no set requirements are known for specific immunocompromised disease states and because assessing nutritional status in patients can be difficult, determining whether the patient is deficient is challenging. Before trying to eliminate the deficiency, the cause of the deficiency should be determined. It may be due to inadequate intake, malabsorption, indigestibility, use by disease processes such as underlying infections, or an inability to correctly metabolize and utilize the nutrient. Being mindful of the age, weight, gender, and total macronutrient intake (carbohydrates, protein, fat) when determining a patient's nutritional status is also important. These factors, in addition to disease state, can change a patient's nutritional requirements. For example, a young child needs nutrients for growth, as well as for disease-fighting processes.

Once the patient's nutritional intake has been normalized, the risk of infection should be considered and nutritional adjustments made to reduce the risk or minimize the effects of infection. Infections stimulate biochemical, metabolic, and hormonal responses that adjust the body's use of nutrients.[10] When infections occur with, or cause, fever and anorexia, they lead to a hypermetabolic state, pushing the requirement for energy and nutrients upwards.[10] Infection can kill the immunocompromised patient and should be prevented if possible.

The last step in developing a nutritional regimen for immunocompromised patients is determining which nutrients can be used to treat the disease itself. Some nutrients are known to benefit patients in certain conditions, and this information should be exploited when prescribing treatment regimens. Nutrients may slow or increase the speed of disease progression, increase the severity of symptoms, or optimize the effects of drug therapies the patient is receiving.

After a brief review of the general effects of nutrition on the immune system, the relationship between nutrition and HIV/AIDS is discussed in detail.

REVIEW OF NUTRITION AND IMMUNITY

Though caloric restriction has been shown to enhance immune function, protein-calorie malnutrition (PCM) is generally detrimental to immunity. When a patient is malnourished, cell-mediated immunity is impaired, and the risk of infection and mortality increases.[42] Specifically, malnourishment has been shown to suppress T helper cell proliferation, limit T suppressor cell counts, generate low CD4/CD8 ratios (T helper/T cytotoxic cell ratios), cause high levels of immunoglobulin E (IgE) to be synthesized, and limit differentiation of T cells.[42] In addition, an immune response elicits increased nutritional requirements by altering cellular and systemic metabolic processes. When these increased needs are not fully met, the situation often leads to malnutrition and a subsequent impairment of response since body stores are used and not replenished.

Before considering the effects of nutrition and malnutrition on the immune system itself, it should be noted that adequate nutritional status is necessary for the maintenance of the physical, first-line barriers in the overall environmental defense system. The body's barriers include epithelial barriers, such as the skin and intestinal mucosa, the enzymes and acidic pH of the stomach, and other mucosal secretions, such as rhinobronchial secretions, vaginal mucus, intestinal mucin, and tears. Micro-nutrients and macronutrients alike help maintain the functions of these systems. Vitamin A and niacin are important to the proper differentiation and proliferation, respectively, of epithelial barriers. Proteins and amino acids are important for the synthesis of gastric enzymes and the enzymes involved in the synthesis of mucosal secretions by glandular cells. The acidic environment of the stomach not only supports protein digestion but also destroys certain microorganisms, thus protecting the body from those invaders. The establishment and maintenance of the normal intestinal microflora is another aspect of the health of the intestinal epithelial barrier.[17] However, a full discussion of the role of probiotics is beyond the scope of this chapter, as the applications of this area of research to the immunocompromised patient are not fully developed.[23]

MACRONUTRIENTS

Feeding of protein can be beneficial to an immune response because the amino acids allow the patient to synthesize defense factors, such as antibody, complement, metal-binding proteins, or lymphocyte cell components. Protein-energy malnutrition (PEM) has been shown to impair delayed-type hypersensitivity and reduce the number of circulating T cells in humans.[42] This is probably due to inadequate amino acid availability for synthesis of transcription factors and the proteins required to synthesize immunoglobulins, complement, and other lymphocyte proteins. Response to PEM depends on specific deficiencies and the extent to which the patient is undernourished.

Supplementation of specific amino acids can improve immune function. The effects of glutamine and arginine supplementation are the most researched. Arginine may decrease protein degradation by stimulating the release of large amounts of anabolic hormones. CD4 (T helper cell) lymphocyte counts and T cell proliferation response to Concanavalin A (Con A) increase in lymphocytes isolated from surgical patients who receive arginine-supplemented nutritional therapy.[18] Arginine has been shown to preserve the thymus, promote T cell proliferation, and increase interleukin(IL)-2 and IL-2 receptor expression by T cells.[2] Furthermore, arginine is a precursor for nitric oxide (NO), a compound which influences leukocyte–endothelial cell adhesion used in microbicidal activity. Glutamine may be important to immune function because it is a preferred energy substrate for rapidly proliferating cells such as those of the immune system and gastrointestinal (GI) tract. Furthermore, glutamine is important because it is a nitrogen donor for protein, purine, and pyrimidine synthesis. Enterocytes need a stable supply of this amino acid in order to maintain the integrity of the gut mucosa. Without glutamine, the gut mucosa will become perforated and allow bacteria and other antigens to invade the circulation, forcing the immune system to deal with these insults. *In vitro* data show that depletion of glutamine in lymphocytes prevents late activation, and deprivation of glutamine in monocytes lowers the number of surface antigens available to initiate phagocytosis.[56] Finally, glutamine is used in the synthesis of glutathione (GSH), an important cellular reductant for mitigating oxidative stress. Clearly, adequate arginine and glutamine intake is essential for immune function; however, this area of research would benefit from further animal or human study.

In addition to protein and amino acids, lipids are another macronutrient that can have profound effects on the immune system. Much current lipid research is focusing on n-6 and n-3 fatty acids (FAs). These FAs are involved in synthesis of prostaglandins (PG) and leukotrienes (LT), important immune signaling molecules. n-3 FAs are considered anti-inflammatory. They mediate their anti-inflammatory effect through decreased T cell proliferation or impaired IL-2 production.[48] Feeding of n-3 FAs increases the content of eicosanoids, such as eicosapentanoic acid (EPA) and docosahexanoic acid (DHA), while decreasing the content of arachidonic acid (AA) in immune cell membranes.[68] The n-6 FAs are considered inflammatory because feeding of these lipids has been shown to increase AA incorporation into membranes of immune cells. Increased amounts of AA allow for increased synthesis and release of PGs and LTs, such as PGE2 and LTB4. PGs and LTs are synthesized by macrophages and mediate inflammatory responses. PGE2 and LTB4 have been shown to stimulate T cell movement across the basement membrane[43] and to have variable effects on T cell differentiation and proliferation, depending on cell maturity.[53] Balance between EPA/DHA and AA concentrations in cell membranes is important to immune function. Current American diets consist of n-6 to n-3 ratios of around 20:1. A ratio of 4:1 or 1:1 is believed to represent the traditional diet and fits well with the idea that smooth functioning of the immune system requires a balance between n-3 and n-6 FAs.[64]

The final macronutrients to consider are nucleotides. Any cells that are rapidly proliferating, such as activated immune cells, require a ready source of nucleotides. Nucleotide-free diets have an immunosuppressive effect on T cell function, and

T cell-dependent immune responses can be enhanced by supplementing a nucleotide-free diet with mononucleotides.[77] Dietary sources of nucleotides, such as ribonucleic acid (RNA), have been shown to be important for immune system function, gut development, and liver health. In situations of stress, nucleotides become essential because the body does not have the resources to synthesize the purines and pyrimidines needed by rapidly proliferating immune, gut, and other cells.[58] Nucleotides are also important in growth and particularly important in growing children. Human infants given nucleotide supplements had increased natural killer (NK) cell activity.[13] In addition, septic or critically ill patients given a nucleotide-containing diet had shorter hospital stays than those fed nucleotide-free diets.[11] Nucleotide-free diets decrease cellular immune function, increase risk of infection, and increase allograft survival. Though these effects have not been adequately demonstrated in humans, the possible value of nucleotides in treating immunocompromised patients should not be overlooked.

MICRONUTRIENTS

Micronutrients also influence immune system function. Vitamins and minerals play important roles in normal cellular function and have specific roles in the various cells of the immune system. Vitamin A, vitamin E, zinc, and selenium are some of the well-studied nutrients known to affect immune system function. However, deficiencies in nearly all vitamins and minerals will have some effect on the immune system of the deficient individual. It is important to note that over-supplementation of many of these compounds can also have detrimental effects on the immune system. These effects may be the result of interactions with other nutrients. For example, excessive intake of zinc can decrease the absorption of copper and lead to copper-deficiency problems. Effects of copper, iron, magnesium, selenium, and zinc on the immune system have been well-characterized and are summarized in Table 11.1. The effects of certain vitamins have also been characterized and are summarized in Table 11.2.

HIV/AIDS

As discussed above, nutrition is important in maintaining immune function. Thus, patients suffering from a disease of immunocompromise, such as HIV/AIDS, are likely to benefit from nutritional intervention. A number of studies have shown that progressive malnutrition occurs in patients with AIDS. Clinical stability is associated with nutritional stability in these patients.[41] In AIDS patients, as in all people, malnutrition occurs as a result of malabsorption, changes in caloric intake, or changes in energy expenditure. Treating the malnutrition seen in HIV/AIDS will benefit the patient's immune and other body systems. As Kotler put it, ". . . malnutrition promotes adverse outcomes. . . nutrition status can be improved through nutritional support, and . . . such improvements improve outcomes."[40] Though most literature reports are based on observational studies which have looked at the effects of HIV/AIDS on nutritional status, research has also been done on the effects of nutritional interventions, either with single micronutrient supplements, multi-micronutrient supplements,

TABLE 11.1
Minerals: Function and Effect of Deficiency, Supplementation,
or Excess on the Immune System

Copper

Functions
 Complement function
 Interacts with iron
 Cell membrane integrity
 Immunoglobulin structure
 Copper–zinc superoxide dismutase
Effects of deficiency
 Increased susceptibility to infection
 Decreased antibody production
 Impaired phagocyte function
 Decreased T cell proliferation, particularly that of T helper cells
 Decreased T cell count
 Decreased production of interleukin-2
 Increased B cell count
 Fewer circulating granulocytes/neutrophils
 Decreased neutrophil respiratory burst in mice
 Impaired T cell function
Effects of excesses
 Decreased phagocyte numbers
 Impaired phagocyte function

Iron

Functions
 Folate metabolism
 Mitochondrial energy production
 Function and migration patterns of T cell subpopulations
 Component of many enzymes
 Formation of reactive oxygen species and radicals during respiratory burst
Effects of Deficiency
 T Cell and phagocyte abnormality
 Decreased cytokine release
 Decreased inflammatory response
 Impaired neutrophil and macrophage cytotoxicity
 Decreased allograft rejection
 Lymphoid tissue atrophy
 Decreased T cell count
 Decreased antibody production
 Inhibits proliferation of T helper 1 (Th1) cells
 Removal of stores decreases T cell proliferation and differentiation
 Chronic inflammation is associated with low serum iron and high levels of iron stores
Supplementation
 High iron correlates with lower interferon-γ and nitric oxide production
Effects of excesses
 Excess intracellular iron stores reduce the phagocytic activity of macrophages and other phagocytic
  immune cells

TABLE 11.1 (CONTINUED)
Minerals: Function and Effect of Deficiency, Supplementation, or Excess on the Immune System

Magnesium

Functions
  Cyclic adenosine monophosphate production
  Complement function
  Cytotoxic T cell lysis of target cells
  Component of enzymes
  Component of adenosine tri- and di-phosphate
Effects of deficiency
  Increased thymic cellularity
  Creation of an oxidative environment
  Decreased acute phase molecules
  Increased numbers of inflammatory cells (especially eosinophils)
  Increased interleukin-1, interleukin-6, tumor necrosis factor α and histamine levels
  Decreased complement activity

Selenium

Functions
  Limits auto-oxidation
  May mediate post-translational modification of important immune system proteins
  Important in alternative reading frames for mRNA based on selenoproteins
  Component of glutathione peroxidase
Effects of deficiency
  Decreased resistance to infection
  Decreased lymphocyte proliferation
  Increased risk of viral infection
  Decreased antibody synthesis
  Decreased cytotoxicity
  Decreased cytokine secretion
  Decreased cell-mediated cytotoxicity
Supplementation
  Increases most immune parameters

Zinc

Functions
  Associated with many enzymes
  T cell function
  Activity of many transcription factors
  Thymocyte development
  Thymic integrity
Effects of deficiency
  Thymic atrophy
  Decreased numbers of T and B cells
  Decreased thymic hormone release
  Decreased T cell response to mitogens
  Decreased T cell maturation

TABLE 11.1 (CONTINUED)
Minerals: Function and Effect of Deficiency, Supplementation, or Excess on the Immune System

Effects of deficiency
  Reduced natural killer cell activity
  Decreased cytolytic T cell response
  Decreased T cell help to antibody production
  Increased corticosteroid production
  Disturbances in calcium/phosphate metabolism
  Enlarged adrenal glands
  Depressed delayed-type hypersensitivity response
  Suboptimal zinc during the *in utero* development of mice causes persistent states of
    immunodeficiency that can be transferred to subsequent generations
  Depressed concentration of serum albumin, prealbumin and transferrin
  Problems with regulation of blood sugar
  May accelerate progression to AIDS
Supplementation
  Increases secretion of thymulin enhancing thymic activity
  Antiviral effects *in vitro*
  Inversely related to susceptibility to apoptosis
  May decrease diarrhea and weight loss in AIDS patients

TABLE 11. 2
Vitamins: Function and Effect of Deficiency, Supplementation, or Excess on the Immune System

Vitamin A

Functions
  Maintains integrity of epithelial cell boundaries
  Supports T helper 2 (Th2) cell development
  Maintains production of mucosal secretions
  Inhibits T cell apoptosis (retinoic acid)
  Regulates T helper 1 cell (Th1) population
  May inhibit viral replication
Effects of deficiency
  Decreased numbers of leukocytes
  Decreased CD4 (T helper cell membrane glycoprotein) count
  Decreased levels of circulating complement
  Impaired T cell function
  Increased severity of infection
  Decreased resistance to immunogenic tumors
  Decreased lymphoid organ weight
  Decreased antigen-specific immunoglobulin G and immunoglobulin E responses
  Induces hypergammaglobulincmia
  May affect vertical transmission of HIV
  May affect progression to AIDS

TABLE 11. 2 (CONTINUED)
Vitamins: Function and Effect of Deficiency, Supplementation, or Excess on the Immune System

Supplementation
  Correlates with fewer respiratory infections
  Increases lymphocyte proliferation
  Lowers neutrophil count
  Improves reticuloendothelial system function
  Increases graft rejection
  Enhances cytotoxic T cell activity
  Increases tumor resistance
Effects of excesses
  Increased antibody and cell-mediated immune responses
  Stimulation of Kupffer cells
  May lead to inhibition of T cell apoptosis

Pyridoxine (B₆)

Functions
  Nucleic acid and protein synthesis in proliferating lymphocytes
  Amino acid metabolism
  Heme synthesis
Effects of deficiency
  Lower lymphoid tissue weights
  Decreased T cell count
  Decreased proliferative response to mitogens
Effects of excesses
  Deficiency in cell-mediated immunity: decreases allograft rejection, interleukin-2 production and delayed-type hypersensitivity responses
  Decreased antibody response
  Decreased antigen-specific secondary responses
Supplementation
  Protects mice from immunosuppressive effects of ultraviolet B radiation

Vitamin B₁₂ (Cobalamin) and Folate

Functions
  Synthesis of thymidylate and thus DNA
  Formation of methionine from homocysteine
Effects of deficiency
  Decreased respiratory burst
  Decreased phagocytosis by peripheral mononuclear cells
  Decreased CD4 (T helper cell) count
  Depressed phagocytic function
  Decreased delayed-type hypersensitivity response
  Decreased T cell proliferation to phytohemagglutinin stimulation
  May accelerate progression to AIDS
  May cause or increase the severity of AIDS dementia complex

TABLE 11. 2 (CONTINUED)
Vitamins: Function and Effect of Deficiency, Supplementation, or Excess on the Immune System

Biotin

Functions
   Carboxylation reactions
Effects of deficiency
   Lower thymic weight
   Decreased lymphocyte proliferation
   Decreased antigen-specific antibody response
   Decreased lymphocyte-mediated suppressor activity

Pantothenic Acid

Functions
   Moves acyl groups from molecule to molecule
Effects of deficiency
   Decreased antibody response

Thiamine (B₁)

Functions
Carbohydrate metabolism
Effects of deficiency
   Increased susceptibility to infections
   Decreased antibody response
   Premature thymic atrophy
   Decreased mobility of polymorphonuclear leukocytes

Riboflavin (B₂)

Functions
   Energy production
Effects of deficiency
   Increased polymorphonuclear leukocytes
   Decreased peripheral blood mononuclear cells
   Lower thymic weight
   Decreased antibody response
   Increased susceptibility to infection

Vitamin C

Functions
   Antioxidant
   Spares selenium by recycling glutathione peroxidase
   Minimizes oxidation of phagocytes and damage to bystander host cells
Effects of deficiency
   Decreased resistance to infection
   Fewer delayed-type hypersensitivity reactions
   Decreased phagocytosis
   Decreased phagocyte mobility
   Reduced skin allograft rejection
   Impaired wound repair

TABLE 11. 2 (CONTINUED)
Vitamins: Function and Effect of Deficiency, Supplementation, or Excess on the Immune System

Supplementation
  Increases antibody response
  Increases delayed-type hypersensitivity response

Vitamin D

Functions
  Promotes immunosuppression
  Influences mineral metabolism, particularly that of calcium and phosphate
  Increases phagocytosis
  Inhibits acquired immune response by decreasing major histocompatibility complex expression
  Stimulates maturation of normal and neoplastic myelomonocytic cells to more differentiated
    monocytes and macrophages
  Decreases tumor growth and the accompanying immunosuppression by blocking release of
    granulocyte/macrophage colony-stimulating factor
  Inhibits CD4 (T helper cell) production of interleukin-2 and interferon-γ
  Inhibits T cell help in antibody response
  Decreases T helper 1 (Th1) cell functions but not T helper 2 cell (Th2) or CD8 (T cytotoxic cell)
    T cell activity

Vitamin E

Functions
  Scavenges radicals
  Limits cell membrane peroxidation
Effects of deficiency
  Decreased leukocyte production
  Unaffected or enhanced natural killer cell activity
  Decreased phagocytosis and chemotaxis by polymorphonuclear leukocytes and macrophages
  Reduced lymphocyte proliferation
Supplementation
  Increases phagocytosis
  Increases lymphocyte proliferation
  Increases antibody levels
  Decreases oxidative damage
  Restores delayed-type hypersensitivity reactions
  Decreases prostaglandin E2 synthesis
  Decreases plasma lipid peroxides
  Restores T helper 1 cell (Th1) activity
  Increases resistance to infection in the elderly
  Restores interleukin-2 and prostaglandin F1a production
  Has antiviral effects *in vitro*

or special nutritional formulas. This section will cover the main relationships between HIV/AIDS and nutrition, beginning with common deficiencies seen in HIV/AIDS.

COMMON DEFICIENCIES

Deficiencies of many vitamins and minerals have been found in HIV/AIDS patients. B vitamin levels decrease over time in these patients.[9] Patients infected with HIV/AIDS have low plasma levels of vitamins B_{12} and B_6[9] and exhibit deficiencies of niacin and vitamin B_6.[6,51] Other vitamins and minerals are also present at abnormal levels in these patients. These include β-carotene,[52] selenium,[52] zinc,[50] vitamin A,[73] and vitamin E.[7] Deficiencies may be caused by a lack of adequate intake, interactions with treatment regimens, malabsorption from an abnormal GI tract, changes in the patient's REE (resting energy expenditure), resulting in altered nutrient metabolism, or losses of the nutrients in diarrhea, vomit, or excessive sweating. Furthermore, the body pools of these nutrients are being used to keep the stress response, viral processes, and immune system running. With body pools mobilized and no fresh supply available, nutrients can be depleted.

CAUSES OF DEFICIENCIES

A major problem in keeping the HIV/AIDS patient well-nourished is the change occurring in the GI tract of the infected person. HIV/AIDS patients have depressed appetites. When a patient's food intake decreases, malabsorption in the small intestine arises as enterocytes atrophy. Enterocyte and villus atrophy even occur in patients with no secondary infections.[75] In addition, HIV/AIDS patients often suffer from diarrhea. Diarrhea increases the speed with which digesta move from the stomach through the small intestine and out of the body. Intestinal cells thus have less time to absorb nutrients, allowing vital nutrients to be lost in the feces. HIV/AIDS patients with opportunistic or secondary infections more commonly have GI tract problems and can experience frequent episodes of diarrhea.[20] Patients in this situation are even more likely to be malnourished. Changes in the GI tract and malabsorption are probable main causes of nutritional deficiencies in HIV/AIDS patients.

As might be expected, people afflicted with HIV/AIDS have unique nutritional needs. Even in populations where dietary intakes of HIV-positive males meet or exceed the recommended daily allowance (RDA), nutritional deficiencies are present. The RDAs were set to meet the nutritional needs of a population of healthy people, not those of individuals afflicted with chronic disease. Nutritional guidelines must be created for the HIV/AIDS patient based on what is known about nutrition as it relates to HIV/AIDS and to the patient himself.

The unique nutritional needs of the HIV/AIDS patient can be attributed to a variety of causes. First, there are the increased needs of an immune system that is in a state of chronic activation. This can drain the patient of his nutritional stores and ultimately cause irreparable damage if nutrients are not supplied in the amounts necessary to sustain heightened levels of immune activity. Besides an increase in nutrient and energy expenditure by the immune system, HIV/AIDS patients experience an overall rise in metabolism. REE has been shown to increase with infection and disease. In general, REE increases by 12.5% for every degree of fever over 37°C.[30] AIDS subjects have been shown to have REEs that are 14% higher than would be predicted by the Harris–Benedict formula.[46] These patients also had higher fat-free-mass REE than controls.[46] In a study by another researcher, AIDS patients were found to have an REE that was 9% higher than that of healthy volunteers with similar body composition, consuming a similar diet.[34] Concentrations of catecholamines, thyroid hormones, cortisol, and tumor necrosis factor (TNF) were similar between the two groups, but experimental subjects had lower norepinephrine levels than controls.

Now that some of the causes of nutritional deficiencies have been discussed, the causes and effects of deficiencies in two minerals that play specific roles in HIV/AIDS will be covered.

SELENIUM DEFICIENCY

A unique nutritional problem is selenium deficiency, rare in healthy individuals but common in HIV patients. Though the exact role of selenium in the body's fight against HIV/AIDS is unknown, it is thought to be involved in protecting the body against oxidative damage, thereby keeping lymphocytes from initiating apoptosis. Selenium deficiency may also result from its use by the HIV virus, since HIV viral deoxyribonucleic acid (DNA) encodes several selenium-dependent enzymes.[74] The synthesis of billions of virus particles each day in the HIV-infected patient, symptomatic or not, has serious effects on normal metabolic processes and therefore increases the patient's requirements for nutrients involved.[51]

ZINC DEFICIENCY

Another nutrient for which the requirement increases in HIV/AIDS patients is zinc. This mineral serves many viral functions. Viral particles must bind zinc ions in T lymphocytes in order to produce pro-viral peptides that will be part of new infectious particles. Integrase, the protein required for the integration of viral DNA into host DNA, contains zinc finger domains, and its activity is optimized by the binding of zinc to those protein structures.[66] In addition, HIV protease is inactivated by binding a sufficient number of Zn ions.[78] The use of zinc by viral processes exemplifies how HIV infection can change nutrient requirements.

OTHER VITAMINS AND MINERALS

Other nutrients are also needed for viral processes. Viral load increases energy requirements. Weight loss has been associated with elevated viral burden.[55] Patients on antiviral therapy, on the other hand, gain weight.[63] HIV/AIDS patients also have an increased need for antioxidants. Antioxidants offset the numbers of oxidative events triggered by viral infection and limit immune cell apoptosis.

Several increased requirements are the result of changes occurring in specific tissues of the HIV/AIDS patient. Vitamin B_{12} requirements may increase when the gut atrophies and less intrinsic factor or intrinsic factor receptor is synthesized.[31] 1,25-dihydroxyvitamin D_3 $(1,25\text{-}(OH)_2\text{-}D_3)$ levels are also lower in symptomatic HIV patients.[32] This may be the result of HIV nephropathy since 25-hydroxyvitamin D_3 $(25(OH)D_3)$ is made into $1,25\text{-}(OH)_2\text{-}D_3$ in the kidney.

The increased nutritional needs of HIV/AIDS patients are demonstrated time and again by studies in which HIV/AIDS patients, whose dietary intakes are adequate for healthy persons, have low serum levels of vitamins and minerals. For example, vitamin B_6 deficiency was seen in patients whose dietary intake appeared to be adequate.[3] Furthermore, HIV/AIDS patients consumed B vitamins at levels that were multiples of the RDA in order to achieve normal plasma levels.[3] Additionally, HIV patients were found to be at risk for and had lower plasma magnesium levels than noninfected controls.[65] Thus, it is clear that the nutrient requirements for HIV/AIDS patients can be substantially higher than those for members of healthy populations.

Nutritional Status as a Predictor of Morbidity/Mortality

Because of the effects of HIV/AIDS on a patient's nutritional status, nutritional intervention must begin early in the disease process, preferably as soon as HIV infection is confirmed. This way the body can have ready the nutrients it needs to fight the infection and make up for nutrients lost to viral processes, diarrhea, and malabsorption. Supplying the HIV-infected patient with the nutrients he needs will help him or her mount a continuous immune response and may delay the immuno-deficient state by enabling the immune system to function at maximum capacity and keep the virus in check.[33] Multivitamin users have a decreased risk of low CD4 counts.[1] In addition, being malnourished increases the risk of opportunistic infection and morbidity in AIDS patients.[42] In a study of 17 HIV-infected patients, nutritional supplementation and counseling were effective in weight gain/maintenance only in patients with no secondary infections.[67] Patients who had secondary infections lost weight while on the same treatment regimen. These examples illustrate that the earlier nutritional intervention is begun, the more likely disease progression will be slowed and complications reduced.

Nutritional Status and Risk of Infection

Impaired nutritional status has been associated with infection in several studies. In a study where 228 AIDS patients were tested for zinc deficiency, 29% had low serum zinc and 21% had marginally low serum zinc levels. Those patients with zinc deficiency had a significantly higher likelihood of bacterial infection.[38] In a 5-year, matched (1:2), case-control study, patients with advanced HIV infection and impaired nutritional status were found to have a greater chance of experiencing HIV-associated enteric salmonellosis.[70] Furthermore, the possibility of atypical symptoms for pulmonary tuberculosis is large for severely malnourished HIV-infected patients.[44] These studies, when combined with general knowledge about nutrition and infection, demonstrate the importance of nutritional status in keeping HIV/AIDS patients free of secondary infections.

IRON AND RISK OF INFECTION

Iron also is rate-limiting for the growth and virulence of some microorganisms that commonly infect immunocompromised patients. These microorganisms have evolved ways of hijacking the host's iron stores, some via secreted siderophores.[36] Infections with *Candida* spp., *Pneumocystis carinii*, and *Mycobacterium* spp. have been associated with high iron stores. For instance, de Monye et al.[19] compared 188 patients with high bone marrow macrophage iron scores to 130 patients with normal to low iron. *Candida*, *Pneumocystis*, and *Mycobacterium* infection were more common in patients with high iron stores (P = 0.006) than those with low stores. Moreover, individuals with high iron stores died sooner.[25] In thalassemia major patients, where iron overload is treated with deferoxamine, patients who had lower ferritin values had a significantly slower progression from HIV-1 seropositivity to AIDS than those with high ferritin values.[60] In mice, iron overload significantly increased the rate of disseminated infection by low-virulence *C. albicans*; however, treatment with deferoxamine resulted in cure and restoration of fungicidal phagocytic activity.[47]

Infections with *Mycobacterium tuberculosis* and *M. avium* result in high morbidity and mortality in AIDS patients. From murine models of these infections, it is apparent that higher iron stores are associated with increased dissemination and higher mortality and that treatment with iron chelators is beneficial.[24] In humans, because deferoxamine is extremely expensive and only available in a form suitable for IV administration, chelation treatment has been used in situations of obvious iron overload, thalassemia major, but not in HIV clinical trials.[76]

From the information presented above, it is apparent that routine iron supplementation is unnecessary, and perhaps, harmful, for the HIV/AIDS patient. However, since responses to erythropoietin are poor in patients with iron deficiency, HIV patients receiving erythropoietin may need iron supplements if found to be truly iron-deficient and not merely anemic.[45]

NUTRITIONAL STATUS AND PROGRESSION TO AIDS

In light of its role in secondary infection, nutritional status can be an important predictor of progression to AIDS. A study by Abrams showed that daily multivitamin use reduced the risk of progression to AIDS by 30%.[1] There was a significant association between mortality and impaired nutritional status in HIV-positive IV drug users. Specifically, deficiencies of vitamin A, vitamin B_{12}, zinc, and selenium were linked to mortality.[4] The status of several specific vitamins and minerals has been associated with risk of progression. Risk of progression to AIDS is significantly lower in patients taking B vitamins at levels that are multiples of those recommended for healthy populations.[72] Two studies have shown vitamin B_{12} status to influence disease progression.[8,59] Thiamin and niacin have each also been shown to slow disease progression.[71] Moderate vitamin A supplementation delays progression and reduces risk of progression by approximately 50%.[71] In addition, HIV patients who had abnormally low levels of $1,25\text{-}(OH)_2\text{-}D_3$ had decreased survival time compared to those HIV patients with adequate amounts of the vitamin D metabolite.[32] Selenium deficiency correlates with progression and mortality in HIV/AIDS patients,[52] as does zinc deficiency. In HIV-positive drug users, relative risk for mortality increased threefold when plasma zinc was low.[4] Graham et al.[27] studied 54 asymptomatic HIV-positive progressors, 54 seropositive non-progressors, and 54 seronegative subjects and found that high serum copper and low serum zinc levels predicted progression to AIDS independently of CD4 count, age, or calorie-adjusted dietary intake of the mineral. Although nutrient status has been linked to disease progression, neither cause nor mechanism is yet understood.

HIV/AIDS Wasting

Another situation where cause and effect has yet to be determined is the relationship between wasting and disease progression. The term *wasting* refers to loss of lean muscle mass. The weight loss may be a result of the hypermetabolism of the disease state,[28] inadequate digestion and absorption by a malfunctioning GI tract,[20] or diminished food intake.[29] These abnormalities may be caused by the disease, nutritional deficiencies, or secondary infections. This loss is strongly associated with disease progression and morbidity.[16] Morbidity, as it relates to wasting, may be a result of the decrease in CD4 cells that occurs when weight decreases.[62] Patients with secondary infections experience more extreme wasting than patients without infections due to decreased caloric intake, not increased REE.[29] Micronutrient deficiencies are also more frequent in patients with wasting,[15] most likely a result of the atrophied gut mucosa. Micronutrient and caloric deficiencies may link wasting to disease progression and morbidity.

Nutritional Status and Vertical Transmission of HIV

Disease transmission is affected by nutritional status. In the case of mother-to-fetus transmission, the nutritional status of the mother is important. When the mother is adequately nourished, disease progression, the possibility and extent of an opportunistic infection, and viral load in the blood, milk, and genital secretions are reduced.[21,22] Furthermore, the GI tract of babies whose mothers were adequately nourished during gestation will be healthy.[21,22] Seropositive mothers who are severely vitamin A deficient are more likely to pass the infection on to their children than seropositive mothers with adequate vitamin A levels.[21]

Effect of Specific Supplements on Immune Status *in vitro* and *in vivo*

The most important micronutrients for an HIV/AIDS patient are those involved in enhancing CD4 or other immune cell counts, facilitating antioxidation, and limiting infection. A number of micronutrients have been demonstrated to affect CD4 or CD8 cell counts. These include vitamins E, B_{12}, and $1,25\text{-}(OH)_2\text{-}D_3$ and minerals iron and zinc. Vitamin E supplementation decreases the CD8 cell count, raising the CD4-to-CD8 cell ratio.[50] In a study of 108 HIV positive men, vitamin B_{12} deficiency was associated with lowered CD4 counts, and increases in vitamin B_{12} serum levels led to higher CD4 counts.[8] The number of CD4 cells circulating in peripheral blood correlates positively with serum $1,25\text{-}(OH)_2\text{-}D_3$ levels.[32] Iron supplementation also increases the number of peripherally circulating lymphocytes.[50] Normalization of serum zinc levels[8] and zinc supplementation[37] led to higher CD4 counts. In addition, patients taking zinc supplements had higher CD3 (cells expressing the signal-transduction element of the T cell receptor) counts and body weights.

Micronutrients are also involved in limiting the amount of oxidative damage experienced by the HIV-positive patient. The disease stimulates apoptosis in lymphocytes. This apoptosis depends upon oxidative signals that are generated not only by the infection itself but also by antioxidant imbalances within the patient.[12] Apoptosis stimulated by oxidative stress is the major cause of T cell death in HIV/AIDS.[26] An adequate intake of antioxidants limits lymphocyte apoptosis and elevates CD4 counts. Levels of the antioxidants β-carotene and vitamin E have been independently correlated with oxidative damage in HIV/AIDS patients.[73] Apoptosis has also been related to the GSH redox status of HIV/AIDS patients.[12] The enzyme glutathione peroxidase (GPx) is a selenium-dependent enzyme that returns GSH to its reduced form. Selenium deficiencies deplete reduced GSH and leave T cells susceptible to the apoptosis induced by oxidative stress.

Some of the same micronutrients involved in limiting oxidative stress are also involved in limiting infection. Selenium supplementation *in vitro* inhibits replication of HIV viral particles.[35] Furthermore, selenoproteins are important in the regulation of viral growth and can be either inhibitory or stimulatory, depending on the redox environment and selenium concentration in the host.[74] Zinc salts inhibit viral activity directly or by modulating the immune system.[61]

The examples above are some of the many nutrients involved in regulating T cell counts, oxidative damage, and infection in HIV-positive persons.

NUTRITIONAL STATUS AND RESPONSE TO MEDICATIONS

Besides using nutrients to control T cell count, apoptosis, and extent of infection, nutrients can also be used to optimize the effectiveness of drugs used to treat HIV/AIDS. Medications may increase nutritional needs as minerals/vitamins may be needed to activate the treatment or to synthesize other molecules required to make treatment effective. Thus, supplementation with those nutrients would ensure more efficient use of the therapy. For example, azidothymidine (AZT) requires a zinc-dependent thymidine kinase for its activation. Of HIV patients receiving AZT treatment, 64% were zinc deficient, while only 24% of the non-AZT-treated population of HIV patients experienced zinc deficiency.[5] Not only are patients on AZT more likely to experience zinc deficiency, but they are also more likely to suffer adverse consequences from a deficiency since their medication will be less potent. In a study where 18 HIV-positive patients with CD4 counts between 250 and 400/mm³ received zinc supplements with AZT treatment, those receiving no supplement had a significantly higher relative risk of infection than those receiving supplements.[49] Clearly, the addition of zinc to antiviral medications can make the medications more effective.

Interestingly, when 28 HIV-positive patients receiving highly active antiretroviral therapy (HAART — a combination antiviral treatment with two nucleoside analogues and a protease inhibitor) took zinc supplements, the risk of opportunistic infection in the supplemented group was no lower than that in the unsupplemented group.[49] However, HAART-treated patients have better zinc absorption and zinc status than AZT-treated patients. Perhaps increased zinc uptake in HAART patients made zinc supplementation unnecessary.

These examples demonstrate the importance of nutritional status in patients being treated with conventional medical therapies.

Supplements in HIV/AIDS

Nutritional supplements can benefit the HIV/AIDS patient. These include vitamin and mineral supplements, total parenteral nutrition (TPN) treatments, as well as enteral nutrition liquid supplements sold in cans. TPN and enteral nutrition formulas have been shown to ameliorate weight loss in HIV/AIDS patients. Enteral feeding is the preferred method of feeding because it maintains the gut epithelium. TPN, on the other hand, has been associated with atrophy of the gut mucosa and GI system. Generally, this atrophy increases the risk of bacterial translocation. Sometimes use of TPN is unavoidable and perhaps even desirable. TPN is the better option when the patient's small intestine is inaccessible or malfunctioning.[40] Treatment with TPN might also be considered when the patient has serious malabsorption or diarrhea problems. In those situations, it is unlikely that the patient will be able to absorb the needed amount of nutrients from an enterally fed formula and will benefit from receiving intravenously delivered nutrition.

Enteral formulas that include functionally significant vitamins and minerals not only allow patients to gain weight but also lead to fewer hospitalizations.[14] When enteral formulas contain n-3 fatty acids in addition to other essential nutrients, such as arginine and nucleotides, weight gain may occur and more EPA is incorporated into cell membranes.[69] Supplements containing antioxidants, n-3 fatty acids, and the medication pentoxifylline (used to improve microcirculation in peripheral artery disease but now under investigation for immunomodulating activity) may inhibit cytokine activity and, therefore, decrease the protein breakdown that normally is promoted by pro-inflammatory cytokines.[40] The addition of fatty acids to nutritional supplements is often debated. A study in which patients received supplements containing either long-chain or medium-chain triglycerides for 6 months showed that both classes of triglycerides allow clinically stable HIV patients to gain weight, mostly in the form of fat.[39]

To minimize fat gain, HIV/AIDS patients on any feeding regimen should be advised to participate in weight-bearing exercise.[57] Exercise will increase retention of lean muscle mass. It may also benefit the patient's psyche, increase production of endogenous growth hormone, and enhance immune system function.[57] A final thought to keep in mind, when considering nutritional supplementation, is that patients who receive nutritional counseling in conjunction with supplements respond better than patients who are given only supplements or counseling.[54,67] However, if only supplementation or nutritional counseling is feasible, either will promote patient health.

The preceding sections provide evidence that, although HIV/AIDS causes malnutrition in those afflicted by the disease, malnutrition, in turn, also significantly affects the outcome of the disease.

ABBREVIATIONS

Abbreviations used: *HIV* Human Immunodeficiency Virus; *AIDS* Acquired Immune Deficiency Syndrome; *REE* resting energy expenditure; *PCM* protein-calorie malnutrition; *CD4* T helper cell; *CD8* T cytotoxic cell; *IgE* immunoglobulin E; *PEM* protein-energy malnutrition; *Con A* concanavalin A; *IL* interleukin; *NO* nitric oxide; *GI* gastrointestinal; *GSH* glutathione; *FA* fatty acid; *PG* prostaglandin; *LT* leukotriene; *EPA* eicosapentanoic acid; *DHA* docosahexanoic acid; *AA* arachidonic acid; *RNA* ribonucleic acid; *NK cell* natural killer cell; *RDA* recommended daily allowance; *FFM* fat-free mass; *TNF* tumor necrosis factor; *DNA* deoxyribonucleic acid; *1,25-(OH)$_2$–D$_3$* 1,25-dihydroxyvitamin D$_3$; *25(OH)D^3* 25-hydroxy vitamin D$_3$; *CD3* cells expressing the signal-transduction element of the T cell receptor; *AZT* azidothymidine; *HAART* highly active antiretroviral therapy; *TPN* total parenteral nutrition.

REFERENCES

1. Abrams B, Duncan D, and Hertz-Picciotto I. A prospective study of dietary intake and acquired immune deficiency syndrome in HIV-seropositive homosexual men. *J. Acquir. Immune Defic. Syndr.* 6: 949–958, 1993.
2. Barbul A. Arginine and immune function. *Nutrition.* 6: 53–58; discussion 9–62, 1990.
3. Baum M, Cassetti L, Bonvehi P, Shor-Posner G, Lu Y, and Sauberlich H. Inadequate dietary intake and altered nutrition status in early HIV-1 infection. *Nutrition.* 10: 16–20, 1994.
4. Baum MK. Role of micronutrients in HIV-infected intravenous drug users. *J. Acquir. Immune Defic. Syndr.* 25: S49–S52, 2000.
5. Baum MK, Javier JJ, Mantero-Atienza E, Beach RS, Fletcher MA, et al. Zidovudine-associated adverse reactions in a longitudinal study of asymptomatic HIV-1-infected homosexual males. *J. Acquir. Immune Defic. Syndr.* 4: 1218–1226, 1991.
6. Baum MK, Mantero-Atienza E, Shor-Posner G, Fletcher MA, Morgan R, et al. Association of vitamin B6 status with parameters of immune function in early HIV-1 infection. *J. Acquir. Immune Defic. Syndr.* 4: 1122–1132, 1991.
7. Baum MK, Shor-Posner G, Bonvehi P, Cassetti I, Lu Y, et al. Influence of HIV infection on vitamin status and requirements. *Ann. N.Y. Acad. Sci.* 669: 165–173; discussion 73–74, 1992.
8. Baum MK, Shor-Posner G, Lu Y, Rosner B, Sauberlich HE, et al. Micronutrients and HIV-1 disease progression. *AIDS.* 9: 1051–1056, 1995.
9. Beach RS, Mantero-Atienza E, Shor-Posner G, Javier JJ, Szapocznik J, et al. Specific nutrient abnormalities in asymptomatic HIV-1 infection. *AIDS.* 6: 701–708, 1992.
10. Beisel WR. Effects of infection on nutritional status and immunity. *Fed. Proc.* 39: 3105–3108, 1980.
11. Bower RH, Cerra FB, Bershadsky B, Licari JJ, Hoyt DB, et al. Early enteral administration of a formula (Impact) supplemented with arginine, nucleotides, and fish oil in intensive care unit patients: results of a multicenter, prospective, randomized, clinical trial. *Crit. Care Med.* 23: 436–449, 1995.
12. Buttke TM and Sandstrom PA. Oxidative stress as a mediator of apoptosis. *Immunol. Today.* 15: 7–10, 1994.

13. Carver JD, Pimentel B, Cox WI, and Barness LA. Dietary nucleotide effects upon immune function in infants. *Pediatrics.* 88: 359–363, 1991.

14. Chlebowski RT, Beall G, Grosvenor M, Lillington L, Weintraub N, et al. Long-term effects of early nutritional support with new enterotropic peptide-based formula vs. standard enteral formula in HIV-infected patients: randomized prospective trial. *Nutrition.* 9: 507–512, 1993.

15. Coodley GO, Coodley MK, Nelson HD, Loveless MO. Micronutrient concentrations in the HIV wasting syndrome. *AIDS.* 7: 1595–1600, 1993.

16. Coodley GO, Loveless MO, and Merrill TM. The HIV wasting syndrome: a review. *J. Acquir. Immune Defic. Syndr.* 7: 681–694, 1994.

17. Dai D and Walker WA. Protective nutrients and bacterial colonization in the immature human gut. *Adv. Pediatr.* 46: 353–382, 1999.

18. Daly JM, Reynolds J, Thom A, Kinsley L, Dietrick-Gallagher M, et al. Immune and metabolic effects of arginine in the surgical patient. *Ann. Surg.* 208: 512–523, 1988.

19. de Monye C, Karcher DS, Boelaert JR, and Gordeuk VR. Bone marrow macrophage iron grade and survival of HIV-seropositive patients. *AIDS.* 13: 375–380, 1999.

20. Dworkin B, Wormser GP, Rosenthal WS, Heier SK, Braunstein M, et al. Gastrointestinal manifestations of the acquired immunodeficiency syndrome: a review of 22 cases. *Am. J. Gastroenterol.* 80: 774–778, 1985.

21. Fawzi W. Nutritional factors and vertical transmission of HIV-1. Epidemiology and potential mechanisms. *Ann. N.Y. Acad. Sci.* 918: 99–114, 2000.

22. Fawzi WW and Hunter DJ. Vitamins in HIV disease progression and vertical transmission. *Epidemiology.* 9: 457–466, 1998.

23. Gibson GR and Fuller R. Aspects of *in vitro* and in vivo research approaches directed toward identifying probiotics and prebiotics for human use. *J. Nutr.* 130: 391S–395S, 2000.

24. Gomes MS, Boelaert JR, and Appelberg R. Role of iron in experimental Mycobacterium avium infection. *J. Clin. Virol.* 20: 117–122, 2001.

25. Gordeuk V, Delanghe JR, Langlois MR, and Boelaert JR. Iron status and the outcome of HIV infection: an overview. *J. Clin. Virol.* 20: 111–115, 2001.

26. Gougeon ML and Montagnier L. Apoptosis in AIDS. *Science.* 260: 1269–1270, 1993.

27. Graham NM, Sorensen D, Odaka N, Brookmeyer R, Chan D, et al. Relationship of serum copper and zinc levels to HIV-1 seropositivity and progression to AIDS. *J. Acquir. Immune Defic. Syndr.* 4: 976–980, 1991.

28. Grunfeld C and Feingold KR. Metabolic disturbances and wasting in the acquired immunodeficiency syndrome. *N. Engl. J. Med.* 327: 329–337, 1992.

29. Grunfeld C, Pang M, Shimizu L, Shigenaga JK, Jensen P, and Feingold KR. Resting energy expenditure, caloric intake, and short-term weight change in human immunodeficiency virus infection and the acquired immunodeficiency syndrome. *Am. J. Clin. Nutr.* 55: 455–460, 1992.

30. Halsted CH. Evaluating malnutrition: what should the physician look for?, in *Nutrition and Immunology*, Gershwin, M, German, J, and Keen, C (Eds.). Totowa, New Jersey: Humana Press, 2000, 15–20.

31. Harriman GR, Smith PD, Horne MK, Fox CH, Koenig S, et al. Vitamin B12 malabsorption in patients with acquired immunodeficiency syndrome. *Arch. Intern. Med.* 149: 2039–2041, 1989.

32. Haug C, Muller F, Aukrust P, and Froland SS. Subnormal serum concentration of 1,25-vitamin D in human immunodeficiency virus infection: correlation with degree of immune deficiency and survival. *J. Infect. Dis.* 169: 889–893, 1994.

33. Hegde HR, Woodman RC, and Sankaran K. Nutrients as modulators of energy in acquired immune deficiency syndrome. *J. Assoc. Physicians. India.* 47: 318–325, 1999.

34. Hommes MJ, Romijn JA, Godfried MH, Schattenkerk JK, Buurman WA, et al. Increased resting energy expenditure in human immunodeficiency virus-infected men. *Metabolism.* 39: 1186–1190, 1990.

35. Hori K, Hatfield D, Maldarelli F, Lee BJ, and Clouse KA. Selenium supplementation suppresses tumor necrosis factor alpha-induced human immunodeficiency virus type 1 replication *in vitro. AIDS Res. Hum. Retroviruses.* 13: 1325–1332, 1997.

36. Howard DH. Acquisition, transport, and storage of iron by pathogenic fungi. *Clin. Microbiol. Rev.* 12: 394–404, 1999.

37. Isa L, Lucchini A, Lodi S, and Giachetti M. Blood zinc status and zinc treatment in human immunodeficiency virus-infected patients. *Int. J. Clin. Lab. Res.* 22: 45–47, 1992.

38. Koch J, Neal EA, Schlott MJ, Garcia-Shelton YL, Chan MF, et al. Zinc levels and infections in hospitalized patients with AIDS. *Nutrition.* 12: 515–518, 1996.

39. Kotler DP. Body composition studies in HIV-infected individuals. *Ann. N.Y. Acad. Sci.* 904: 546–552, 2000.

40. Kotler DP. Nutritional alterations associated with HIV infection. *J. Acquir. Immune Defic. Syndr.* 25: S81–S87, 2000.

41. Kotler DP, Tierney AR, Brenner SK, Couture S, Wang J, and Pierson RN, Jr. Preservation of short-term energy balance in clinically stable patients with AIDS. *Am. J. Clin. Nutr.* 51: 7–13, 1990.

42. Lehmann S. Immune function and nutrition. The clinical role of the intravenous nurse. *J. Intraven. Nurs.* 14: 406–420, 1991.

43. Leppert D, Hauser SL, Kishiyama JL, An S, Zeng L, and Goetzl EJ. Stimulation of matrix metalloproteinase-dependent migration of T cells by eicosanoids. *FASEB J.* 9: 1473–1481, 1995.

44. Madebo T, Nysaeter G, and Lindtjorn B. HIV infection and malnutrition change the clinical and radiological features of pulmonary tuberculosis. *Scand. J. Infect. Dis.* 29: 355–359, 1997.

45. Matzkies FK, Cullen P, von Eckardstein A, Nofer JR, Rahn KH, and Schaefer RM. Diagnosis of iron deficiency in patients infected with human immunodeficiency virus. *Haematologica.* 85: 871–873, 2000.

46. Melchior JC, Salmon D, Rigaud D, Leport C, Bouvet E, et al. Resting energy expenditure is increased in stable, malnourished HIV-infected patients. *Am. J. Clin. Nutr.* 53: 437–441, 1991.

47. Mencacci A, Cenci E, Boelaert JR, Bucci P, Mosci P, et al. Iron overload alters innate and T helper cell responses to *Candida albicans* in mice. *J. Infect. Dis.* 175: 1467–1476, 1997.

48. Meydani SN, Lichtenstein AH, Cornwall S, Meydani M, Goldin BR, et al. Immunologic effects of national cholesterol education panel step-2 diets with and without fish-derived N-3 fatty acid enrichment. *J. Clin. Invest.* 92: 105–113, 1993.

49. Mocchegiani E, Muzzioli M, Gaetti R, Veccia S, Viticchi C, and Scalise G. Contribution of zinc to reduce CD4+ risk factor for 'severe' infection relapse in aging: parallelism with HIV. *Int. J. Immunopharmacol.* 21: 271–281, 1999.

50. Moseson M, Zeleniuch-Jacquotte A, Belsito DV, Shore RE, Marmor M, and Pasternack B. The potential role of nutritional factors in the induction of immunologic abnormalities in HIV-positive homosexual men. *J. Acquir. Immune Defic. Syndr.* 2: 235–247, 1989.

51. Murray MF. Niacin as a potential AIDS preventive factor. *Med. Hypotheses.* 53: 375–379, 1999.

52. Patrick L. Nutrients and HIV: part one — beta carotene and selenium. *Altern. Med. Rev.* 4: 403–13, 1999.

53. Payan DG and Goetzl EJ. Recognition of leukotriene B4 by a unique subset of human T-lymphocytes. *J. Allergy Clin. Immunol.* 74: 403–406, 1984.

54. Rabeneck L, Palmer A, Knowles JB, Seidehamel RJ, Harris CL, et al. A randomized controlled trial evaluating nutrition counseling with or without oral supplementation in malnourished HIV-infected patients. *J. Am. Diet. Assoc.* 98: 434–438, 1998.

55. Rivera S, Briggs W, Qian D, and Sattler FR. Levels of HIV RNA are quantitatively related to prior weight loss in HIV-associated wasting. *J. Acquir. Immune Defic. Syndr. Hum. Retrovirol.* 17: 411–418, 1998.

56. Roth E, Spittler A, and Oehler R. Glutamine: effects on the immune system, protein balance and intestinal functions. *Wien Klin. Wochenschr.* 108: 669–676, 1996.

57. Roubenoff R. Acquired immunodeficiency syndrome wasting, functional performance, and quality of life. *Am. J. Manag. Care.* 6: 1003–1016, 2000.

58. Rudolph FB and Van Buren CT. The metabolic effects of enterally administered ribonucleic acids. *Curr. Opin. Clin. Nutr. Metab. Care.* 1: 527–530, 1998.

59. Rule SA, Hooker M, Costello C, Luck W, and Hoffbrand AV. Serum vitamin B12 and transcobalamin levels in early HIV disease. *Am. J. Hematol.* 47: 167–171, 1994.

60. Salhi Y, Costagliola D, Rebulla P, Dessi C, Karagiorga M, et al. Serum ferritin, desferrioxamine, and evolution of HIV-1 infection in thalassemic patients. *J. Acquir. Immune Defic. Syndr. Hum. Retrovirol.* 18: 473–478, 1998.

61. Sergio W. Zinc salts that may be effective against the AIDS virus HIV. *Med. Hypotheses.* 26: 251–253, 1988.

62. Sharkey SJ, Sharkey KA, Sutherland LR, and Church DL. Nutritional status and food intake in human immunodeficiency virus infection. GI/HIV Study Group. *J. Acquir. Immune Defic. Syndr.* 5: 1091–1098, 1992.

63. Shepp DH and Ashraf A. Effect of didanosine on human immunodeficiency virus viremia and antigenemia in patients with advanced disease: correlation with clinical response. *J. Infect. Dis.* 167: 30–35, 1993.

64. Simopoulos AP. Essential fatty acids in health and chronic disease. *Am. J. Clin. Nutr.* 70: 560S-569S, 1999.

65. Skurnick JH, Bogden JD, Baker H, Kemp FW, Sheffet A, et al. Micronutrient profiles in HIV-1-infected heterosexual adults. *J. Acquir. Immune Defic. Syndr. Hum. Retrovirol.* 12: 75–83, 1996.

66. South TL, Kim B, Hare DR, and Summers MF. Zinc fingers and molecular recognition. Structure and nucleic acid binding studies of an HIV zinc finger-like domain. *Biochem. Pharmacol.* 40: 123–129, 1990.

67. Stack JA, Bell SJ, Burke PA, and Forse RA. High-energy, high-protein, oral, liquid, nutrition supplementation in patients with HIV infection: effect on weight status in relation to incidence of secondary infection. *J. Am. Diet. Assoc.* 96: 337–341, 1996.

68. Surette ME, Whelan J, Lu G, Hardard'ottir I, and Kinsella JE. Dietary n - 3 polyunsaturated fatty acids modify Syrian hamster platelet and macrophage phospholipid fatty acyl composition and eicosanoid synthesis: a controlled study. *Biochim. Biophys. Acta.* 1255: 185–191, 1995.

69. Suttmann U, Ockenga J, Schneider H, Selberg O, Schlesinger A, et al. Weight gain and increased concentrations of receptor proteins for tumor necrosis factor after patients with symptomatic HIV infection received fortified nutrition support. *J. Am. Diet. Assoc.* 96: 565–569, 1996.

70. Tacconelli E, Tumbarello M, Ventura G, Leone F, Cauda R, and Ortona L. Risk factors, nutritional status, and quality of life in HIV-infected patients with enteric salmonellosis. *Ital. J. Gastroenterol. Hepatol.* 30: 167–172, 1998.

71. Tang AM, Graham NM, Kirby AJ, McCall LD, Willett WC, and Saah AJ. Dietary micronutrient intake and risk of progression to acquired immunodeficiency syndrome (AIDS) in human immunodeficiency virus type 1 (HIV-1)-infected homosexual men. *Am. J. Epidemiol.* 138: 937–951, 1993.

72. Tang AM, Graham NM, and Saah AJ. Effects of micronutrient intake on survival in human immunodeficiency virus type 1 infection. *Am. J. Epidemiol.* 143: 1244–1256, 1996.

73. Tang AM and Smit E. Selected vitamins in HIV infection: a review. *AIDS Patient Care STDS.* 12: 263–273, 1998.

74. Taylor EW, Cox AG, Zhao L, Ruzicka JA, Bhat AA, et al. Nutrition, HIV, and drug abuse: the molecular basis of a unique role for selenium. *J. Acquir. Immune Defic. Syndr.* 25: S53–S61, 2000.

75. Ullrich R, Zeitz M, Heise W, L'Age M, Hoffken G, and Riecken EO. Small intestinal structure and function in patients infected with human immunodeficiency virus (HIV): evidence for HIV-induced enteropathy. *Ann. Intern. Med.* 111: 15–21, 1989.

76. van Asbeck BS, Georgiou NA, van der Bruggen T, Oudshoorn M, Nottet HS, and Marx JJ. Anti-HIV effect of iron chelators: different mechanisms involved. *J. Clin. Virol.* 20: 141–147, 2001.

77. Yamauchi K, Adjei AA, Ameho CK, Chan YC, Kulkarni AD, et al. A nucleoside-nucleotide mixture and its components increase lymphoproliferative and delayed hypersensitivity responses in mice. *J. Nutr.* 126: 1571–1577, 1996.

78. Zhang ZY, Reardon IM, Hui JO, O'Connell KL, Poorman RA, et al. Zinc inhibition of renin and the protease from human immunodeficiency virus type 1. *Biochemistry.* 30: 8717–8721, 1991.

12 Food Intake Management in Patients with Psychiatric Disorders

Alexander R. Lucas, Diane L. Olson, and F. Karen Olson

CONTENTS

This chapter provides practical guidelines to physicians managing the dietary requirements of patients with psychiatric disorders throughout the life span. Nutritional considerations are discussed, as are interactions between diet and the medications used to treat these disorders.

0-8493-0945-X/03/$0.00+$1.50
© 2003 by CRC Press LLC

FEEDING DISORDERS OF INFANCY
AND EARLY CHILDHOOD

Pica is the eating of nonfood substances beyond the age of infancy, when mouthing objects indiscriminately is common. It occurs most often in young children, many of whom are mentally retarded or developmentally impaired. Pica also is seen in children who are unsupervised and neglected. When it occurs, a pediatric evaluation is warranted to evaluate the child for nutritional deficiencies or the presence of other complications. Pica may cause anemia, intestinal parasitic infestations, lead poisoning, or malnutrition. When present, those conditions require specific treatment. The family and social environment should be evaluated to determine whether there is adequate supervision or neglect. In the treatment of pica, preventive steps should be taken to ensure that the child receives adequate nutrition and that caretakers provide a safe, lead-free, nurturing environment.[4]

One group of feeding disorders of infancy and early childhood involves a persistent failure to eat adequately, reflected by weight loss or failure to gain weight. Referred to as *failure to thrive* or *psychosocial dwarfism*, these conditions can be due to neurologic, metabolic, or endocrine causes, to difficulties in the parent–child relationship, or to a combination of both.

Chatoor[4] has developed treatment techniques to help parents deal with the subgroup of feeding disorders she identifies as *infantile anorexia*. This disorder is characterized by food refusal in infants and leads to acute or chronic malnutrition. The treatment techniques educate parents regarding their infant's temperament and their own anxieties about feeding and provide guidelines that allow parents to regulate their infant's food intake on the basis of the infant's hunger and satiety.

NUTRITIONAL GUIDELINES

The feeding relationship is the complex of interactions that occur between parent and child as they engage in food selection, ingestion, and regulation behaviors. Feeding is a reciprocal process that depends on the capabilities of parents and children. Optimal feeding interactions require that parents be emotionally healthy, sensitive, and responsive and that children be able to achieve a minimal level of communication and stability.[22]

A division of responsibility in feeding is recommended. Parents are responsible for what children are offered to eat. Children are responsible for deciding how much and even whether to eat what parents put before them. Parents must choose appropriate food, provide structured meals and snacks after the first year, and ensure a pleasant eating environment.[22] According to a prevalent myth arising from the pioneering research of pediatrician Clara Davis, infants and children have the innate ability to select a balanced, nutritious diet. Children in Dr. Davis' studies, however, were provided only nutritious foods and did not have access to foods of low nutrient density.[28]

Feeding in a developmentally appropriate fashion is vital. Food selection and methods of feeding should be in concert with the child's neuromuscular development and feeding skills. This is discussed in detail in Satter's book, *Child of Mine: Feeding with Love and Good Sense.*[23]

Assessment of nutrition in a child with failure to grow and gain weight starts with a detailed dietary history to review feeding during infancy, breast-feeding patterns, accuracy of infant formula preparation, age at introduction to solid foods, and age at weaning. The history also elicits information on the eating patterns and habits of the child and family, food allergies, excessive losses of nutrients due to vomiting or diarrhea, food likes and dislikes, understanding of child-size portions, and the family's views of the feeding behaviors and problems. Because of misconceptions about nutrition, some families excessively limit food choices and the caloric density of foods. The food intake record should be evaluated for adequacy of calories, protein, and other key nutrients for which deficiencies may exist.[20d]

Most feeding problems can be prevented by routine support and education about feeding and parenting. This includes education in the child's nutritional needs, ways to meet these needs with age-appropriate and developmentally appropriate foods, and the roles of the parent and child in the feeding process.[20d] An ongoing supportive rapport with the family is often necessary for resolution of feeding and growth problems.[24]

ENCOPRESIS

Encopresis is characterized by fecal soiling after the age of 4 years. It may occur in children who have never been toilet trained or in children who begin soiling after a period of bowel control. It may be voluntary (intentional) or involuntary. Organic disorders of the gastrointestinal tract should first be ruled out by a general medical evaluation. The most common nonorganic form of encopresis involves fecal retention and overflow incontinence.[18]

Treatment of pediatric encopresis entails removal of the impaction, use of stool softeners or laxatives, and a program of sitting on the toilet at regular times each day. A nonconstipating, high-fiber diet is prescribed.[25] Voluntary soiling and soiling unresponsive to this regimen may require psychiatric consultation.

NUTRITIONAL GUIDELINES

A balanced diet that includes whole grains, fruits, and vegetables is recommended as part of the treatment for constipation and encopresis in children.[2,20e] As yet, no randomized controlled studies with children have been published that demonstrate a positive effect by increasing intakes of fiber,[2,5] fluid, or nonabsorbable carbohydrates, such as sorbitol-containing fruit juices.[2]

Nutritional therapy begins with a thorough review of the child's food and fluid intake. This helps the dietitian identify current problems and formulate dietary recommendations that take into account growth, development, and activity needs.[20e] If the child's diet is low in fiber, a gradual increase in dietary fiber from a varied diet decreases the likelihood of gastrointestinal side effects. Recommendations must reflect family food preferences and be implemented in a noncoercive manner.

ATTENTION DEFICIT–HYPERACTIVITY DISORDER

Increasingly recognized in preschool and school-age children, attention deficit–hyperactivity disorder (ADHD) is characterized by a chronic pattern of inattention or hyperactivity and impulsiveness that interferes with social and academic functioning. Most often these children are treated with stimulant medications that improve attention and reduce hyperactivity. The most commonly prescribed stimulants are methylphenidate (Ritalin®; see Appendix 1 for list of drugs and manufacturers) and dextroamphetamine (Dexedrine®). Several preparations have a longer duration of action and can be taken once per day. Whether they have fewer side effects is not yet known. Side effects commonly seen include insomnia, anorexia, stomach pains, and weight loss. Dosage adjustments and timing of the administration may minimize adverse effects. Nonetheless, many children receiving these medications experience marked weight loss during the months after medication is begun. Therefore, monitoring of weight and growth is mandatory during treatment. Early studies suggested that stimulants suppressed growth, but subsequent studies showed that catch-up growth occurred later, even when the medications were continued, and that ultimate height was not impaired. Physicians and parents should closely monitor each child's growth and provide individualized nutritional intervention when weight and height lag behind growth norms.

NUTRITIONAL GUIDELINES

Children receiving stimulant medication may present with various feeding problems, including being picky about foods eaten, dawdling at mealtime, refusing school lunches, or being uninterested in eating. Parents' concerns regarding the child's refusal to eat may lead to bribery, indulgence, or threats in order to improve food intake. As a result, some children may consume large amounts of sweets or other low-nutrient foods.[16]

Appetite suppression due to stimulants is typically dose related and worse at the beginning of treatment. It can, however, persist throughout the treatment period.[1] The medication's effect on appetite is usually minimal or absent 4 to 6 hours after administration.[16] Children medicated before their school day begins typically experience a reduced appetite at lunchtime. Their appetite returns when the medication wears off, and they may have a robust appetite after school and before bedtime.[1] If the child's appetite is suppressed, the following actions should be considered:[1]

- Split the morning stimulant dose so that one half is given approximately 30 minutes before the other half
- Encourage temporary trials off medication (e.g., the weekend or summer)
- Consider a brief trial off medication to see if medication is still needed
- Change to a different stimulant medication or a nonstimulant medication
- Administer a lower dose of the same medication
- Change from a long-acting to a short-acting preparation

Nutrition therapy begins with a detailed history of the child's intake. This includes information on the type and amount of food eaten, when and where food is eaten, and appetite at different times of day. Also gathered is information on food likes and dislikes, vitamin, mineral, herbal, and nutritional supplements used, and the family's views of feeding behaviors and problems. Stimulant medication type, dosage, and administration must be known.

The child's usual intake should be evaluated for adequacy of calorie, protein, vitamin, and mineral sufficiency. Guidelines detailing the amount, type, and frequency of foods needed each day should be developed and reviewed with the child and the parents. Breakfast should be consumed before the morning stimulant dose. Intake, especially of energy- and nutrient-dense foods, should be optimized when the child's appetite is best. The family's meal schedule must be considered.

Dietary management of ADHD is a controversial topic that has become more complicated over time. Current research into the interaction of diet and behavior is focused primarily on the roles that nutritional deficiency, malnutrition, metabolic disturbances, food allergies, and food sensitivities play in the biochemical basis of behavior.[3] No research evidence or physiologic explanation supports an independent role of sugar[3] or aspartame[3,31] in causing or exacerbating ADHD. At present, evidence supporting the role of herbs or homeopathy in the specific treatment of ADHD is inconsistent or lacking. Although a daily multivitamin and mineral supplement may be an inexpensive way to meet the recommended daily allowance of nutritional requirements,[1,3] the use of megadoses of any micronutrient in the treatment of children with ADHD is unsupported.[1]

EATING DISORDERS

Eating disorders most often begin during puberty or later in adolescence. They are much more common in girls than in boys. Girls are affected during the time of rapid growth and acquisition of fat. Eating disorders also occur in young adulthood and less commonly later in life. The two major forms of eating disorder are anorexia nervosa (AN) and bulimia nervosa (BN).

AN is characterized by self-imposed food restriction, endocrine dysfunction manifested by amenorrhea in females, and a distorted, psychopathologic attitude toward eating and weight. The result is a marked weight loss of at least 15% of the previous body weight or failure to gain weight during the growth phase of adolescence. Although onset typically occurs in puberty, AN can be premenarchal or occur later in life. Approximately 1 in 10 patients with AN is male.

BN is characterized by frequent binge eating and purging associated with loss of control over eating and a persistent concern about body shape and weight. It occurs most frequently in young women.

Depending on the severity of the illness, treatment of eating disorders ranges from minimal intervention consisting of educational counseling about the illness to extensive psychiatric treatment, including hospitalization. In mild cases, counseling about adolescent growth, normal nutrition, and the consequences of starvation, binging, and purging may be sufficient. The severe forms of the illness are best treated by someone experienced in treating eating disorders. Often the illnesses take a chronic course, requiring close, supportive treatment for a long time.

NUTRITIONAL GUIDELINES

Fundamental principles regarding the nutritional aspects of eating disorders apply in all cases, mild or severe, inpatient or outpatient. The goal of dietary treatment in AN and BN is to help the patient reestablish normal eating patterns. The techniques for accomplishing this occur in the following stages:[20b]

- Taking the dietary history
- Analyzing the caloric content of the initial diet
- Designing the dietary plan
- Executing the dietary plan
- Establishing weight-gain expectations
- Planning for weight maintenance

The initial dietary history provides information on the extent of dietary restriction in AN and the deviations in eating pattern in BN. The frankness with which the history is revealed indicates how cooperative with treatment recommendations the patient is likely to be. An estimate of the basal energy requirement calculated from the patient's weight and height is the starting point for a meal plan. When a patient with AN has been undernourished for a considerable time, basal energy expenditure is likely reduced. It is important to start with relatively small meals intended to stabilize the weight. In BN it is most important to stop the damaging behaviors of vomiting, diuretic abuse, or laxative abuse before stabilizing the patient's weight with the meal plan. Patients with BN are usually at an acceptable weight but experience dramatic weight fluctuations. An individualized meal plan is designed for both conditions and requires close monitoring by a dietitian, through the use of a food diary. In AN, a gradual progression in the meal plan is recommended to bring the patient's weight to an acceptable level. In BN, elimination of damaging behaviors lessens the wide fluctuations in weight that are typical. A meal plan based on caloric needs should stabilize the patient's weight if these behaviors are curtailed. Finally, a weight maintenance plan is designed once an acceptable weight and healthy eating practices are established.

Details of the dietary management of these eating disorders can be found in the *Mayo Clinic Diet Manual: A Handbook of Nutrition Practices*[20b] and in the paper by Huse and Lucas[11] on the dietary treatment of AN.

MAJOR PSYCHIATRIC DISORDERS

Nutritional problems in the management of patients with psychiatric disorders chiefly concern changes in eating habits inherent to the illness and weight changes associated with psychotropic medications. From a practical standpoint, most of the issues concern weight gain associated with medication treatment. Sachs and Guille[21] pointed out that nearly 25% of all cases of obesity in patients receiving psychotropic medication are drug related. This section briefly describes the major psychiatric disorders and discusses the medications frequently associated with weight gain or other issues requiring nutritional consideration.

MOOD DISORDERS

Depressive Disorders

Patients with major depressive disorder have a clinical course characterized by one or more episodes of severe depression. These episodes typically include loss of appetite and weight, sleep disturbance, decreased psychomotor activity, decreased energy, feelings of worthlessness or guilt, difficulty in thinking or making decisions, recurrent thoughts of death, or suicidal ideation. Loss of appetite is a major symptom, as are loss of interest in one's surroundings and withdrawal from social relationships. Often the recurrences take a chronic course. Although these disorders may begin in childhood, they are more common in adulthood.

Patients with dysthymic disorder have a chronically depressed mood of lesser severity than that seen in major depressive disorder. During periods of depressed mood, there may be poor appetite or overeating, insomnia or hypersomnia, low energy or fatigue, low self-esteem, poor concentration or difficulty making decisions, and feelings of hopelessness.

The anorexia seen in both major depressive disorder and dysthymic disorder is manifested by loss of interest in eating and its associated pleasure. It is more profound in major depression. Depressed patients tend to ignore mealtime and say that they do not feel hungry. Weight loss generally amounts to less than 15% of body weight. When depression is severe enough to have caused weight loss and other symptoms, treatment with antidepressant medication is instituted. This usually restores appetite and interest in eating. If substantial weight loss occurred before treatment was started, special attention to the restoration of adequate nutrition may be required. In such cases, a dietary history may reveal important nutritional deficiencies, and the assistance of a dietitian may be necessary to educate the patient about appropriate meal choices. Small frequent meals of calorically dense foods are the dietary recommendation. Nutritional beverage supplements may be needed; they are both convenient and calorically dense. Ongoing follow-up care with a dietitian is recommended until consistent weight restoration is achieved.

Antidepressant medications consist of the tricyclic antidepressants (TCAs), the monoamine oxidase inhibitors (MAOIs), the selective serotonin reuptake inhibitors (SSRIs), and atypical antidepressants.

The antidepressant properties of TCAs (Table 12.1) were first discovered in 1957,[19] and for much of the past 40 years they have been the mainstay in the pharmacologic treatment of depression. Treatment with TCAs is often associated with weight gain. Initially, this is a beneficial occurrence and may be due in part to the return of normal appetite as the symptoms of depression are alleviated. However, it may become problematic if excessive carbohydrate craving occurs after normal weight is restored. Many patients with chronic depression require maintenance treatment with medications, and excessive weight gain may not become apparent

TABLE 12.1
Tricyclic Antidepressants

Amitriptyline (Elavil®)
Amoxapine (Asendin®)
Clomipramine (Anafranil®)
Desipramine (Norpramin®, Pertofrane®)
Doxepin (Adapin®, Sinequan®)
Imipramine (Tofranil®)
Nortriptyline (Aventil®, Pamelor®)
Protriptyline (Vivactil®)
Trimipramine (Surmontil®)

TABLE 12.2
Monoamine Oxidase Inhibitors

Irreversible nonselective
    Isocarboxazid (Marplan®)
    Phenelzine (Nardil®)
    Tranylcypromine (Parnate®)
Irreversible selective
    Selegiline (Eldepryl®)

until much time has passed. Amitriptyline and imipramine are the TCAs most likely to be associated with weight gain. Individual differences in patient response are common, however; some patients may lose weight when taking a TCA. Many patients stop taking their medication because of weight gain, even though they experience symptomatic improvement of their depression while on medication.[21] Desipramine and protriptyline are the TCAs least likely to be associated with weight gain. Patients consuming high-fiber diets experience a decrease in the serum levels of antidepressant; these lowered serum levels may interfere with the clinical efficacy of the medication. The TCAs have side effects that must be weighed against the benefits. In addition to potentially troubling cholinergic effects, the TCAs have cardiovascular effects. Orthostatic hypotension and tachycardia are expected accompaniments of treatment with TCAs. Serious cardiac toxicity caused by delay in conduction time and risk of heart block may occur at higher doses.[19]

Introduced more than 40 years ago, the MAOIs (Table 12.2) received considerable use in the treatment of depression because they are often effective in patients unresponsive to TCAs. The older group of irreversible nonselective MAOIs must be used with great caution, however, because they interact with foods high in tyramine and may cause a hypertensive crisis. Reversible selective inhibitors of monoamine oxidase types A and B that can be used without dietary restrictions have been developed but are not available in the U.S. Selegiline, an irreversible selective inhibitor of monoamine oxidase type B is marketed in the U.S. but prescribed primarily in the treatment of parkinsonism.[14] If selegiline is prescribed in daily doses of 20 mg or more, tyramine should be restricted.

In patients receiving the older MAOIs, as little as 6 mg of tyramine in the daily diet induces increased blood pressure, and 25 mg of tyramine may induce a life-threatening hypertensive crisis.[20c] Before MAOIs are prescribed, the ability of the patient to follow a restrictive diet must be considered carefully and the diet reviewed with the patient. Of lesser concern, weight gain commonly is seen with the use of MAOIs, as it is with other antidepressant medications.

Since the advent of SSRIs (Table 12.3) in 1988,[13] the TCAs have been used much less frequently because of their cardiotoxic effects and other disturbing side effects. Thus the SSRIs are safer and better tolerated than the TCAs and MAOIs.

Moreover, the SSRIs are equal in efficacy. MAOIs are now used rarely; they are indicated only when depression is resistant to other medications or electroconvulsive treatment.

In comparison with MAOIs and TCAs, SSRIs are remarkably devoid of side effects and toxicity. They have become, by far, the most widely prescribed antidepressant drugs. According to Sussman and Ginsberg,[29] more than 80% of new prescriptions for brand-name antidepressants are written for the four most popular SSRIs. Early clinical trial data did not predict that the SSRIs would induce weight gain. In fact, fluoxetine, particularly in high doses, was believed to be associated with some anorexia and weight loss. Accumulating clinical evidence, however, indicated that many patients on maintenance medication develop excessive weight gain, just as they do with the other antidepressants. The patients who initially lose some weight tend to regain it; over the long term, patients tend to gain weight. Whether SSRIs affect body weight by altering caloric intake or by modifying energy expenditure is not well understood.[29]

Other antidepressants, differing in chemical structure from the TCAs, MAOIs, and SSRIs, include bupropion (Wellbutrin®), mirtazapine (Remeron®), nefazodone (Serzone®), trazodone (Desyrel®), and venlafaxine (Effexor®). Except for bupropion, these medications have been reported to be associated with weight gain. Bupropion is a novel antidepressant compound that causes appetite suppression and may lead to weight loss during treatment.[8,15] It is contraindicated in patients with bulimia nervosa because of seizures that have occurred in patients with this disorder.

**TABLE 12.3
Selective Serotonin
Reuptake Inhibitors**

Citalopram (Celexa®)
Fluoxetine (Prozac®)
Fluvoxamine (Luvox®)
Paroxetine (Paxil®)
Sertraline (Zoloft®)

Bipolar Disorders

Bipolar I disorder is characterized by the occurrence of one or more manic episodes. In a manic episode, there is a distinct period of abnormally and persistently elevated, expansive, or irritable mood. The mood disturbance is accompanied by inflated self-esteem or grandiosity, decreased need for sleep, pressure of speech, flight of ideas, distractibility, increased involvement in goal-directed activities or psychomotor agitation, and excessive involvement in pleasurable activities that have a high potential for painful consequences. Bipolar II disorder has a clinical course that is characterized by recurrent major depressive episodes with hypomanic episodes. In a hypomanic episode, there is a distinct period of abnormally and persistently elevated, expansive, or irritable mood of briefer duration and lesser severity than that seen in a manic episode.

Cyclothymic Disorder

Patients with cyclothymic disorder have a chronic, fluctuating mood disturbance involving numerous periods of hypomanic symptoms and numerous periods of depressive symptoms. Lithium is indicated in the treatment of mania, bipolar disorders, and cyclothymic disorder. Long-term lithium treatment reduces the frequency, severity, and duration of manic and depressive episodes. Because of its metabolic effects, monitoring of thyroid and renal function is necessary during lithium therapy.[12] One third to two thirds of patients treated with lithium gain weight. Some patients have a pronounced weight gain that is dose dependent.[21] This creates a therapeutic dilemma, because the drug dose that is low enough to avoid weight gain is usually subtherapeutic. Weight gain is a major reason why patients discontinue taking lithium. Lithium impairs glucose tolerance and may increase sensitivity to insulin. It also inhibits the effects of the antidiuretic hormone, resulting in polyuria and polydipsia. If dietary sodium is restricted, lithium excretion decreases and lithium toxicity may occur. Various mechanisms have been suggested for lithium-induced weight gain, including increased appetite, altered metabolism, lithium-induced hypothyroidism, increased fluid retention, and endocrine changes.[21]

Mood stabilizers (Table 12.4) include a group of anticonvulsant drugs that are frequently prescribed for bipolar disorder. Weight gain has been reported with each of these drugs except topiramate. Again, weight gain is the most frequent reason for lack of patient compliance.[21]

TABLE 12.4
Mood Stabilizers

Carbamazepine (Tegretol®)
Divalproex (Depakote®)
Gabapentin (Neurontin®)
Lamotrigine (Lamictal®)
Topiramate (Topamax®)
Valproic acid (Depakene®)
Vigabatrin (Sabril®)

Nutritional Guidelines for Tyramine-Restricted Diet

Tyramine is the amine of the amino acid tyrosine, which occurs in certain proteins and aged foods. Dietary tyramine is deaminated by monoamine oxidase in the gastrointestinal tract and liver. When MAOIs are taken, this deaminative oxidation does not occur. Tyramine, consequently, is taken up in the brain, potentially causing severe hypertension. Patients taking MAOIs, therefore, should restrict their dietary tyramine.

The tyramine-restricted diet limits the intake of foods that contain tyramine naturally or through aging. Because blood pressure has been reported to increase with as little as 6 mg of tyramine per day (d), a conservative intake limit would be 5 mg/d.[17] Aging can contribute to tyramine content through the breakdown of protein. Therefore, only fresh foods or freshly prepared foods should be eaten. Any protein food item that has been aged, unrefrigerated, or stored for a long period should be avoided. It is recommended that patients meet with a dietitian for guidelines before starting MAOIs.

In the past, as many as 70 food items have been restricted for patients taking MAOI medications.[7] Concern about noncompliance has led to the avoidance of a highly effective treatment option. The exclusion of many food items has often been based on inadequate scientific evidence, which has led to reanalysis.[26,30] However, areas of controversy still exist; for example, in cheese the level of tyramine content can vary according to the location of a cheese sample in the cheese wheel (outside surface vs. deeper into the body). Some investigators have allowed processed cheese to be included in the diet, with the amount limited to 30 g or 1 oz/d.[30] Aged cheeses are universally restricted. Bottled beers and wine, both red and white, have been found to have low amounts of tyramine and can be consumed in moderation. In light of a tyramine analysis and two well-documented case reports, all tap (draft) beers should be absolutely restricted.[26] Moderate intake is defined as 12 to 24 oz bottled beer or 4 oz wine per day. Gardner et al.[7] suggest limiting nonalcoholic beer, as with other bottled beer.

The tyramine content of aged or cured meats can be excessively high, but is variable. Of greatest concern are the air-dried sausages, such as pepperoni, salami, summer sausage, and mortadella. Patients should be advised to store all meats under proper refrigeration. Meats should be eaten in a timely manner or discarded if there is any suspicion of spoilage or improper storage. It is believed that many of the earlier food restrictions stemmed from the effects of eating spoiled foods. Studies suggested that spoilage contributed to an increase in the tyramine content of chicken livers and pickled herring. Fresh chicken livers contain very small amounts of tyramine, and fresh herring flesh contains no detectable levels.[27,30]

Marmite has consistently been identified with a high tyramine content, while other yeast extracts contain no appreciable amounts of tyramine.[27]

Soy sauce contains marked, yet variable, amounts of tyramine. Some brands contain as much as 941 mg/L, with 10 mL providing more than 6 mg tyramine.[7]

Table 12.5 lists foods to be avoided entirely, those to be consumed in limited amounts, and those whose consumption does not need restriction. Because some

---

TABLE 12.5
Dietary Restrictions on the Basis of Tyramine Content

Entirely restricted
  All aged cheeses
  Salad dressings with cheese
  Tap (draft) beer
  Yeast extracts, e.g., marmite
  Sauerkraut, broad beans (fava)
  Sausages, e.g., dry, summer, pepperoni, salami, mortadella
Restricted to limited amounts
  Processed cheese (1 oz per day)
  Bottled beer and nonalcoholic beer (12 to 24 oz per day)[a]
  Red or white wines (4 oz per day)[a]
Unrestricted
  Fresh and mild cheeses, e.g., ricotta, cottage cheese, cream cheese, farmer's cheese
  Fresh smoked fish
  Pickled herring if properly refrigerated
  Liver if fresh and properly refrigerated
  Other yeast extracts, e.g., baker's yeast, brewer's yeast
  Soy milk
  Banana pulp, avocado, raspberries, chocolate

[a] Unless contraindicated with other medications the patient is taking.

---

foods contain variable amounts of tyramine, it is prudent that a registered dietitian educate patients before MAOI therapy is initiated. After discontinuation of the MAOI, the dietary restrictions should be continued for 2 weeks.

PSYCHOTIC DISORDERS

Psychotic disorders, including schizophrenia and its subtypes, are characterized by illogical thinking, bizarre behavior, hallucinations, and delusions. Patients with schizophrenia may have delusions about food, resulting in idiosyncratic or bizarre eating habits. Because of their delusions, such patients may avoid eating or eat excessively. Treatment with antipsychotic medication (Table 12.6) restores rational thinking and eliminates delusions. Eating habits usually return to normal without specific nutritional intervention.

However, treatment with antipsychotic medication is often associated with fluid retention and weight gain, especially when these medications are used chronically. Because many psychotic disorders are chronic conditions, fluid retention and weight gain may become serious problems. Weight gain is greatest with clozapine, olanzapine, and thioridazine. Haloperidol, risperidone, chlorpromazine, and sertindole are intermediate in inducing weight gain. Ziprasidone and fluphenazine are associated with less weight gain. Molindone is an exception, in that weight gain has not been reported with its use.[6,21]

TABLE 12.6
Antipsychotic Medications

Chlorpromazine (Thorazine®)
Clozapine (Clozaril®)
Fluphenazine (Permitil®, Prolixin®)
Haloperidol (Haldol®)
Molindone (Moban®)
Nefazodone (Serzone®)
Olanzapine (Zyprexa®)
Risperidone (Risperdal®)
Sertindole (Serlect®)
Thioridazine (Mellaril®)
Ziprasidone (Geodon®)

The newer antipsychotic medications, clozapine, olanzapine, and risperidone, have advantages over the conventional neuroleptic medications in that they are less likely to cause extrapyramidal symptoms, tardive dyskinesia, and other adverse effects. However, weight gain remains a problem with all of these medications.

Sachs and Guille[21] advised that patients be informed of the risk of weight gain associated with the use of psychotropic medications. Several management options are available, including decreasing the drug dose, switching to another medication, and implementing a weight loss program. Dietary advice before the start of treatment may prevent marked weight gain.

For all of these medications, antidepressants, mood stabilizers, and antipsychotics, the mechanisms underlying weight gain are not well understood. Improvement in the underlying psychiatric condition accounts for only a small portion of the weight change, and, in those instances, the change is a restoration of weight to a normal level. Increased appetite, fluid retention, changes in metabolism, and endocrine mechanisms have been implicated but not studied thoroughly. Surprisingly, changes in the patient's activity level tend not to be implicated. It would seem that the antipsychotic medications and lithium, which profoundly reduce hyperactivity and agitation, cause a decrease in activity levels sufficient to bring about a marked weight gain over time. Considering the high prevalence of obesity in the U.S., it is also conceivable that patients receiving psychotropic medication would have gained weight in the course of time had they not been receiving medication. Undoubtedly, in an individual patient, numerous factors conspire to cause unwanted weight gain, including the effects of the medication, the patient's eating habits, and the patient's exercise habits.

NUTRITIONAL GUIDELINES FOR WEIGHT CONTROL

Dietary education is a necessary component of treatment. To avoid excessive weight gain, patients must be taught appropriate food choices and portion control. Consultation with a dietitian is particularly helpful to staff who plan meals for chronically psychotic patients in group homes and sheltered living situations.

Preventing or minimizing weight gain is the best means of managing the problem. Before patients begin treatment, they should be informed of the potential for marked weight gain. They should be educated about appropriate caloric intake and portion control; a registered dietitian can be very helpful in providing guidelines. If marked weight gain does occur, a weight management program with a registered dietitian is often indicated. This is likely to be more effective when initiated at the onset of treatment.[10] Rapport must be established between the patient and the dietitian. This is followed by individualized assessment of the weight loss need, determination of appropriate weight, the setting of realistic goals, and lifestyle changes for improved health. Treatment involves dietary changes, exercise, and behavioral or psychological modifications. At the initial appointment, the dietitian takes a dietary history, discusses a healthy weight range, and helps the patient set realistic weight loss goals. The rate of weight loss established should be moderate, for example, 1 to 1.5 lb per week. Caloric needs can be determined by estimating the resting energy expenditure using the Mayo Clinic nomogram or the Harris–Benedict equation.[20a] Treatment should be adapted to the individual needs of the patient. Meals, frequency of meals, snacks, and restaurant dining all must be considered in establishing a plan that the patient is likely to follow. The general prescription for the diet is 15 to 20% protein, 25 to 30% fat, and 55% complex carbohydrates. If the daily caloric level planned is a maximum of 1200 calories, a daily generic multivitamin is recommended. Other supplements, such as calcium, may be recommended as needed. Regular exercise is encouraged as part of a healthy lifestyle. Increases in activity should be made gradually. The physician must consider the patient's degree of deconditioning before prescribing an exercise program.

Behavioral therapy may be needed to institute healthy lifestyle changes. Weight control can be a great challenge; a combination of appropriate meal planning, regular exercise, and behavioral modification produces the best results. Physicians can help to minimize the undesired consequences associated with psychotropic medications by choosing drugs and dosages least likely to cause weight gain.[9]

REFERENCES

1. Adesman AR and Morgan AM. Management of stimulant medications in children with attention-deficit/hyperactivity disorder. *Pediatr. Clin. North Am.* 46: 945–963, 1999.
2. Baker SS, Liptak GS, Colletti RB, Croffie JM, Di Lorenzo C, Ector W, and Nurko S. Constipation in infants and children: evaluation and treatment. A medical position statement of the North American Society for Pediatric Gastroenterology and Nutrition. *J. Pediatr. Gastroenterol. Nutr.* 29: 612–626, 1999.
3. Baumgaertel A. Alternative and controversial treatments for attention-deficit/hyperactivity disorder. *Pediatr. Clin. North Am.* 46: 977–992, 1999.
4. Chatoor I. Feeding and eating disorders of infancy and early childhood, in *Kaplan & Sadock's Comprehensive Textbook of Psychiatry.* 7th ed., vol. 2, Sadock BJ and Sadock VA (Eds.) Philadelphia: Lippincott Williams & Wilkins, 2000, 2704–2710.
5. Di Lorenzo C. Childhood constipation: finally some hard data about hard stools! *J. Pediatr.* 136: 4–7, 2000.
6. Ganguli R. Weight gain associated with antipsychotic drugs. *J. Clin. Psychiatry.* 60 Suppl 21: 20–24, 1999.
7. Gardner DM, Shulman KI, Walker SE, and Tailor SA. The making of a user friendly MAOI diet. *J. Clin. Psychiatry.* 57: 99–104, 1996.
8. Golden RN and Nicholas LM. Bupropion, in *Kaplan & Sadock's Comprehensive Textbook of Psychiatry.* 7th ed., vol. 2, Sadock BJ and Sadock VA (Eds.) Philadelphia: Lippincott Williams & Wilkins, 2000, 2324–2328.
9. Greenberg I, Chan S, and Blackburn GL. Nonpharmacologic and pharmacologic management of weight gain. *J. Clin. Psychiatry.* 60 Suppl 21: 31–36, 1999.
10. Holt RA and Maunder EMW. Is lithium-induced weight gain prevented by providing healthy eating advice at the commencement of lithium therapy? *J. Hum. Nutr. Dietetics.* 9: 127–134, 1996.
11. Huse DM and Lucas AR. Dietary treatment of anorexia nervosa. *J. Am. Diet. Assoc.* 83: 687–690, 1983.
12. Jefferson JW and Greist JH. Lithium, in *Kaplan & Sadock's Comprehensive Textbook of Psychiatry.* 7th ed., vol. 2, Sadock BJ and Sadock VA (Eds.) Philadelphia: Lippincott Williams & Wilkins, 2000, 2377–2390.
13. Kelsey JE and Nemeroff CB. Selective serotonin reuptake inhibitors, in *Kaplan & Sadock's Comprehensive Textbook of Psychiatry.* 7th ed., vol. 2, Sadock BJ and Sadock VA (Eds.) Philadelphia: Lippincott Williams & Wilkins, 2000, 2432–2435.
14. Kennedy SH, McKenna KF, and Baker GB. Monoamine oxidase inhibitors, in *Kaplan & Sadock's Comprehensive Textbook of Psychiatry.* 7th ed., vol. 2, Sadock BJ and Sadock VA (Eds.) Philadelphia: Lippincott Williams & Wilkins, 2000, 2397–2407.
15. Lasslo-Meeks M. Weight gain liabilities of psychotropic and seizure disorder medications. *Scan's Pulse* 20: No. 1 Winter, 2001, 10–13 (American Dietetic Association).
16. Lucas B. Nutrition for school-age children, in *Nutrition in Infancy and Childhood.* 6th ed., Trahms CM and Pipes PL (Eds.) New York: WCB/McGraw-Hill, 1997, 282–304.
17. McCabe BJ. Dietary tyramine and other pressor amines in MAOI regimens: a review. *J. Am. Diet. Assoc.* 86: 1059–1064, 1986.
18. Mikkelsen EJ. Elimination disorders, in *Kaplan & Sadock's Comprehensive Textbook of Psychiatry.* 7th ed., vol. 2, Sadock BJ and Sadock VA (Eds.) Philadelphia: Lippincott Williams & Wilkins, 2000, 2720–2728.

19. Nelson JG. Tricyclics and tetracyclics, in *Kaplan & Sadock's Comprehensive Textbook of Psychiatry*. 7th ed., vol. 2, Sadock BJ and Sadock VA (Eds.) Philadelphia: Lippincott Williams & Wilkins, 2000, 2491–2502.

20. Nelson JK, Moxness KE, Jensen MD, and Gastineau CF. *Mayo Clinic Diet Manual: A Handbook of Nutrition Practices*. 7th ed., St. Louis: Mosby, 1994, p. *a*, 185–195; *b*, 303–311; *c*, 311–313; *d*, 459–462; *e*, 547–550.

21. Sachs GS and Guille C. Weight gain associated with use of psychotropic medications. *J. Clin. Psychiatry*. 60 Suppl 21: 16–19, 1999.

22. Satter E. Feeding dynamics: helping children to eat well. *J. Pediatr. Health Care*. 9: 178–184, 1995.

23. Satter E. *Child of Mine: Feeding With Love and Good Sense*. 3rd ed., Palo Alto, Bull Publishing Company, 2000.

24. Schechter M. Weight loss/failure to thrive. *Pediatr. Rev.* 21: 238–239, 2000.

25. Schmitt BD. Encopresis, in *Primary Pediatric Care*. 3rd ed., Hoekelman RA. et al. (Eds.) St. Louis: Mosby, 1997, 722–726.

26. Shulman KI, Tailor SA, Walker SE, and Gardner DM. Tap (draft) beer and monoamine oxidase inhibitor dietary restrictions. *Can. J. Psychiatry*. 42: 310–312, 1997.

27. Shulman KI, Walker SE, MacKenzie S, and Knowles S. Dietary restriction, tyramine, and the use of monoamine oxidase inhibitors. *J. Clin. Psychopharmacol.* 9: 397–402, 1989.

28. Story M and Brown JE. Do young children instinctively know what to eat? The studies of Clara Davis revisited. *N. Engl. J. Med.* 316: 103–106, 1987.

29. Sussman N and Ginsberg D. Rethinking side effects of the selective serotonin reuptake inhibitors: sexual dysfunction and weight gain. *Psychiatric Ann.* 28: 89–97, 1998.

30. Walker SE, Shulman KI, Tailor SA, and Gardner D. Tyramine content of previously restricted foods in monoamine oxidase inhibitor diets. *J. Clin. Psychopharmacol.* 16: 383–388, 1996.

31. Wolraich ML, Lindgren SD, Stumbo PJ, Steglink LD, Appelbaum MI, and Kiritsy MC. Effects of diets high in sucrose or aspartame on the behavior and cognitive performance of children. *N. Engl. J. Med.* 330: 301–307, 1994.

APPENDIX 1
Drugs Listed by Nonproprietary Name (Proprietary Name and Manufacturer Information in Parentheses)

Amitriptyline (Elavil®, Astra Zeneca, Wilmington, DE)
Amoxapine (Asendin®, ESI Lederle, Philadelphia, PA)
Bupropion (Wellbutrin®, Glaxo Wellcome, Research Triangle Park, NC)
Carbamazepine (Tegretol®, Novartis, East Hanover, NJ)
Chlorpromazine (Thorazine®, SmithKline Beecham, Philadelphia, PA)
Citalopram (Celexa®, Forest, St. Louis, MO)
Clomipramine (Anafranil®, Novartis, East Hanover, NJ)
Clozapine (Clozaril®, Novartis, East Hanover, NJ)
Desipramine (Norpramin®, Aventis Behring, King of Prussia, PA; Pertofrane®, Rorer, Fort Washington, PA)
Dextroamphetamine (Dexedrine®, SmithKline Beecham, Philadelphia, PA)
Divalproex (Depakote®, Abbott, Abbott Park, IL)
Doxepin (Adapin®, Lotus Biochemical, Bristol, TN; Sinequan®, Pfizer, New York, NY)
Fluoxetine (Prozac®, Eli Lilly, Indianapolis, IN)
Fluphenazine (Permitil®, Schering, Kenilworth, NJ; Prolixin®, Apothecon, Princeton, NJ)
Fluvoxamine (Luvox®, Solvay, Marietta, GA)
Gabapentin (Neurontin®, Parke-Davis, Morris Plains, NJ)
Haloperidol (Haldol®, Ortho-McNeil, Raritan, NJ)
Imipramine (Tofranil®, Novartis, East Hanover, NJ)
Isocarboxazid (Marplan®, Roche, Nutley, NJ)
Lamotrigine (Lamictal®, Glaxo-Wellcome, Research Triangle Park, NC)
Methylphenidate (Ritalin®, Novartis, East Hanover, NJ)
Mirtazapine (Remeron®, Organon, West Orange, NJ)
Molindone (Moban®, Endo, Chadds Ford, PA)
Nefazodone (Serzone®, Bristol-Myers Squibb, Princeton, NJ)
Nortriptyline (Aventil®, Eli Lilly, Indianapolis, IN; Pamelor®, Novartis, East Hanover, NJ)
Olanzapine (Zyprexa®, Eli Lilly, Indianapolis, IN)
Paroxetine (Paxil®, SmithKline Beecham, Philadelphia, PA)
Phenelzine (Nardil®, Parke-Davis, Morris Plains, NJ)
Protriptyline (Vivactil®, Merck, West Point, PA)
Risperidone (Risperdal®, Janssen, Titusville, NJ)
Selegiline (Eldepryl®, Somerset, Tampa, FL)
Sertindole (Serlect®, Abbott, Abbott Park, IL)
Sertraline (Zoloft®, Pfizer, New York, NY)
Thioridazine (Mellaril®, Novartis, East Hanover, NJ)
Topiramate (Topamax®, Ortho-McNeil, Raritan, NJ)
Tranylcypromine (Parnate®, SmithKline Beecham, Philadelphia, PA)
Trazodone (Desyrel®, Apothecon, Princeton, NJ)
Trimipramine (Surmontil®, Wyeth-Ayerst, Philadelphia, PA)
Valproic acid (Depakene®, Abbott, Abbott Park, IL)
Venlafaxine (Effexor®, Wyeth-Ayerst, Philadelphia, PA)
Vigabatrin (Sabril®, Aventis, Parsippany, NJ)
Ziprasidone (Geodon®, Pfizer, New York, NY)

13 Nutrition and Alcoholism

Khursheed P. Navder and Charles S. Lieber

CONTENTS

0-8493-0945-X/03/$0.00+$1.50
© 2003 by CRC Press LLC

INTRODUCTION

Alcoholism, an addiction to alcohol resulting in heavy and frequent drinking, is a major public health issue, with a cost to society exceeding that of all other drugs of abuse combined. Yet it is attracting very little concern, neither among the public nor health professionals, because the perception is that not much can be done about it. Alcoholism affects virtually all organs of the body, with the most severe functional and structural alterations occurring in the liver and brain. Alcohol also alters the utilization of many nutrients and interacts with the effects and toxicity of commonly used medications. New insights into the pathophysiology of alcohol-induced disorders allow for earlier recognition and more successful efforts at prevention and treatment.

Ethanol — to use its chemical name — differs from other alcohols in being both a beverage ingredient and an intoxicant. Ethanol also differs from other alcohols in being a palatable source of energy as well as of euphoria. It is a small, un-ionized molecule that is completely miscible with water and is also somewhat fat soluble. It is produced by the breakdown of sugars and starches by a fungus (yeast) in a process called fermentation. Beer is made by fermenting barley hops. Wine comes from the fermentation of grapes or other fruits, hard liquors from the fermentation of starches derived from various grains or from potatoes. Our own intestinal bacteria produce some alcohol.

NUTRITIONAL VALUE OF ALCOHOLIC BEVERAGES

Alcoholic beverages contain water, ethanol, variable amounts of carbohydrate, and little else of nutritive value. The carbohydrate content varies greatly: whiskey, cognac, and vodka have virtually none; red and dry white wines have 2 to 10 g/L; beer and dry sherry, 30 g/L; and sweetened white and port wines, as much as 120 g/L. The protein, vitamin and mineral content of these beverages is extremely low.

Alcoholic beverages differ in their alcohol content. Regular beer contains about 4% alcohol. Light beers have nearly as much alcohol (3%) but contain fewer calories (Table 13.1). Wine coolers are low in alcohol content (3.5 to 6%) compared to regular wine (11 to 12%) but are high in calories. The alcohol content of distilled spirits such as whiskey, rum, gin, or brandy is more variable. It is measured in "proof," the definition of which varies from one country to another. In the U.S., 1 proof equals 0.5% alcohol. Based on national consumption data, the estimated contribution of alcohol to the average American diet is 4.5% of total calories. An example of the calculation of calories for an alcoholic beverage (bourbon) is given in Table 13.1.

TABLE 13.1
Estimation of Calories from Alcohol

For example, One Drink (1.5 oz each) of 80-Proof Bourbon

1. Sp. Gr. × % pure ethanol (proof/2) × ml = g
   0.8 × 0.40 × 45 = 14.4 g
2. g × 7 Kcal/g = kcal
   14.4 × 7 = 101 kcal

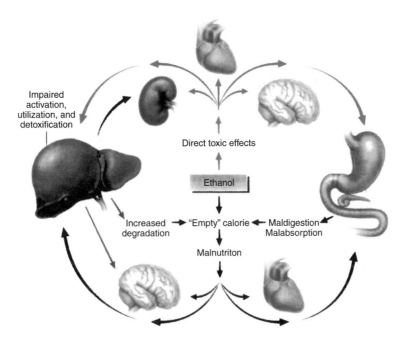

FIGURE 13.1 Organ damage in the alcoholic. The direct toxic effects of ethanol on various organs (gray arrows) interact with malnutrition (black arrows) due to dietary deficiencies or to maldigestion, malabsorption, increased degradation, or impaired activation of nutrients, as well as altered hepatic detoxification. (From Lieber CS. Medical disorders of alcoholism. *N. Eng. J. Med.* 333:1058–1065, 1995. With permission.)

ALCOHOL AND MALNUTRITION

Alcoholism is a major cause of malnutrition. There are three main reasons for this: 1) alcohol interferes with central mechanisms that regulate food intake, causing food intake to decrease; 2) alcohol is rich in energy (7.1 kcal/g) but, like pure sugar or fat, the calories are "empty," devoid of nutrients. The more alcohol one drinks, the less likely one is to eat enough food to obtain adequate nutrients, even if total energy intake is sufficient. Thus, chronic alcohol abuse causes primary malnutrition by displacing other nutrients in the diet. 3) Gastrointestinal and liver complications associated with alcoholism interfere with digestion, absorption, metabolism, and activation of nutrients, and thereby cause secondary malnutrition (Figure 13.1).

ALCOHOL AND BODY WEIGHT

It is important to note that although ethanol is rich in energy (7.1 kcal/g), chronic consumption does not produce the expected gain in body weight. Excessive alcohol intake may provide less energy than an equicaloric amount of carbohydrate. This energy deficit can be attributed, in part, to damaged mitochondria and the resulting poor coupling of oxidation of fat with energy production. The microsomal pathway (vide infra) that oxidizes ethanol could also be contributing, because it produces heat rather than ATP, and thereby does not conserve chemical energy (Figure 13.2). Because of these energy considerations, alcoholics with high total caloric intake do not have a corresponding weight gain, even though their physical activity levels are comparable to those of the nonalcohol-consuming population.[3]

ABSORPTION AND METABOLISM

ABSORPTION

Unlike foods, which require time for digestion, alcohol needs no digestion and is quickly absorbed from the intestine. The presence of food in the stomach, particularly fatty foods, delays gastric emptying, keeping the alcohol in the stomach longer. Therefore, eating snacks when drinking alcoholic beverages slows alcohol absorption. Carbonated beverages, on the other hand, speed up passage of the stomach's contents into the small intestine, where absorption is most rapid.

Only 2 to 10% of ethanol absorbed is eliminated through the kidneys and lungs; the rest is metabolized in the body, principally in the liver. A small amount of ethanol is also metabolized by gastric alcohol dehydrogenase isozymes in the stomach (first first-pass metabolism [FPM]). This FPM explains why, for a given dose of ethanol, blood levels are higher after intravenous than after oral administration. FPM is partly lost in the alcoholic, together with decreased gastric ADH activity. Premenopausal women have less of this stomach enzyme than men, which explains, in part, why women become more intoxicated on less alcohol than men.

METABOLISM

Hepatocytes are the primary cells in the body that can oxidize alcohol at an appreciable rate. This relative organ specificity for the liver, coupled with the high energy content of ethanol and the lack of effective feedback control of its rate of hepatic metabolism, results in a displacement, by ethanol, of up to 90% of the liver's normal metabolic substrates.

The hepatocyte contains three main pathways for ethanol metabolism, each located in a different subcellular compartment: 1) the alcohol dehydrogenase (ADH) pathway of the cytosol or the soluble fraction of the cell; 2) the microsomal ethanol oxidizing system (MEOS), located in the endoplasmic reticulum; and 3) catalase, located in the peroxisomes. Each of these pathways produces specific metabolic and toxic disturbances, and all three result in the production of acetaldehyde (CH_3CHO), a highly toxic metabolite. ADH appears to be the only active enzyme at low ethanol concentrations, but at high concentrations (> 10 mM), MEOS may account for 40% of ethanol metabolism. Because the role of catalase is normally small, it will not be discussed further here.

The ADH Pathway

The oxidation of ethanol via the ADH pathway results in the production of acetaldehyde (CH_3CHO) with loss of H, which reduces NAD to NADH. This is the

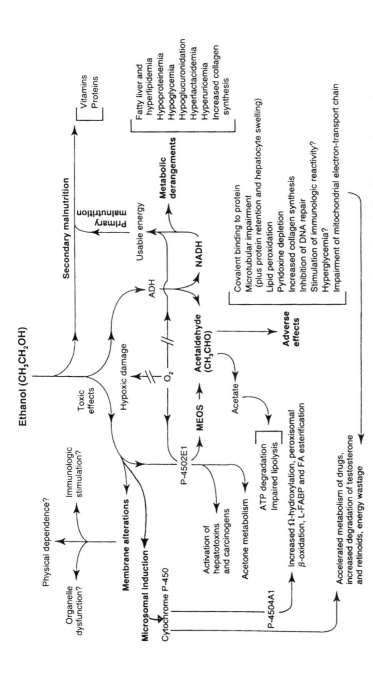

FIGURE 13.2 Toxic and metabolic disorders caused by ethanol abuse. Most of the toxic and metabolic disorders that result from long-term ethanol abuse can be explained on the basis of the metabolism of ethanol by alcohol dehydrogenase (ADH) and the associated generation of reduced NADH or by the microsomal ethanol-oxidizing system (MEOS), with the induction of cytochrome P-4502E1 and other microsomal enzymes, as well as by the toxic effects of acetaldehyde produced by both pathways, primarily in the liver. FA denotes fatty acids, and L-FABP denotes liver fatty-acid-binding protein. (From Lieber CS. Medical disorders of alcoholism. *N. Eng. J. Med.* 333:1058–1065, 1995. With permission.)

rate-limiting step in the metabolism of alcohol. Approximately 13 to 14 g of ethanol are metabolized per hour (the amount in a typical drink), a rate reached when blood alcohol is approximately 10 mg/dl. The large amounts of reducing equivalents generated overwhelm the hepatocyte's ability to maintain redox homeostasis, and a number of metabolic disorders ensue (Figure 13.2). The increased NADH also promotes fatty acid synthesis and opposes lipid oxidation, resulting in fat accumulation.

Microsomal Ethanol Oxidizing System (MEOS)

This pathway also converts a portion of ethanol to acetaldehyde, using cytochrome P4502E1 (CYP2E1). Like other microsomal oxidizing systems, this system is inducible, i.e., it increases in activity after chronic consumption of substantial amounts of substrate. This induction contributes to the metabolic tolerance to ethanol that develops in the alcoholic. This tolerance, however, should not be confused with resistance to disease, and it is important to note that even though an individual can tolerate large amounts, the harmful effects persist and are, in fact, increased.

THE PROCESS OF INTOXICATION

If taken on an empty stomach, some alcohol gets into the blood and reaches the brain within minutes. When consumed with food, absorption takes more time (vida supra). In the brain, alcohol acts as a depressant and as a sedative (producing a sense of calm), as an anesthetic (producing numbness), and finally as a hypnotic (producing sleep). However, its initial effect is a stimulating one, producing a feeling of well being, even of euphoria.

The first brain functions to be depressed are those in the frontal lobe, which controls judgment and behavior. The drinker will usually be more talkative and outgoing, perhaps more aggressive. With higher levels of alcohol in the blood, the parts of the brain that control speech, vision, and voluntary muscles will be affected. Reaction time, depth perception, night vision, and voluntary muscle movement will be impaired. A drinker at this stage is legally intoxicated. In some states, this corresponds to a blood concentration of 0.08%, and in others 0.10%. Many states have adopted a "zero-tolerance" policy for drivers under age 21, using 0.02% as the limit.

If drinking continues and the blood alcohol concentration reaches 0.4%, brain function will be depressed to the state of unconsciousness. At a blood-alcohol concentration of about 0.5%, paralysis of the breathing functions can result in death. This seldom occurs, however, because the alcohol is either vomited or the drinker passes out before a fatal dose can be taken. Occasionally, however, rapid drinking leads to death, because enough alcohol is consumed before the drinker loses consciousness so that the blood alcohol concentration continues to rise to the fatal level. People who drink and take medications that also depress the central nervous system, such as tranquilizers or barbiturates, may suffer respiratory failure at lower levels of alcohol.

ALCOHOLIC LIVER DISEASE

CLINICAL PRESENTATION

Alcoholic liver disease (ALD) occurs in three forms — fatty liver (steatosis), alcoholic hepatitis, and cirrhosis.

Fatty Liver (Steatosis)

Steatosis is seen in more than 80% of binge and heavy drinkers. Hepatocytes are the principal cells in the body that can oxidize alcohol at an appreciable rate, and disposing of the alcohol takes precedence over other liver functions. As a result, dietary protein and fat are not metabolized as readily, and instead, are stored in the liver. The stored protein can absorb 10 times its volume of water and thus causes the liver to swell. Unmetabolized fat deposited in the liver produces a fatty liver, which further impairs the organ's function. After only a weekend of heavy drinking, fatty deposits are evident. A fatty liver is the first step in the organ's deterioration and is characteristic of heavy drinkers. Symptoms include malaise, anorexia, vomiting, weakness, and tenderness of an enlarged liver. Usual laboratory results include elevated aspartate aminotransferase (AST). Fatty liver is treated with abstinence from alcohol and a nutritious diet.

Alcoholic Hepatitis

With continued alcohol abuse, some heavy drinkers develop inflammation of the liver, called *alcoholic hepatitis* (AH). AH is less common than fatty liver, occurring in approximately 10 to 20% of chronic abusers. Severe alcoholic hepatitis can be fatal or can lead to chronic liver disease with cirrhosis. Alcoholic hepatitis symptoms include fatigue, weakness, anorexia, fever, jaundice, and hepatomegaly. Laboratory results show an elevated AST level, hyperbilirubinemia, hypoalbuminemia, and a prolonged prothrombin time.

Alcoholic Cirrhosis

Alcoholic cirrhosis (Laennec's cirrhosis) develops in approximately 10% of chronic alcoholics. The incidence of cirrhosis is significantly increased in men who consume more than 40 to 60 g/day (d) of ethanol for 10 years, while women are already at increased risk by consuming 20 g/d. Among alcoholics who drink approximately 180 g of alcohol daily for 8 years, about one third have biopsy-confirmed cirrhosis or precirrhotic lesions. The symptoms of fatty liver and alcoholic hepatitis continue. Hepatoma (primary liver cancer) develops in 5% or more of the patients with Laennec's cirrhosis. Prognosis of alcoholic cirrhosis improves if alcohol consumption ceases. As the disease progresses, the increasing fibrotic scar tissue impairs blood circulation through the liver, and *portal hypertension* follows. The impaired portal circulation with increasing venous pressure may lead to *esophageal varices*. Increased intrahepatic pressure combined with a decrease in plasma proteins also results in accumulation of large quantities of fluid in the abdomen — *ascites*. Blood clotting mechanisms are impaired because factors such as prothrombin and fibrinogen are not adequately produced. There is excess tissue catabolism and negative nitrogen balance.

Nutritional Management and Anabolic Steroids for ALD

To reverse the catabolic process, both nutritional and anabolic steroid therapy are effective. A high-calorie, high-protein diet has been shown to improve hepatic function and reduce mortality (see Chapter 3). In one study, this was achieved by providing a regular diet plus supplements of 60 g/d of protein and 1600 kcal/d for the first 30 days and followed by supplements of 45 g/d of protein and 1200 kcal/d for the next 60 days.[11] Although protein intake with this treatment was high — exceeding 100 g/d — encephalopathy was usually not exacerbated.

Mendenhall et al.[11] have reported that adding anabolic steroids (oxandrolone, 80 mg/d for 30 days, then 40 mg/d for 60 days) to nutritional therapy for moderately malnourished patients improved recovery of liver functions. It also decreased protein-calorie malnutrition and lowered mortality. For oxandrolone to be effective, however, adequate nutrition is absolutely necessary. Presumably, this is because both energy and protein building blocks (amino acids) are essential for the increased anabolism associated with the steroid. Using amino acid mixtures supplemented with BCAA was also advocated in the early 1980s, but there is little evidence of additional benefit from their use.

Nutritional therapy also includes micronutrients. Thiamine deficiency is presumed in alcoholics, and, unless the deficiency has been excluded, 50 mg of thiamine per day should be given. Riboflavin and pyridoxine should be routinely administered at dosages contained in standard multivitamin preparations. Adequate folic acid replacement can be attained with the usual diet. Additional replacement is optional unless deficiency is severe. Vitamin A replacement should only be given for well-documented deficiency, and to patients whose abstinence from alcohol is assured. Zinc, magnesium, and iron deficiencies that are clearly diagnosed must be corrected.

The most important goal in the treatment of ALD is abstinence; it prolongs survival — even in patients with symptoms of portal hypertension or hepatic insufficiency.

Emerging Therapies

Understanding of the pathogenesis and treatment of ALD has increased considerably in recent years.

PPC Therapy

One of the promising approaches to preventing ALD involves using a mixture of polyunsaturated phosphatidylcholine (PPC), specifically its major species, dilinoleoylphosphatidylcholine (DLPC). An early event in the development of ALD is a decrease in hepatic levels of phospholipids, especially phosphatidylcholines. This phospholipid deficiency also involves mitochondrial membranes and correlates with morphologic abnormalities and impaired function.

In a study in nonhuman primates, administration of PPC corrected the ethanol-induced decrease in phopholipid levels and prevented progression to alcohol-induced septal fibrosis and cirrhosis.[8] A possible explanation for the beneficial effect of PPC is the ability of liver cells to directly incorporate it into cell membranes; thus it can directly influence membrane structure and function. PPC also acts as an antioxidant[9] and promotes collagen breakdown.[8] Human clinical trials are under way for this promising compound with the potential for preventing progression to cirrhosis.

S-adenosylmethionine (SAMe)

Another approach being investigated is correction of SAMe deficiency.[6] SAMe is produced endogenously from methionine and adenosine triphosphate by SAMe synthetase and is essential in three metabolic pathways: transmethylation, transsulfuration, and aminopropylation.

In transmethylation reactions, it serves as the methyl group donor. It is essential in the synthesis, activation, and metabolism of hormones, nucleic acids, proteins, phopholipids, and some drugs.

In the transsulfuration pathway, it leads to the formation of glutathione, which serves as an important cellular antioxidant. Glutathione is crucial in endogenous detoxification processes, including that of acetaldehyde.

In patients with cirrhosis, SAMe synthetase levels are decreased, leading to reduced levels of SAMe and sulfated compounds including glutathione. SAMe deficiency resulting from decreased SAMe synthetase activity can be countered by exogenous SAMe administration.

In experimental models, SAMe replenishes methionine in its active form and is effective in the early stages of alcoholic liver disease of baboons.[7] Human trials are currently under way. Preliminary results show decreased mortality in alcoholic cirrhosis.[10]

Silymarin

This is the active component of milk thistle, *Silybum marianum*. Experimentally, silymarin exerts hepatoprotective actions through free radical scavenging and immunomodulatory effects. In clinical trials, it appears to improve results of liver function tests. However, the role of silymarin in the treatment of alcoholic liver disease remains to be determined.[2,13]

OTHER ORGAN TOXICITIES

ACUTE VS. CHRONIC: HARMFUL EFFECTS

The problems of someone who occasionally becomes drunk differ from those who go on drinking binges at regular intervals.

"Acute" Harmful Effects of Alcohol Intoxication

Occasional excess drinking can cause nausea, vomiting, and hangovers (especially in inexperienced drinkers). The acute neurologic effects of alcohol intoxication are dose-related, progressing from euphoria, relief from anxiety and removal of inhibitions to ataxia, impairment of vision, judgment, reasoning, and muscle control. Massive amounts of alcohol ingested over a short period of time can be fatal (when the respiratory and circulatory centers of the brain are anesthetized). Every year, one encounters news stories of the tragic deaths from excessive alcohol consumption at fraternity initiation rites or other gatherings of young people.

"Chronic" Harmful Effects of Alcohol Excess

Chronic excessive drinking can adversely affect virtually all tissues of the body. Alcoholics have a mortality rate and suicide rate two times greater than average, and an accident rate seven times greater than average. It is well established that dire consequences of alcohol are seen not only in liver (vide supra), but also in other organ systems in the body, including the brain, the cardiovascular system, and the gastrointestinal tract, as well as with sexual function and in pregnancy.

THE BRAIN

Extensive long-term use of alcohol destroys the neurons, producing a decline of intellectual function, memory loss, and inability to concentrate. The brain of a heavy drinker ages prematurely. Alcohol also affects the brain's normal production of neurotransmitter serotonin and endorphins, the body's natural opiates.

Chronic alcohol use can reduce blood flow and oxygen to the brain and may result in blacking out. During a blackout, the drinker functions as usual and seems to be aware of the surroundings but later cannot remember anything that happened. Not all heavy drinkers have blackouts, and whether or when they occur is unpredictable.

Wernicke's Encephalopathy: This type of alcoholic dementia may also occur among heavy drinkers. It is sometimes difficult to differentiate from Alzheimer's and other forms of dementia, but x-rays of the head often show a characteristic shrinking of the brain. It is accompanied by nystagmus, abducens, and conjugate gaze palsies, ataxia of gait, and mental disturbance (global confusion, apathy, drowsiness). It is caused by thiamine deficiency. One manifestation is Korsakoff psychosis, in which recent retentive memory is impaired out of proportion to other cognitive functions.

THE CARDIOVASCULAR SYSTEM

Alcohol can affect the heart and blood vessels in many ways, ranging from dilation of peripheral blood vessels, producing flushing and warmth, to damage of the heart muscle, which can result in sudden death. Even a single episode of heavy drinking may produce arrhythmia (irregular heartbeat), and those who abuse alcohol have an increased risk of hypertension and heart muscle disease (i.e., cardiomyopathy in

---

TABLE 13.2
Drinking in Moderation

Moderation is defined as no more than one drink per day for women and no more than two drinks per day for men (based on differences between the sexes in both weight and metabolism).

A *drink* is any alcoholic beverage that delivers approximately 14.4 g of *pure ethanol.*

Count as a drink:
    12 oz of regular beer — 4.5% alcohol by volume (13 g alcohol; 92 kcal)
    5 oz of wine — 12% alcohol by volume (14.4 g alcohol; 102 kcal)
    1.5 oz of distilled spirits — 80 proof, 40% alcohol (14.4 g alcohol; 102 kcal)

---

which the heart becomes enlarged and cannot pump effectively, resulting in a backup of blood into tissues and organs).

It has been claimed that drinking in moderation (Table 13.2) may lower the risk for coronary heart disease (mainly among men over 45 and women over 55), but to what extent the protection is primarily due to alcohol rather than to some other associated factor, such as lifestyle, is not yet settled.

THE GASTROINTESTINAL TRACT

Alcohol is an irritant to the entire digestive tract, damaging the mucosal linings of the mouth, esophagus, and stomach. Especially in heavy smokers, there is strong evidence of a link between alcohol and cancers of the gastrointestinal tract, the mouth, the pharynx, the larynx, and the stomach.

In the stomach, alcohol increases the output of hydrochloric acid, irritating the lining and producing heartburn and nausea, and promoting ulcers in both the stomach and duodenum. Peptic ulcers are more common among heavy drinkers. Alcohol increases nutritional deficiencies (primary or secondary), with altered absorption and utilization of vitamins. It damages the intestinal absorption of nutrients like lactose, folic acid, and vitamin B12. Alcohol results in severe thiamin deficiency, characterized by visual disorders, ataxia, confusion, and coma — Wernicke's encephalopathy (vide supra).

Alcohol also causes inflammation of the pancreas and impaired production of pancreatic enzymes needed for the digestion of protein and fat.

HORMONES

Alcohol and Sexual Function

In small doses, alcohol releases inhibitions and increases self-confidence, which is responsible for its reputation as a sex stimulant. In larger doses, however, sexual performance suffers and interest in sex declines. Alcoholic men frequently become impotent. In women, heavy drinking may lead to ovarian malfunction and the cessation of menstruation.

Alcohol and Pregnancy

Heavy drinking during pregnancy — especially during the first trimester — can have devastating effects on the fetus. The baby may be born with a condition known as fetal alcohol syndrome. These babies are below normal in weight, height, and head circumference. They have characteristic face malformations and frequently suffer from poor motor development and mental retardation. The abnormalities are irreversible.

Scientific investigation has not yet determined what level of alcohol consumption is safe during pregnancy, or whether there is a safe level at all. The issue remains unresolved, but many doctors advise their pregnant patients to avoid all alcohol, including the alcohol in drugs such as cough syrups.

ALCOHOL AND DRUG INTERACTIONS

The presence of alcohol in the blood alters the way in which the body handles other drugs. One group of enzymes in the liver that is responsible for metabolizing alcohol (cytochromes P450) is also responsible for breaking down drugs (vide supra). Since these enzymes become more active when chronically exposed to alcohol, drugs will be removed more rapidly from the system in drinkers than in nondrinkers. Thus, generally, a drug will have less of an effect in a drinker because it will wear off more quickly than intended. However, if the person is drinking heavily at the time the drug is taken, the alcohol may compete successfully with the drug, which will remain in the body relatively unchanged and, if the drug dose is repeated, high and potentially dangerous levels may build up.

Of the 50 most frequently prescribed drugs, more than half contain ingredients that react adversely with alcohol. The effects of depressants of the central nervous system — including narcotics, barbiturates, sedatives, some sleeping pills, and codeine-containing pain medications — are potentiated when alcohol is present. Beer and wine taken along with monoamine oxidase inhibitors — a group of antidepressants — produce headache, extreme rises in blood pressure, rapid heartbeat, nausea, and vomiting. Heavy use of aspirin combined with heavy alcohol use can lead to bleeding in the stomach. In a heavy drinker, acetaminophen, present in many over-the-counter analgesics and often taken to alleviate symptoms of hangovers (such as headaches), may become toxic to the liver in amounts usually considered safe.

Whether alcohol use is heavy or light, care must be exercised when taking other drugs. For example, some gastric acid inhibitors (H2-blockers) may increase blood alcohol levels. The patient needs to inform the doctor about the level of alcohol consumption so that any needed adjustment in drug dosages can be made. Any change in drinking habits should be reported for the same reason. At least 500 drugs, some prescription, and some sold over the counter, contain alcohol. The majority are vitamin preparations, decongestants, and cough medicines. Some cold remedies contain as much as 25% alcohol. As with other alcohol consumption, medications containing 10% or more alcohol can stimulate gastric secretion and provoke erosions of the gastrointestinal lining.

ALCOHOL ABUSE (ALCOHOLISM) AND ITS TREATMENT

Definitions of alcoholism vary widely, and there is disagreement among health professionals about what causes the disease. One commonly accepted definition is regular drinking to the extent that it affects a person's health, home, or work life. By any definition, alcoholism is a major health and social problem, touching people without regard to age, sex, race, or socioeconomic status. In some urban areas of the U.S., cirrhosis, a complication of alcoholism, is the fifth most frequent cause of death among people 25 to 54 years of age. Once the alcoholic has developed cirrhosis but continues to drink, the outcome is dismal. In a prospective survey of 280 alcoholics, more than half of those with cirrhosis and two thirds of those with cirrhosis and alcoholic hepatitis died within 48 months after enrollment.[1] This outcome is worse than that of many cancers, yet causes much less concern, in part, because there is a prevailing, pervasive but erroneous perception that not much can be done about it.

Two major approaches have been used in the treatment of alcoholics: 1) correction of the medical, nutritional, and psychological problems and 2) alleviating the dependency on alcohol. These traditional approaches, though helpful, often come too late to return the liver to normal. More direct approaches, such as those focusing on prevention, utilizing biochemical markers, such as γ-glutamyltranspeptidase and carbohydrate-deficient transferrin, and screening for signs of medical complications, are helpful to overcome denial (which is common among alcoholics) and facilitate early detection. Naltrexone[12,14] and other pharmacologic agents against craving are now emerging. Along with correction of nutritional deficiencies, supplementation with "supernutrients," such as S-adenosylmethionine (SAMe) and polyunsaturated lecithin,[5] have been shown to offset some of the adverse manifestations, and are now being tested in humans. By combining these approaches, one multiplies the chances to alleviate the suffering of the alcoholic and the public health impact of alcoholism.

ACKNOWLEDGMENT

Skillful typing of the manuscript by Ms. Y. Rodriguez as well as the editing assistance by Ms. F. DeMara is gratefully acknowledged.

REFERENCES

1. Chedid A, Mendenhall CL, Gartside P, French SW, Chen T, and Rabin L. Prognostic factors in alcoholic liver disease. *Am. J. Gastroenterol.* 86:210–216, 1991.
2. Ferenci P, Dragosics B, Dittrich H, Frank H, Benda L, Lochs H, Meryn S, Base W, and Schneider B. Randomized control trial of silymarin treatment in patients with cirrhosis of the liver. *J. Hepatol.* 9:105–113, 1989.
3. Lieber CS. Perspectives: do alcohol calories count? *Am. J. Clin. Nutr.* 54:976–982, 1991.
4. Lieber CS. Medical disorders of alcoholism. *N. Engl. J. Med.* 333:1058–1065, 1995.

5. Lieber CS. Alcohol: its metabolism and interaction with nutrients. *Ann. Rev. Nutrition.* 20:395–430, 2000.

6. Lieber CS. Ademthionine and alcohol, in *S-Adenosylmethionine in the Treatment of Liver Disease.* Lieber, CS (Ed.). UTET (Italy), 2001, 45–60.

7. Lieber CS, Casini A, DeCarli LM, Kim C, Lowe N, Sasaki R, and Leo MA. S-adenosyl-L-methionine attenuates alcohol-induced liver injury in the baboon. *Hepatology.* 11:165–172, 1990.

8. Lieber CS, Robins SJ, Li J, DeCarli LM, Mak KM, Fasulo JM, and Leo M.A. Phosphatidylcholine protects against fibrosis and cirrhosis in the baboon. *Gastroenterology.* 106:152–159, 1994.

9. Lieber CS, Leo MA, Aleynik SI, Aleynik MK, and DeCarli LM. Polyenylphosphatidyl-choline decreases alcohol-induced oxidative stress in the baboon. *Alcoholism: Clin. Exp. Res.* 21:375–379, 1997.

10. Mato JM, Cámara J, Fernández de Paz J, Caballeria L, Coll S, Caballero A, Buey-Garcia L, Beltran J, Benita V, Caballeria J, Sola R, Otero-Moreno R, Felix B, Duce-Marin A, Correa JA, Pares A, Barro E, Magaz-Garcia I, Puerta JL, Moreno J, Boissard G, Ortiz P, and Rodes J. S-Adenosylmethionine in alcoholic liver cirrhosis: a randomized, placebo-controlled, double-blind, multicenter clinical trial. *J. Hepatology.* 30:1081–1089, 1999.

11. Mendenhall C, Moritz T, Roselle GA, Morgan TR, Nemchausky BA, Tamburro CH, McClain CJ, Marsano LS, and Allen JI. A study of oral nutrition support with oxandrolone in malnourished patients with alcoholic hepatitis: results of a Department of Veterans Affairs Cooperative Study. *Hepatology.* 17:564–576, 1993.

12. O'Malley SS, Jaffe AJ, Chang G, Schottenfeld RS, Meyer RE, and Rounsaville B. Naltrexone and coping skills therapy for alcohol dependence. *Arch. Gen. Psychiatry.* 49:881–887, 1992.

13. Trinchet JC, Coste T, Levy VG, Vivet F, Duchatelle V, Legendre C, Gotheil C, Beaugrand M. Treatment of alcoholic hepatitis with silymarin: a double-blind comparative study in 116 patients. *Gastroenterol. Clin. Biol.* 13:120–124, 1989.

14. Volpicelli JR, Alterman AI, Hayashida M, and O'Brien CP. Naltrexone in the treatment of alcohol dependence. *Arch. Gen. Psychiatry.* 49:876–880, 1992.

Index

A

Accent, 188
Acetaldehyde, 310–312
Acetaminophen, 48, 318
Achlorhydria, 112
Acidosis, 10, 159, 250, 258
Acne, 110
Acromegaly, 76
ACTH, 108–112, 243
Acute phase protein synthetic response, 241
ADAH, 294–295
Adenoma, 100
Adenosine triphosphate, 315
Adipose tissue, 200–202, 229
Adrenal gland, 96–98, 109–112
Adrenocorticotropic hormone (ACTH), 108–112, 243
Adrenoleukodystrophy, 113
Adrenomyeloneuropathy, 113
Advil, 48
Aerobic exercise, 31, 77, 284
Agranulocytosis, 104
Alanine, 247
Albumin
    cytokine impact on, 241
    in IBD, 52
    levels, in renal disease, 162–166, 171
    as malnutrition indicator, 2
    in nephrotic syndrome, 158
Alcoholic beverages
    absorption of, 310
    antioxidant properties, 179–180
    for cardiovascular health, 27–28, 33–34
    metabolism of, 310–312
    nutritional value of, 308
    for vision health, 188
Alcoholism, 308–319
    blood pressure level factor, 27–28
    cataract factor, 179
    cirrhosis cause, see Cirrhosis
    fatty liver cause, 57
    fetal alcohol syndrome, 106
    GERD factor, 46
    hepatitis cause, 55
    hyperinsulinemia factor, 76
    hyperuricemia factor, 148

pancreatitis cause, 66
PUD factor, 48–49
Aldosterone, 236, 243
Alendronate, 130–131, 135
Alkaline phosphatase, 9, 136, 138
Allopurinol, 150
Aloe, 87
Alpha-lipoic acid, 83
Aluminum hydroxide, 108, 162
Amenorrhea, 295
American Heart Association Dietary Guidelines, 34
Amidarone, 105
Amide, 50
Amino acid, 95–96, 229, 232–234, 237
    in animal protein, 61
    chromium and, 81–82
    loss in dialysis, 157
    profile, in renal disease patients, 157
    supplements for
        diabetes, 83
        premature infants, 8
    uptake in sepsis, 249–250
    in vegetable protein, 61
Amino acid dialysate, 171
Amino-aciduria, 63, 119
Aminoglycoside antibiotics, 13
Ammonia, 59–61, 234, 247
Amyloidosis, 113
Anaphylactic shock, 115
Anemia from
    celiac disease, 50
    cirrhosis, 58
    cystic fibrosis, 13
    hiatal hernia, 47
    hyperthyroidism, 104
    hypocorticalism, 114
    pica, 292
    PUD, 48
    Wilson's disease, 63
Anorexia
    cancer, 201, 209
    cirrhosis, 58–59
    cystic fibrosis, 12
    cytokines, 240–241
    depression, 297
    hepatitis, 56

321

E

F

Folate, 275
  deficiency, 3
    alcoholism, 314
    celiac disease, 50
    cirrhosis, 58
    IBD, 52
    scleroderma, 144
  homocysteine level factor, 29, 34
  malabsorption from steatorrhea, 103
  supplements for
    cirrhosis, 59
    renal disease, 161
Foley catheter, 261–262
Food allergy, 115
Foremilk, 10
Fructose, 117
Fungal infection, 113
Furosemide, 133, 149

G

Galactose, 115
Galactosemia, 115, 118–121, 125
Gallbladder, 65
Gangrene, 119
Gastric inhibitory peptide, 111
Gastritis, 48
Gastroesophageal reflux disease (GERD), 12, 46–47
Gastrointestinal (GI) tract, 46–67, 143–145
  alcohol impact on, 317
  HIV/AIDS problems, 278
  nutritional therapy, 254
  stress response in, 242–243
  TPN impact on, 258–259
Gastroparesis, 156
Gastrostomy, 12–14, 170, 207–208, 261–262
GERD, 12, 46–47
Gerson diet, 215
GI tract, *see* Gastrointestinal (GI) tract
Giardia lamblia, 123
Ginseng, 84
Glaucoma, 187
Gliadin, 49–51
Glomerular filtration rate, 156
Glucagon, 67, 96–99, 229, 239–240, 244
Glucocorticoid, 229
  disease, 108–114
  hyperinsulinemia, 76
  injury response, 240, 243–244
  osteoporosis, 134–135
Glucomannan, 87
Glucosamine sulfate, 141–142
Glucose, 95–96, 229, 237
  increase in, 249

  intolerance in, 70–88; *see also* Insulin
    resistance syndrome
    cancer, 204
    cirrhosis, 59
    hemochromatosis, 64
    refeeding syndrome, 103
    renal disease, 157
    Syndrome X, 74
  oxidation of, 231
  supplements for premature infants, 7, 10
  in TPN formulas, 255–256
Glucosuria, 12
Glutamine
  for GI function, 259–260
  in gluten, 50
  in human breast milk, 10
  immune system impact, 270
  in sepsis, 250
  supplements for bone marrow transplant, 212–213
  in trauma, 247
Glutathione, 13, 83, 180, 270, 315
Gluten-induced enteropathy, *see* Celiac disease
Glycemic index, 79–80
Glycerol, 95
Glycine, 157
Glycosuria, 63, 119
Glycosylation, 73
Glycyrrhizic acid, 112, 115
Goiter, 101, 105–108
Goitrogens, 105–106
Gonadotropin, 135
Gonads, 99
Gout, 145–150
Gram-negative infection, 250
Graves' disease, 100–101, 104
Growth hormone, 229
  for cystic fibrosis, 14
  disruption of, in chronic inflammation, 167
  in feeding solutions, 260
  for hypophosphatemic rickets, 138
  injury response, 240
  metabolism of, 96–99
  in renal disease, 159–160
Growth retardation
  in BPD, 14–15
  chronic malnutrition indicator, 2–4
  in cystic fibrosis, 11–12
  from galactosemia, 119
  from IBD, 52
  in premature infants, 6
Guar gum, 80, 85
Gums, 3
Gymnema sylvestre, 84

H

H2 histamine blockers, 112
H2 receptor antagonists, 48, 66
HAART, 283–284
Haberman feeder, 9–10
Hair, 3, 64, 107
Harris–Benedict equation, 58
Hashimoto's thyroiditis, 105
HCG, 100
HCVH, 145
HDL cholesterol level, 29–31
  cardiovascular disease factor, 23–26
  in diabetes, 78–79
  raising, by diet, 33
  in renal disease, 158–159
Heart disease, *see* Cardiac dysfunction;
        Cardiovascular disease
Helicobacter pylori, 48
Hematochromatosis, 73
Hemorrhage, 3, 7, 47–48
Hemochromatosis, 58, 64
Hemodialysis, *see* Dialysis
Hepatic-Aid, 61
Hepatitis, 55–57, 104, 313
Hepatomegaly, 64
Hepatosplenomegaly, 62
Hiatal hernia, 46–47
Hibiscus sabdariffa, 87
High-density lipoprotein cholesterol, *see* HDL
        cholesterol
Highly active antiretroviral therapy (HAART),
        283–284
Hindmilk, 10
Histidine, 157
HIV/AIDS, 113, 210, 271–284
Home central venous hyperalimentation (HCVH),
        145
Homocysteine, 24, 29, 34
Human chorionic gonadotrophin (HCG), 100
Human parathyroid hormone 1–34, synthetic, 131
Hydrochlorothiazide, 133
Hydroxytryptophan, 62
Hyperbilirubinemia, 313
Hypercalcemia, 138
Hypercalciuria, 135, 138
Hypercholesterolemia, 30–31, 86, 107, 158
Hypercorticalism, 108–114
Hyperemesis gravidarum, 100
Hyperglycemia, 70–75
  botanical supplement interaction, 84
  carbohydrate impact, 79–80
  lens fiber mitochondria destruction, 180
  in premature infants, 7
  TPN complication, 258

in trauma, 246
Hypergonadotropic hypogonadism, 119
Hyperinsulinemia, 31, 73–78, 104
Hyperkalemia, 113
Hyperkeratosis, 3
Hyperlipidemia, 75, 87, 149, 158
Hypermetabolism, 246, 249
Hyperparathyroidism, 136–139, 157–159
Hyperphosphatemia, 160
Hyperpigmentation, 62
Hyperprolactinemia, 106
Hypertension, 23–33
  Cushing's syndrome, 109
  diabetes factor, 72–76
  garlic for, 86
  from ginseng, 84
  gout link, 149
  magnesium deficiency, 81
  Syndrome X, 74
  tyramine overload, 300
Hyperthyroidism, 100–105
Hypertriglyceridemia, 24, 78–79, 149, 202, 258
Hyperuricemia, 145–150
Hypoalbuminemia, 67, 158, 313
Hypocalcemia, 135–138, 160, 214
Hypochloremic alkalosis, 119
Hypocorticalism, 98, 108–114
Hypoferremia, 242
Hypoglycemia
  chromium supplements for, 81
  in cirrhosis, 59
  diabetes complication, 74–75
  from *Gymnema sylvestre*, 84
  hypocorticalism, 112–114
  from hypothyroidism, 104
  in LGA infants, 6
  in myxedema coma, 106
  in pancreatitis, 67
  in premature infants, 7
  in sepsis, 249
  TPN complication, 258
Hypogonadism, 64, 135
Hypokalemia
  Cushing's syndrome, 109
  hypothyroidism, 104
  IBD, 53
  in premature infants, 10
  from TPN, 256, 258
Hypokalemic alkalosis, 110
Hypolactasia, 116–118
Hypomagnesemia, 4, 13, 81, 214, 258
Hyponatremia, 10, 106, 113
Hypoparathyroidism, 63
Hypophosphatemia, 4, 137–138, 231, 256, 258
Hypoprothrombinemia, 104